DATA FOR RADIOACTIVE WASTE MANAGEMENT AND NUCLEAR APPLICATIONS

DATA FOR RADIOACTIVE WASTE MANAGEMENT AND NUCLEAR APPLICATIONS

DONALD C. STEWART

Associate Director (Retired)
Chemistry Division
Argonne National Laboratory

A WILEY-INTERSCIENCE PUBLICATION

JOHN WILEY & SONS

New York • Chichester • Brisbane • Toronto • Singapore

Library of Congress Cataloging in Publication Data:

Stewart, Donald Charles, 1912–
Data for radioactive waste management and nuclear applications.

"A Wiley-Interscience publication."
Bibliography: p.
Includes index.
1. Radioactive wastes. 2. Ionizing radiation.
I. Title.

TD898.S75 1985 621.48′38 84-22910
ISBN 0-471-88627-0

Printed in the United States of America

10 9 8 7 6 5 4 3 2 1

PREFACE

At the time of my retirement from Argonne National Laboratory, I was offered a consultantship in connection with the radioactive wastes stored at the Western New York Nuclear Service Center in West Valley, New York. While carrying out this assignment, I found myself spending an inordinate amount of time digging out odd bits of data from the extensive and widely scattered nuclear literature. This book is a result of that experience; it is an effort to bring together in a single reference useful information on radioactive waste management so that others can avoid some of the frustrations I encountered. The material should be useful to responsible government officials, project managers, those preparing environmental impact statements and similar documents, and as a supplementary teaching aid. The range of information should also be valuable in other areas of nuclear application (e.g., neutron activation) besides radioactive waste management.

Anyone in this field soon recognizes that radioactive waste management is a two-significant-figure technology at best, since it is only rarely that all of the needed background information regarding a specific waste is known to better than a few percent. Data in this book are generally given to three or four significant figures, but with due appreciation for the fact that this is a definite gilding of the lily.

I have benefited from the wisdom and knowledge of colleagues and individuals in many organizations in preparing the book. Particular thanks go to Paul Fields, Kevin Flynn, Charles Luner, Ruth Sjoblom, Laverne Trevorrow, and Seymour Vogler of Argonne National Laboratory, Argonne, Illinois, M. E. Anderson and H. W. Kirby of the Mound Facility, Monsanto Research Corporation, Miamisburg, Ohio; Norman Holden, Sol Pearlstein, and Frances Scheffel of the Brookhaven Nuclear Data Center, Upton, New York; Betty McGill of the Radiation Shielding Information Center, Oak Ridge, Tennessee; B. T. Kenna of Sandia Laboratory, Albuquerque, New Mexico; and Richard Heckman of Lawrence Livermore Laboratory, Livermore, California. I also wish to thank Dr. Theodore P. Hoffman, Chemistry Editor for Wiley-Interscience, for his patience with an author who very badly overshot his manuscript deadline.

D. C. STEWART

Watsonville, California
January 1985

CONTENTS

PART 1. PHYSICAL DATA 1

Chapter 1. Nuclides of the Lighter Elements 3

1.1. The Fission Products 3
1.1.1. Decay Chains 3
1.1.2. Chain Yields 26
1.1.3. Fission-Product Properties 27
1.1.3.1. Useful Conversion Equations 40
1.2. Hardware Activation Products 42
1.3. Other Selected Isotopes 43
1.4. Miscellaneous Cross Sections and Other Data 43
1.4.1. Thermal and Resonance Integral *n*-Gamma Reaction Cross Sections 43
1.4.2. Reactor Cross Sections 51
1.4.3. Fast Neutron Reactions 51
1.4.4. The Photoneutron Reaction 55
1.5. Rules of Thumb 55

Chapter 2. Nuclides of the Heavier Elements 57

2.1. Buildup Chains 57
2.1.1. The Primary Buildup Chain 59

2.1.2. The ^{241}Am-^{242}Cm Secondary Loop 61
2.1.3. Neptunium-237 61
2.1.4. Plutonium-238 61
2.1.5. Uranium-233 and Uranium-232 61
2.1.6. Plutonium-236 62
2.2. Radiometric Properties of the Very Heavy Nuclides 62
2.2.1. Neutron Emission 74
2.3. Naturally Occurring Heavy Element Radioactivities 75
2.3.1. In High-Level Waste 75
2.3.2. The Four Chains 77
2.3.3. Nuclide Properties 79
2.3.3.1. Nuclear Cross Sections 80
2.3.3.2. Helium Production 81
2.3.4. Natural Uranium and Thorium 83
2.4. Californium-252 87
2.4.1. Cf-252–General Properties 87
2.4.2. Fission-Product Yields 88
2.4.3. Exposures and Doses 88
2.5. (Alpha, n) Neutron Sources 88

Chapter 3. ORIGEN: Calculation of Waste Compositions **94**

3.1. The Computational Problem 94
3.2. Buildup of Active Fission Products 97
3.3. Buildup of Inert Fission Products 97
3.4. Buildup of Actinides 103
3.5. Buildup of Activation Products 105
3.6. Activity Levels During Cooling 108

PART 2. CHEMICAL DATA **115**

Chapter 4. General Tables 117

4.1. The Elements 117
4.2. Standard Oxidation Potentials 117
4.3. Weight Conversion Factors 120
4.4. Properties of Selected Compounds 121
4.5. Solubility Constants 121
4.6. Data for Process Chemicals 134
4.6.1. Common Acids and Bases 134
4.6.2. Extractants and Solvents 134
4.6.3. Ion Exchangers 136

PART 3. RADIOACTIVE WASTES 139

Chapter 5. High-Level Liquid Wastes 141

5.1. Types and Compositions of HLLW 141
5.2. Solidification of HLLW 145
5.2.1. Comparison of Forms 148
5.2.2. Product Descriptions 148
5.2.3. Calcines 158
5.2.4. Glasses 162
5.3. Leachability 174
5.4. Element Volatility 179

Chapter 6. Non-High-Level Wastes 186

6.1. Background and Literature 186
6.2. Low-Level Wastes (LLW) 186
6.2.1. 10CFR20 Regulations 191
6.3. TRU Wastes 191
6.4. Intermediate-Level Wastes 193
6.5. Immobilization of Non-HLW 194

Chapter 7. Packaged Radioactive Wastes 199

7.1. Centerline and Surface Temperatures 199
7.2. Canister Midpoint Exposure Rates 201
7.3. Waste Container Sizes 209
7.4. Corrosion of Canister Materials 209

Chapter 8. Repository Data 210

8.1. Background 210
8.2. Host Rock Properties 212
8.2.1. Permeability and Porosity 212
8.2.2. Thermal Properties 212
8.2.3. Shear Strength 215
8.2.4. Specific Rock Types 215
8.2.4.1. Basalts 215
8.2.4.2. Granites 215
8.2.4.3. Volcanic Tuffs 216
8.3. Nuclide Migration Through Rock 221

PART 4. DATA FOR OPERATIONS **229**

Chapter 9. Shielding **231**

9.1. The Shielding Literature 231
9.2. Shielding Terminology 231
9.3. Shielding Against Alphas and Betas 232
9.4. Shielding Against Gamma Radiation 235
9.5. Shielding Against Neutrons 237
9.6. Shielding Materials 243

Chapter 10. Health Physics **244**

10.1. Background and Literature 244
10.2. Health Physics Terms 245
10.2.1. SI Units 246
10.3. Protection Standards 248

Chapter 11. Radiation Damage **254**

11.1. Background and Literature 254
11.2. Oils and Lubricants 255
11.3. Elastomers and Plastics 256
11.4. Electronic Components 258

Chapter 12. Nuclear Criticality **260**

12.1. Background and Literature 260
12.2. Criticality Terminology 260
12.3. Critical Mass Data 262

Chapter 13. Decontamination **264**

13.1. Background and Literature 264
13.2. Decontamination Procedures 264

PART 5. MISCELLANEOUS DATA **271**

Chapter 14. General Information **273**

14.1. Radioactive Decay 273
14.2. Neutron Activation 276
14.3. Conversions and Equalities 277
14.4. Miscellaneous Data 280

References **281**

Index **293**

DATA FOR RADIOACTIVE WASTE MANAGEMENT AND NUCLEAR APPLICATIONS

PART ONE

PHYSICAL DATA

Bismuth-209 (100% natural abundance) is the heaviest stable nuclide found in nature, everything heavier is radioactive. An arbitrary decision was thus made here to consider all elements of atomic number 83 (Bi) and below as "light," and 84 (Po) and above as "heavy." Chapter 1 accordingly deals with the physical properties of the light nuclides, as defined, and is chiefly concerned with the fission products and with reactor hardware activation products of interest to waste management.

Elements heavier than uranium, the last of the "natural" elements, are most readily produced in quantity by successive neutron captures in a reactor. These buildup chains are given in Chapter 2, as are data on the radiometric properties of the isotopes thus produced.

The interval of the periodic table between lead (Z, 82) and uranium (Z, 92) includes most of the radioactive species found in nature. This region is covered in Section 2.3 since the naturally occurring radioactive decay chains can become part of the waste-management problem in some situations.

Section 2.4 presents data relative to Californium-252. This spontaneously fissioning isotope is a compact and highly useful source of neutrons, and has waste-management implications. Section 2.5 relates to alpha-*n* reactions.

ONE

NUCLIDES OF THE LIGHTER ELEMENTS

1.1 THE FISSION PRODUCTS

1.1.1 Decay Chains

Table 1.1 is a simplified presentation of the radioactive decay chains produced by neutron fission of heavy element targets. The data are taken from Rose and Burrows as given in Report BNL-NCS-5045.[1] This two-volume document is in turn derived from the ENDF/B-IV files maintained by the National Nuclear Data Center at Brookhaven National Laboratory, acting as coordinator for the Nuclear Data Network, a group of American nuclear research centers. (ENDF is the abbreviation for Evaluated Nuclear Data Files.)

COMMENTS ON TABLE 1.1

(a) The format used has been modified from that given in BNL-NCS-5045 in the interest of conserving space, and the full chains as given in that report are truncated in all save the tritium case for the same reason. The first members in the other chains are invariably very short lived and thus of no practical interest to waste-management specialists. [Of the several hundred nuclides thus ignored, only ^{83}Se(70s), ^{111}Rh(63s), ^{113}Pd(90s), ^{131}Sn(63s) and ^{147}Ce(70s) have half-

TABLE 1.1 Simplified Fission Product Decay Chains

Mass Number	Decay Chain(s)	
3	$^{3}H \xrightarrow{12.33y} {}^{3}He(St.)$	
72	$Pr.(3) \longrightarrow {}^{72}Zn \xrightarrow{46.5h} {}^{72}Ga \xrightarrow{14.1h} {}^{72}Ge(St.)$	
73	$Pr.(3) \longrightarrow {}^{73}Zn \xrightarrow{23.5s} {}^{73}Ga \xrightarrow{4.88h} {}^{73m}Ge \xrightarrow{0.53s} {}^{73}Ge(St.)$	
74	$Pr.(3) \longrightarrow {}^{74}Zn \xrightarrow{98s} {}^{74}Ga \xrightarrow{8.2m} {}^{74}Ge(St.)$	
75	$Pr.(3) \longrightarrow {}^{75}Zn \xrightarrow{9.0s} {}^{75}Ga\ (1.9m)$; $^{75}Ga \xrightarrow{4\%} {}^{75m}Ge \xrightarrow{49s} {}^{75}Ge$; $^{75}Ga \xrightarrow{96\%} {}^{75}Ge \xrightarrow{82.8m} {}^{75}As(St.)$	
76	$Pr.(3) \longrightarrow {}^{76}Ga \xrightarrow{27s} {}^{76}Ge(St.)$ $^{76}As \xrightarrow{26.3h} {}^{76}Se(St.)$	(Two chains)

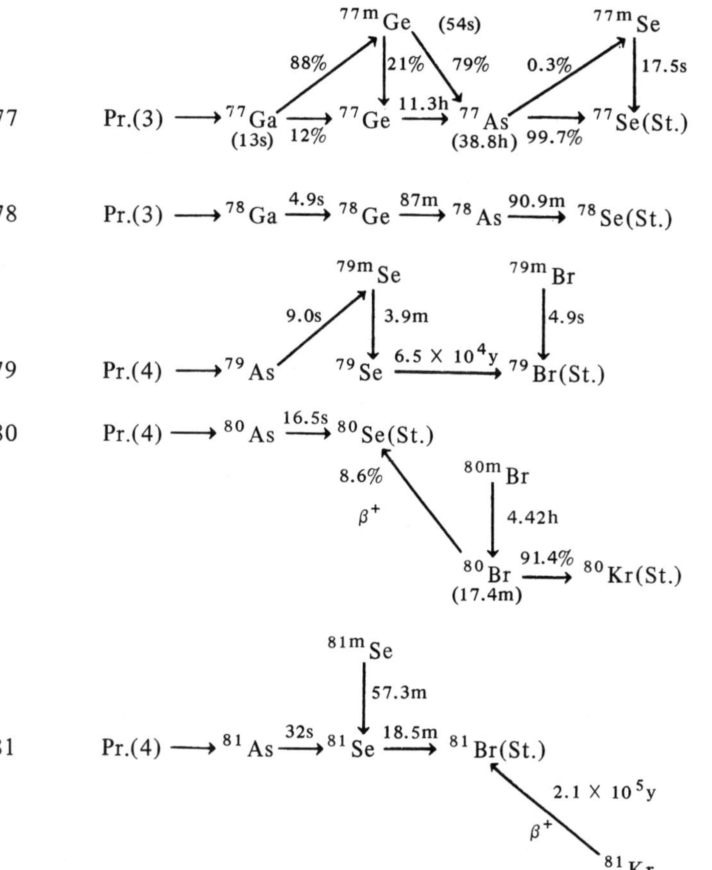

77 Pr.(3) → ^{77}Ga (13s) → 12% → ^{77}Ge → 11.3h → ^{77}As (38.8h) → 99.7% → ^{77}Se(St.)
^{77}Ga → 88% → ^{77m}Ge (54s)
^{77m}Ge → 21% → ^{77}Ge
^{77m}Ge → 79% → ^{77}As
^{77}As → 0.3% → ^{77m}Se
^{77m}Se → 17.5s → ^{77}Se(St.)
78 Pr.(3) → ^{78}Ga → 4.9s → ^{78}Ge → 87m → ^{78}As → 90.9m → ^{78}Se(St.)
79 Pr.(4) → ^{79}As → 9.0s → ^{79m}Se
^{79m}Se → 3.9m → ^{79}Se → 6.5×10^4y → ^{79}Br(St.)
^{79m}Br → 4.9s → ^{79}Br(St.)
80 Pr.(4) → ^{80}As → 16.5s → ^{80}Se(St.)
^{80m}Br → 4.42h → ^{80}Br (17.4m)
^{80}Br → 91.4% → ^{80}Kr(St.)
^{80}Br → 8.6% β^+ → ^{80}Se(St.)
81 Pr.(4) → ^{81}As → 32s → ^{81}Se → 18.5m → ^{81}Br(St.)
^{81m}Se → 57.3m → ^{81}Se
^{81}Kr → β^+ 2.1×10^5y → ^{81}Br(St.)

TABLE 1.1 (Continued)

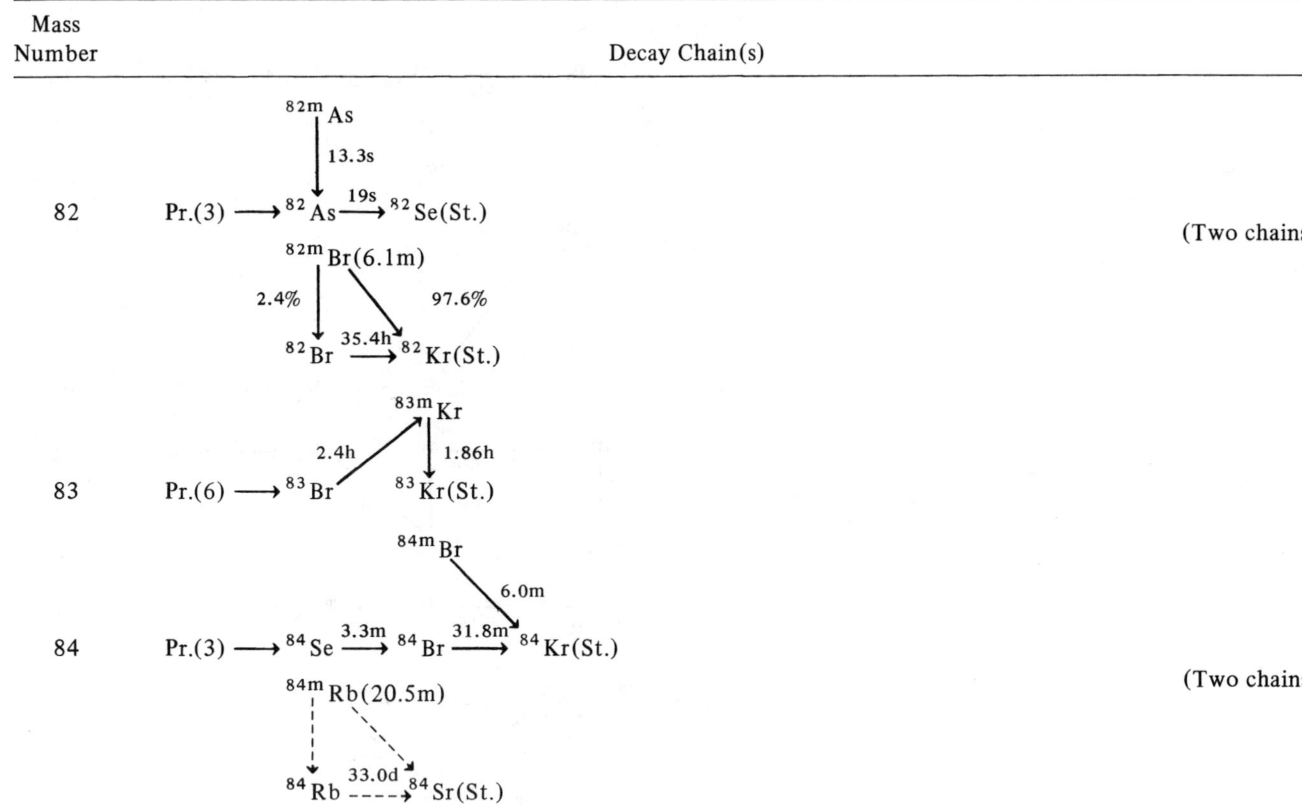

Mass Number	Decay Chain(s)	
82	^{82m}As $\xrightarrow{13.3s}$ ^{82}As; Pr.(3) $\longrightarrow$ ^{82}As $\xrightarrow{19s}$ ^{82}Se(St.); ^{82m}Br(6.1m) $\xrightarrow{2.4\%}$ ^{82}Br $\xrightarrow{35.4h}$ ^{82}Kr(St.); ^{82m}Br(6.1m) $\xrightarrow{97.6\%}$ ^{82}Kr(St.)	(Two chains)
83	Pr.(6) $\longrightarrow$ ^{83}Br $\xrightarrow{2.4h}$ ^{83m}Kr $\xrightarrow{1.86h}$ ^{83}Kr(St.)	
84	^{84m}Br $\xrightarrow{6.0m}$ ^{84}Kr(St.); Pr.(3) $\longrightarrow$ ^{84}Se $\xrightarrow{3.3m}$ ^{84}Br $\xrightarrow{31.8m}$ ^{84}Kr(St.); ^{84m}Rb(20.5m) $\dashrightarrow$ ^{84}Rb $\xrightarrow{33.0d}$ ^{84}Sr(St.); ^{84m}Rb(20.5m) $\dashrightarrow$ ^{84}Sr(St.)	(Two chains)

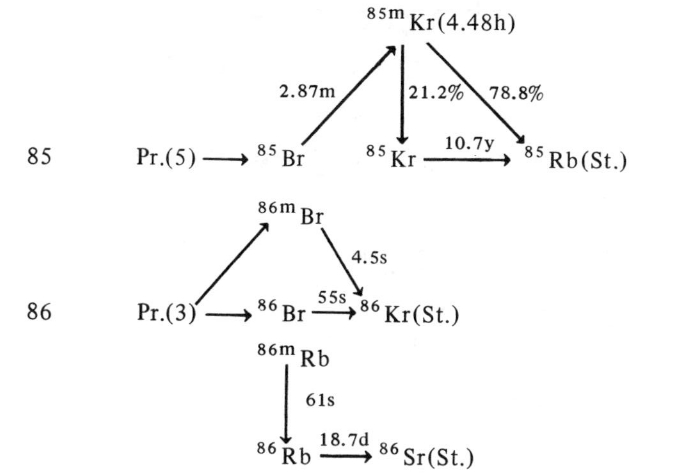

85 $\text{Pr.}(5) \longrightarrow {}^{85}\text{Br} \xrightarrow{2.87\text{m}} {}^{85\text{m}}\text{Kr}(4.48\text{h})$; ${}^{85\text{m}}\text{Kr} \xrightarrow{21.2\%} {}^{85}\text{Kr} \xrightarrow{10.7\text{y}} {}^{85}\text{Rb(St.)}$; ${}^{85\text{m}}\text{Kr} \xrightarrow{78.8\%} {}^{85}\text{Rb(St.)}$

86 $\text{Pr.}(3) \longrightarrow {}^{86}\text{Br} \xrightarrow{55\text{s}} {}^{86}\text{Kr(St.)}$; $\text{Pr.}(3) \longrightarrow {}^{86\text{m}}\text{Br} \xrightarrow{4.5\text{s}} {}^{86}\text{Kr(St.)}$

${}^{86\text{m}}\text{Rb} \xrightarrow{61\text{s}} {}^{86}\text{Rb} \xrightarrow{18.7\text{d}} {}^{86}\text{Sr(St.)}$

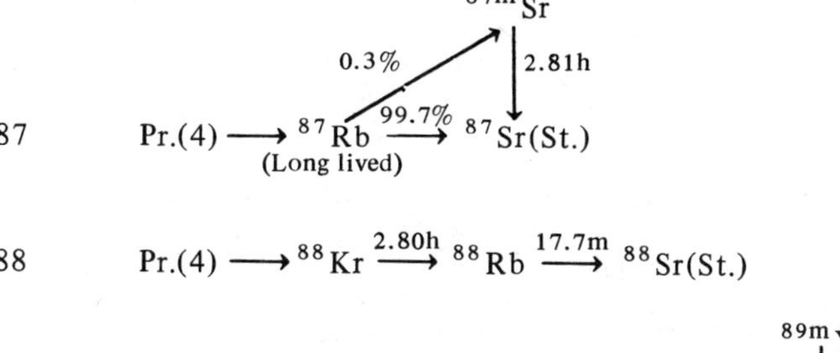

87 $\text{Pr.}(4) \longrightarrow {}^{87}\text{Rb} \xrightarrow{99.7\%} {}^{87}\text{Sr(St.)}$; ${}^{87}\text{Rb} \xrightarrow{0.3\%} {}^{87\text{m}}\text{Sr} \xrightarrow{2.81\text{h}} {}^{87}\text{Sr(St.)}$

(Long lived)

88 $\text{Pr.}(4) \longrightarrow {}^{88}\text{Kr} \xrightarrow{2.80\text{h}} {}^{88}\text{Rb} \xrightarrow{17.7\text{m}} {}^{88}\text{Sr(St.)}$

89 $\text{Pr.}(3) \longrightarrow {}^{89}\text{Kr} \xrightarrow{3.16\text{m}} {}^{89}\text{Rb} \xrightarrow{15.2\text{m}} {}^{89}\text{Sr} \xrightarrow{52.0\text{d}} {}^{89}\text{Y(St.)}$; ${}^{89\text{m}}\text{Y} \xrightarrow{15.7\text{s}} {}^{89}\text{Y(St.)}$

(Two chains)

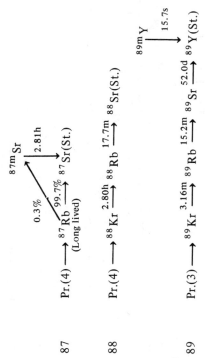

TABLE 1.1 (Continued)

Mass Number	Decay Chain(s)	
90	$\text{Pr.}(4) \longrightarrow {}^{90m}\text{Rb}(4.28\text{m})$; ${}^{90m}\text{Rb} \xrightarrow{5\%} {}^{90}\text{Rb}$; ${}^{90m}\text{Rb} \xrightarrow{95\%} {}^{90}\text{Sr}$ $\text{Pr.}(4) \longrightarrow {}^{90}\text{Rb} \xrightarrow{2.70\text{m}} {}^{90}\text{Sr} \xrightarrow{28.1\text{y}} {}^{90}\text{Y} \xrightarrow{64.1\text{h}} {}^{90}\text{Zr(St.)}$ ${}^{90}\text{Sr} \longrightarrow {}^{90m}\text{Y}(3.1\text{h})$; ${}^{90m}\text{Y} \xrightarrow{99.6\%} {}^{90}\text{Y}$; ${}^{90m}\text{Y} \xrightarrow{0.4\%} {}^{90m}\text{Zr} \xrightarrow{0.83\text{s}} {}^{90}\text{Zr(St.)}$	
91	$\text{Pr.}(3) \longrightarrow {}^{91}\text{Rb} \xrightarrow{58\text{s}} {}^{91}\text{Sr}\,(9.48\text{h}) \xrightarrow{42.4\%} {}^{91}\text{Y} \xrightarrow{58.6\text{d}} {}^{91}\text{Zr(St.)}$ ${}^{91}\text{Sr} \xrightarrow{57.6\%} {}^{91m}\text{Y} \xrightarrow{49.7\text{m}} {}^{91}\text{Y}$	
92	$\text{Pr.}(4) \longrightarrow {}^{92}\text{Sr} \xrightarrow{2.71\text{h}} {}^{92}\text{Y} \xrightarrow{3.53\text{h}} {}^{92}\text{Zr(St.)}$	
93	$\text{Pr.}(4) \dashrightarrow {}^{93}\text{Sr} \xrightarrow{7.5\text{m}} {}^{93}\text{Y} \xrightarrow{10.2\text{h}} {}^{93}\text{Zr}\,(9.5 \times 10^5\text{y}) \xrightarrow{5\%} {}^{93}\text{Nb(St.)}$ ${}^{93m}\text{Y} \xrightarrow{7.3\text{m}} {}^{93}\text{Y}$ ${}^{93}\text{Zr} \xrightarrow{95\%} {}^{93m}\text{Nb} \xrightarrow{12.0\text{y}} {}^{93}\text{Nb(St.)}$	
94	$\text{Pr.}(3) \longrightarrow {}^{94}\text{Sr} \xrightarrow{76\text{s}} {}^{94}\text{Y} \xrightarrow{19\text{m}} {}^{94}\text{Zr(St.)}$ ${}^{94m}\text{Nb}(6.3\text{m}) \xrightarrow{98.8\%} {}^{94}\text{Nb} \xrightarrow{2 \times 10^4\text{y}} {}^{94}\text{Mo(St.)}$ ${}^{94m}\text{Nb} \xrightarrow{1.2\%} {}^{94}\text{Mo(St.)}$	(Two chains)

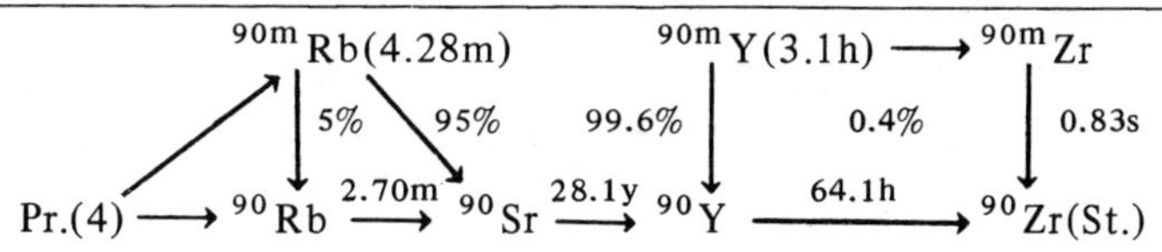

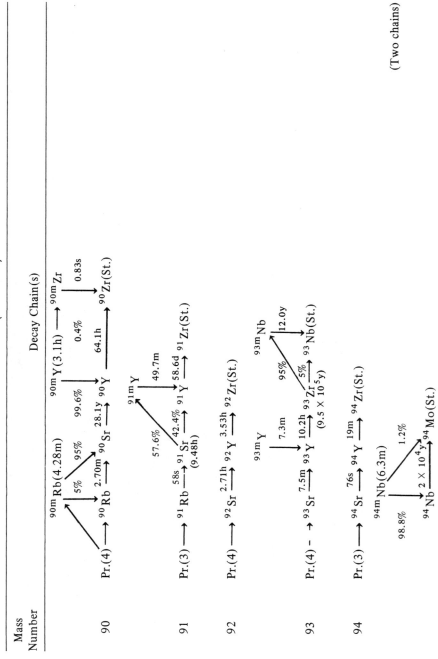

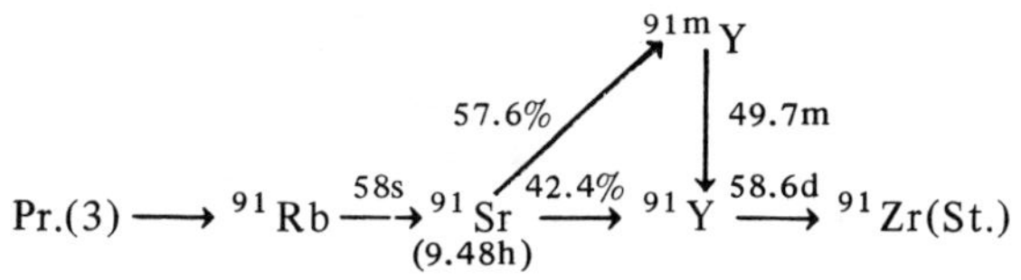

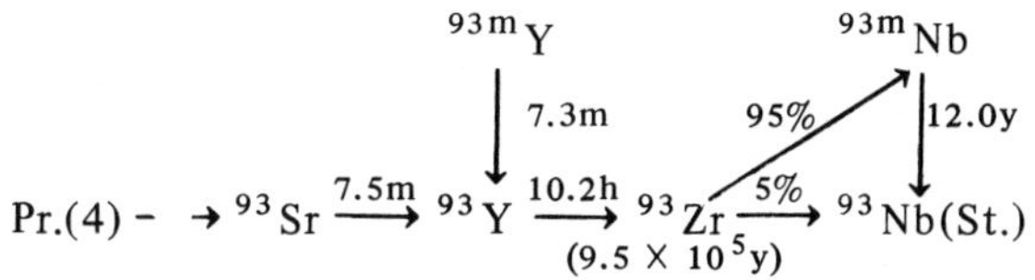

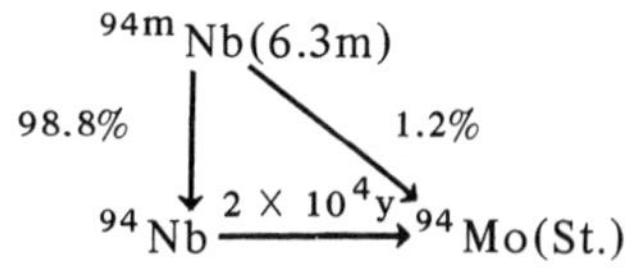

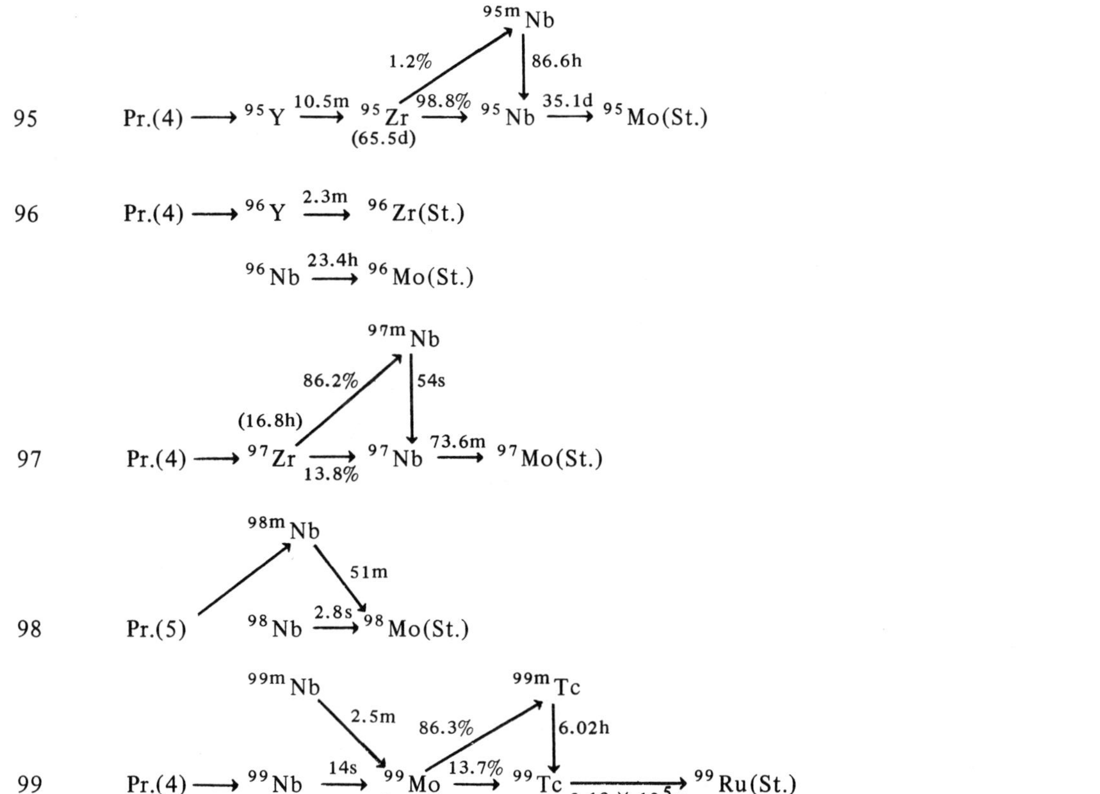
95
Pr.(4)
^{95}Y
10.5m
^{95}Zr
(65.5d)
1.2%
98.8%
^{95m}Nb
86.6h
^{95}Nb
35.1d
^{95}Mo(St.)
96
Pr.(4)
^{96}Y
2.3m
^{96}Zr(St.)
^{96}Nb
23.4h
^{96}Mo(St.)
(Two chains)
97
Pr.(4)
^{97}Zr
(16.8h)
86.2%
13.8%
^{97m}Nb
54s
^{97}Nb
73.6m
^{97}Mo(St.)
98
Pr.(5)
^{98m}Nb
51m
^{98}Nb
2.8s
^{98}Mo(St.)
99
Pr.(4)
^{99}Nb
14s
^{99m}Nb
2.5m
^{99}Mo
(66.0h)
86.3%
13.7%
^{99m}Tc
6.02h
^{99}Tc
2.13 × 10^{5}y
^{99}Ru(St.)

TABLE 1.1 (Continued)

Mass Number	Decay Chain(s)	
100	Pr.(6) $\longrightarrow {}^{100}Mo(St.)$	(Two chains)
	${}^{100}Tc \xrightarrow{16s} {}^{100}Ru(St.)$	
101	Pr.(5) $\longrightarrow {}^{101}Mo \xrightarrow{14.6m} {}^{101}Tc \xrightarrow{14.2m} {}^{101}Ru(St.)$	
102	Pr.(4) $\longrightarrow {}^{102}Mo \xrightarrow{11.1m} {}^{102}Tc \xrightarrow{5.3s} {}^{102}Ru(St.)$; ${}^{102m}Tc \xrightarrow{4.3m} {}^{102}Ru(St.)$	
103	Pr.(6) $\longrightarrow {}^{103}Ru \xrightarrow{39.6d} {}^{103m}Rh \xrightarrow{56.0m} {}^{103}Rh(St.)$	
104	Pr.(4) $\longrightarrow {}^{104}Mo \xrightarrow{96s} {}^{104}Tc \xrightarrow{18.0m} {}^{104}Ru(St.)$; ${}^{104m}Rh(4.35m) \xrightarrow{99.8\%} {}^{104}Rh$ (42s); ${}^{104m}Rh(4.35m) \xrightarrow{0.2\%} {}^{104}Pd(St.)$; ${}^{104}Rh \xrightarrow{99.9\%} {}^{104}Pd(St.)$; ${}^{104}Rh \xrightarrow{\beta^+,\ 0.1\%} {}^{104}Ru(St.)$	

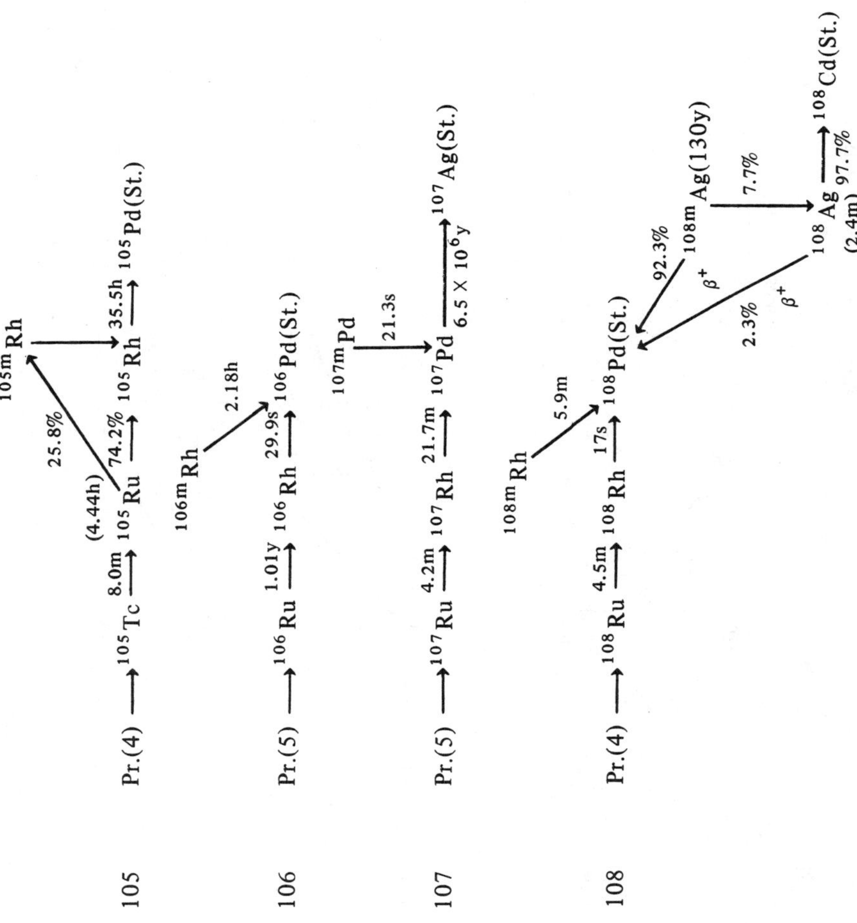
105 Pr.(4) ⟶ 105Tc 8.0m ⟶ 105Ru (4.44h) 74.2% ⟶ 105Rh 35.5h ⟶ 105Pd(St.)
25.8%
105mRh
106 Pr.(5) ⟶ 106Ru 1.01y ⟶ 106Rh 29.9s ⟶ 106Pd(St.)
106mRh 2.18h
107 Pr.(5) ⟶ 107Ru 4.2m ⟶ 107Rh 21.7m ⟶ 107Pd 6.5 × 10^6y ⟶ 107Ag(St.)
107mPd 21.3s
108 Pr.(4) ⟶ 108Ru 4.5m ⟶ 108Rh 17s ⟶ 108Pd(St.)
108mRh 5.9m
108mAg(130y)
92.3%
β+
7.7%
108Ag (2.4m) 97.7% ⟶ 108Cd(St.)
2.3%
β+

TABLE 1.1 (Continued)

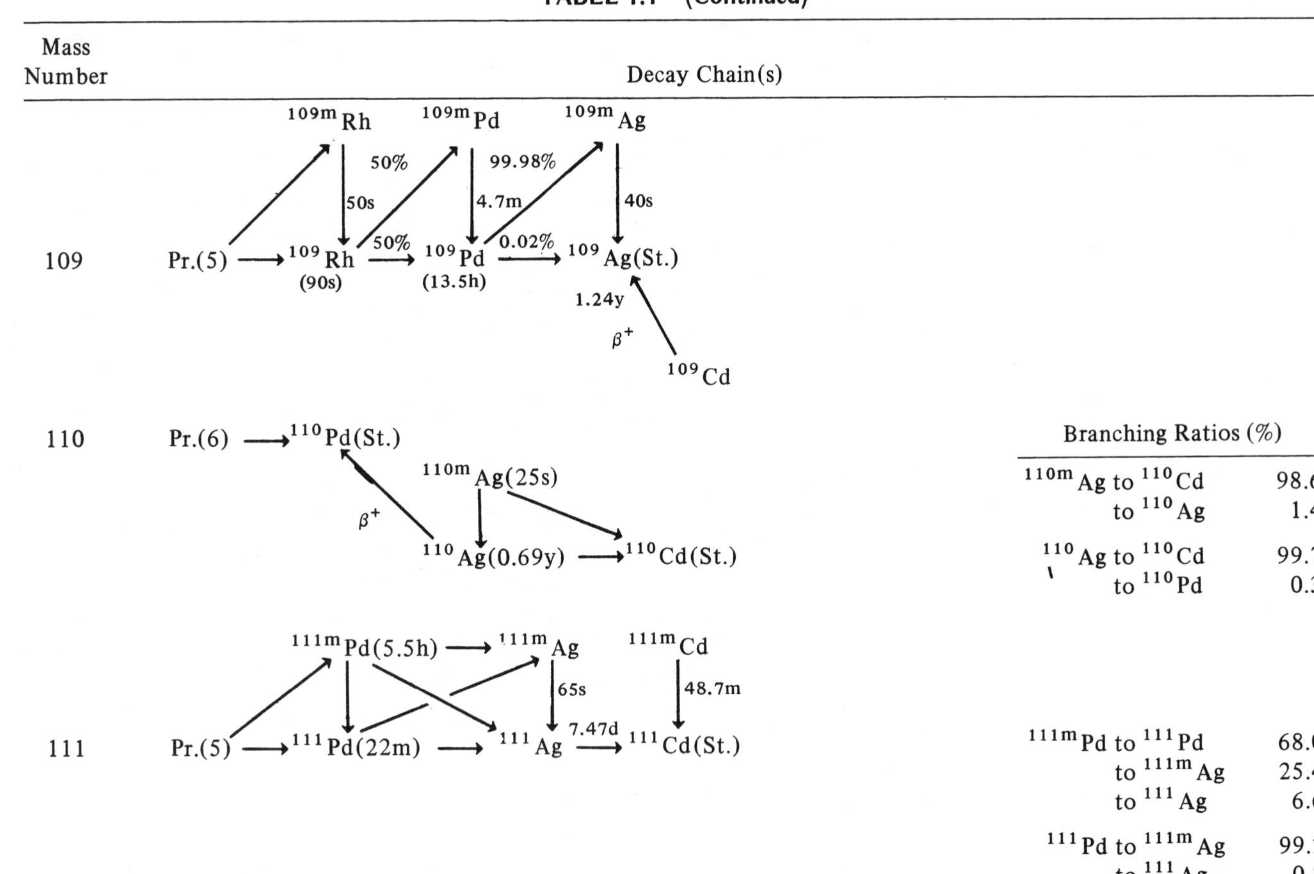

Branching Ratios (%)	
^{110m}Ag to ^{110}Cd	98.6
to ^{110}Ag	1.4
^{110}Ag to ^{110}Cd	99.7
to ^{110}Pd	0.3
^{111m}Pd to ^{111}Pd	68.0
to ^{111m}Ag	25.4
to ^{111}Ag	6.6
^{111}Pd to ^{111m}Ag	99.2
to ^{111}Ag	0.8

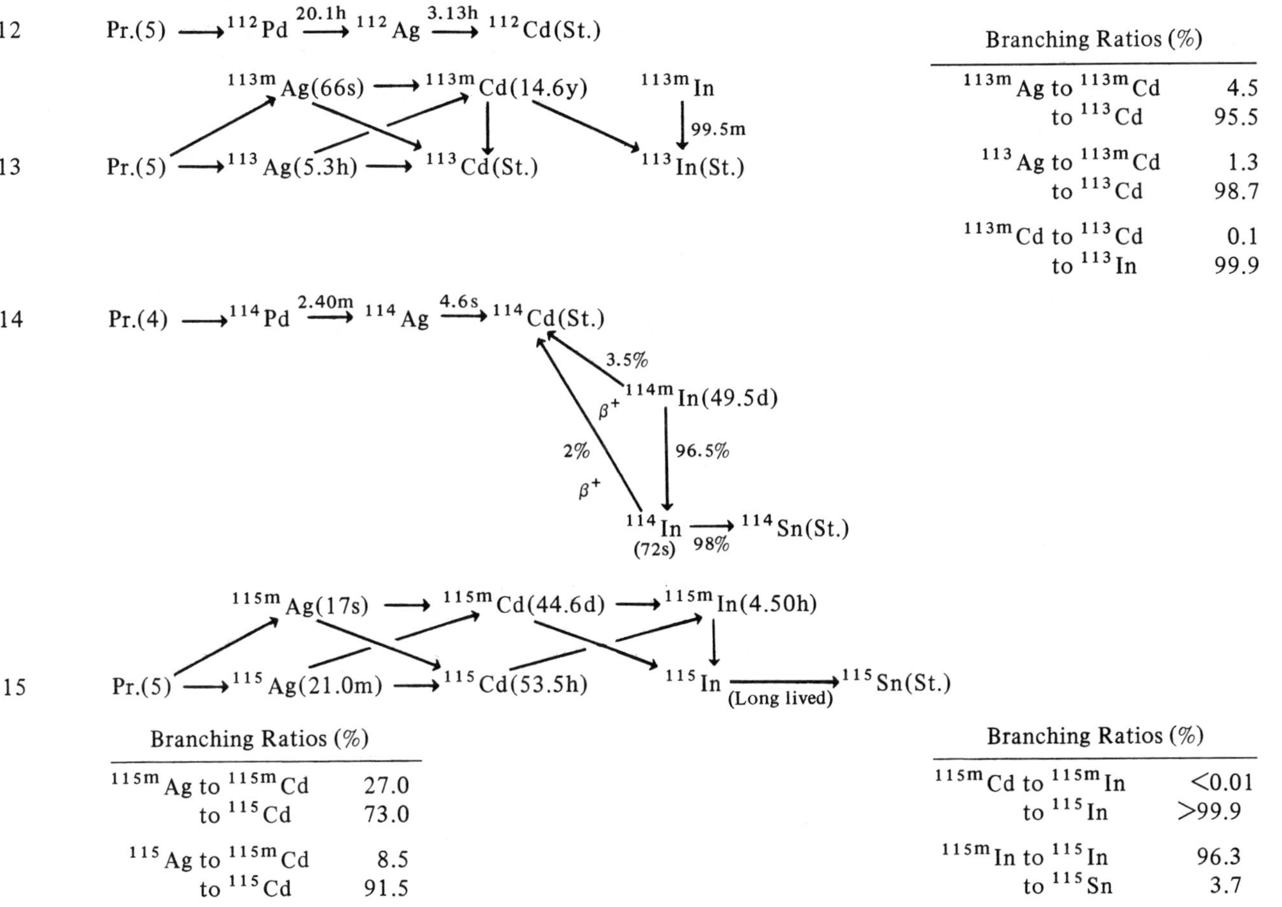

Branching Ratios (%)	
^{113m}Ag to ^{113m}Cd	4.5
to ^{113}Cd	95.5
^{113}Ag to ^{113m}Cd	1.3
to ^{113}Cd	98.7
^{113m}Cd to ^{113}Cd	0.1
to ^{113}In	99.9

Branching Ratios (%)	
^{115m}Ag to ^{115m}Cd	27.0
to ^{115}Cd	73.0
^{115}Ag to ^{115m}Cd	8.5
to ^{115}Cd	91.5

Branching Ratios (%)	
^{115m}Cd to ^{115m}In	<0.01
to ^{115}In	>99.9
^{115m}In to ^{115}In	96.3
to ^{115}Sn	3.7

TABLE 1.1 (Continued)

Mass Number	Decay Chain(s)
116	(Two chains)
117	

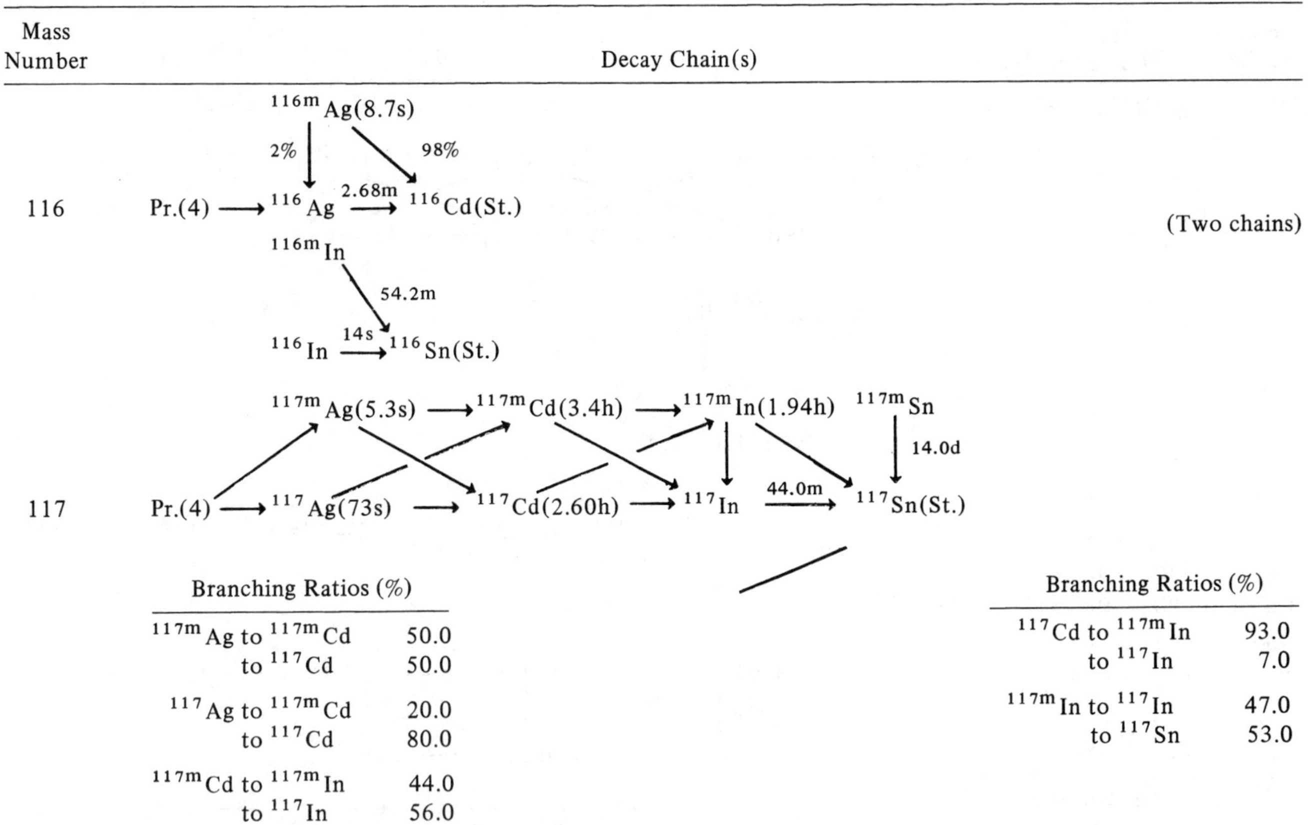

Branching Ratios (%)	
^{117m}Ag to ^{117m}Cd	50.0
to ^{117}Cd	50.0
^{117}Ag to ^{117m}Cd	20.0
to ^{117}Cd	80.0
^{117m}Cd to ^{117m}In	44.0
to ^{117}In	56.0

Branching Ratios (%)	
^{117}Cd to ^{117m}In	93.0
to ^{117}In	7.0
^{117m}In to ^{117}In	47.0
to ^{117}Sn	53.0

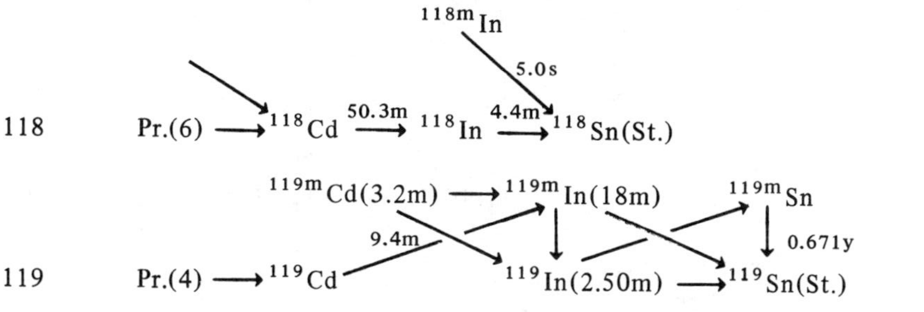

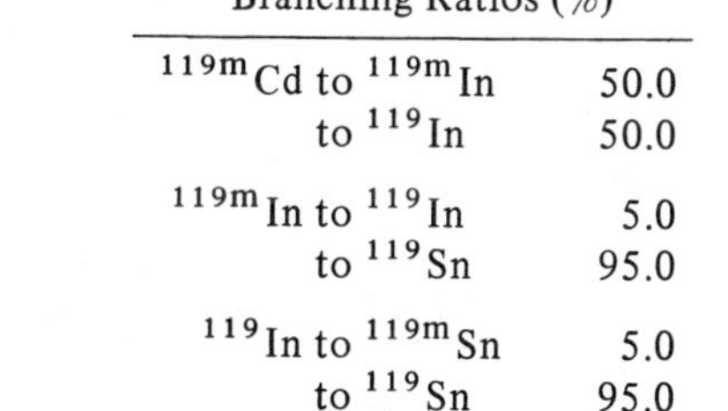

Branching Ratios (%)	
^{119m}Cd to ^{119m}In	50.0
to ^{119}In	50.0
^{119m}In to ^{119}In	5.0
to ^{119}Sn	95.0
^{119}In to ^{119m}Sn	5.0
to ^{119}Sn	95.0

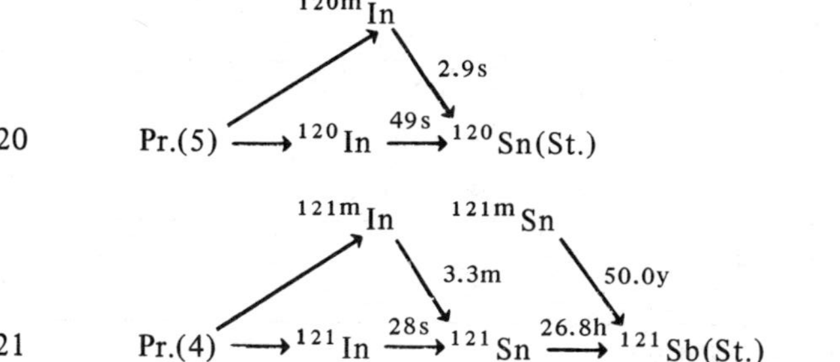

TABLE 1.1 (Continued)

Mass Number	Decay Chain(s)
122	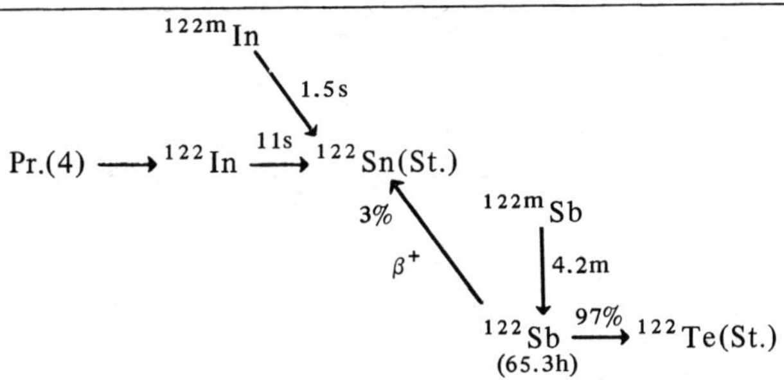 Pr.(4) ⟶ $^{122}In \xrightarrow{11s} {}^{122}Sn(St.)$; $^{122m}In \xrightarrow{1.5s} {}^{122}Sn(St.)$; $^{122m}Sb \xrightarrow{4.2m} {}^{122}Sb$ (65.3h); $^{122}Sb \xrightarrow{97\%} {}^{122}Te(St.)$; $^{122}Sb \xrightarrow{3\%\ \beta^+} {}^{122}Sn(St.)$
123	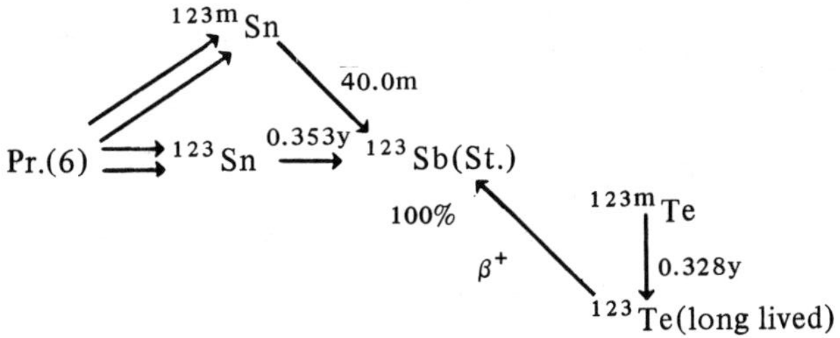 Pr.(6) ⟶ $^{123m}Sn \xrightarrow{40.0m} {}^{123}Sb(St.)$; Pr.(6) ⟶ $^{123}Sn \xrightarrow{0.353y} {}^{123}Sb(St.)$; $^{123m}Te \xrightarrow{0.328y} {}^{123}Te$(long lived); $^{123}Te \xrightarrow{100\%\ \beta^+} {}^{123}Sb(St.)$
124	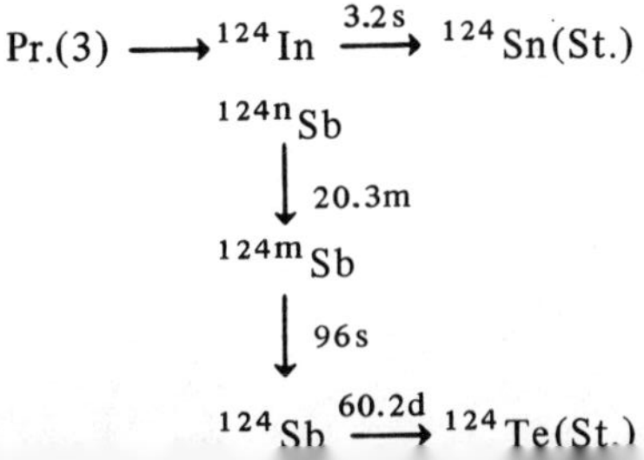 Pr.(3) ⟶ $^{124}In \xrightarrow{3.2s} {}^{124}Sn(St.)$; $^{124n}Sb \xrightarrow{20.3m} {}^{124m}Sb \xrightarrow{96s} {}^{124}Sb \xrightarrow{60.2d} {}^{124}Te(St.)$ (Two chains)

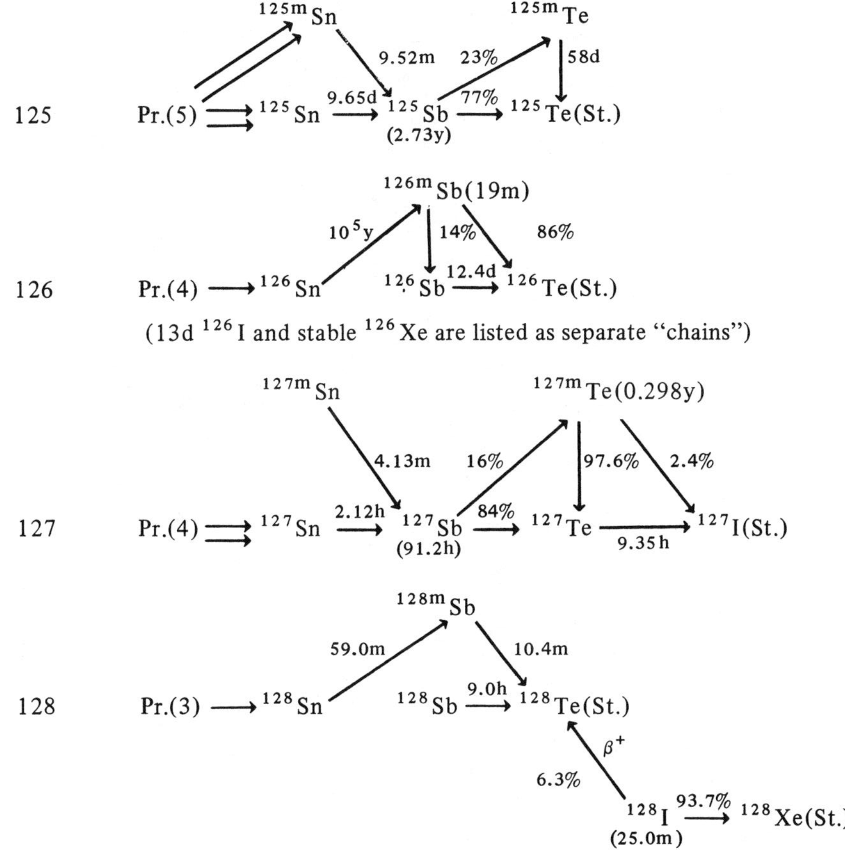

125
Pr.(5)
125mSn
125Sn
9.52m
9.65d
125Sb
(2.73y)
23%
77%
125mTe
58d
125Te(St.)
126
Pr.(4)
126Sn
10^5y
126mSb(19m)
14%
86%
126Sb
12.4d
126Te(St.)
(13d 126I and stable 126Xe are listed as separate "chains")
127
Pr.(4)
127mSn
127Sn
4.13m
2.12h
127Sb
(91.2h)
16%
84%
127mTe(0.298y)
97.6%
2.4%
127Te
9.35h
127I(St.)
128
Pr.(3)
128Sn
59.0m
128mSb
10.4m
128Sb
9.0h
128Te(St.)
β+
6.3%
128I
(25.0m)
93.7%
128Xe(St.)

TABLE 1.1 (Continued)

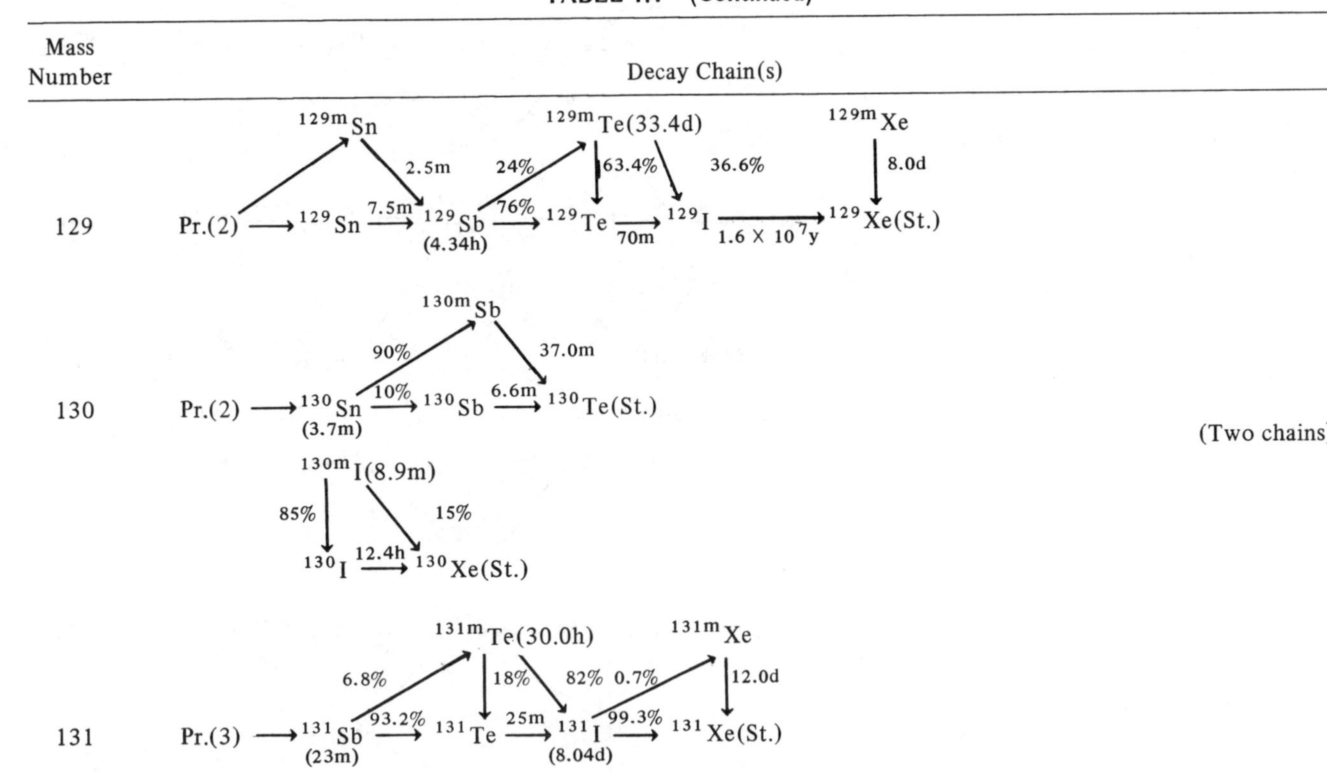

132

133

134

135

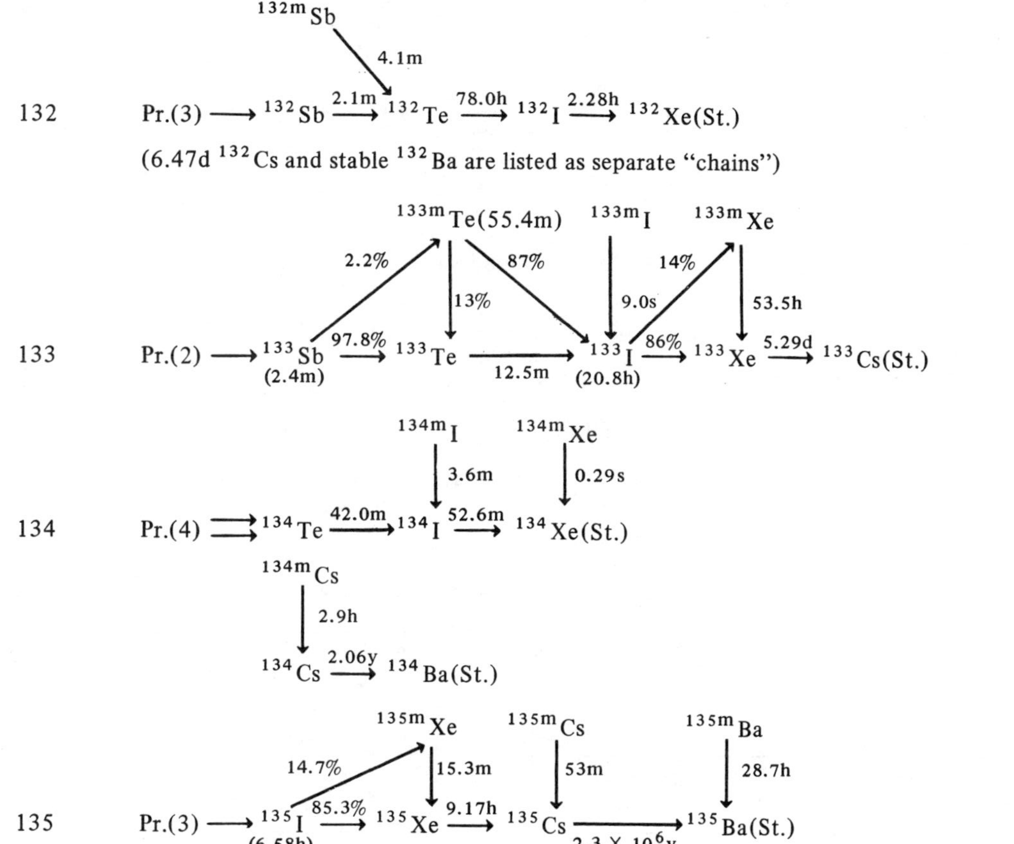

(Two chains)

TABLE 1.1 (Continued)

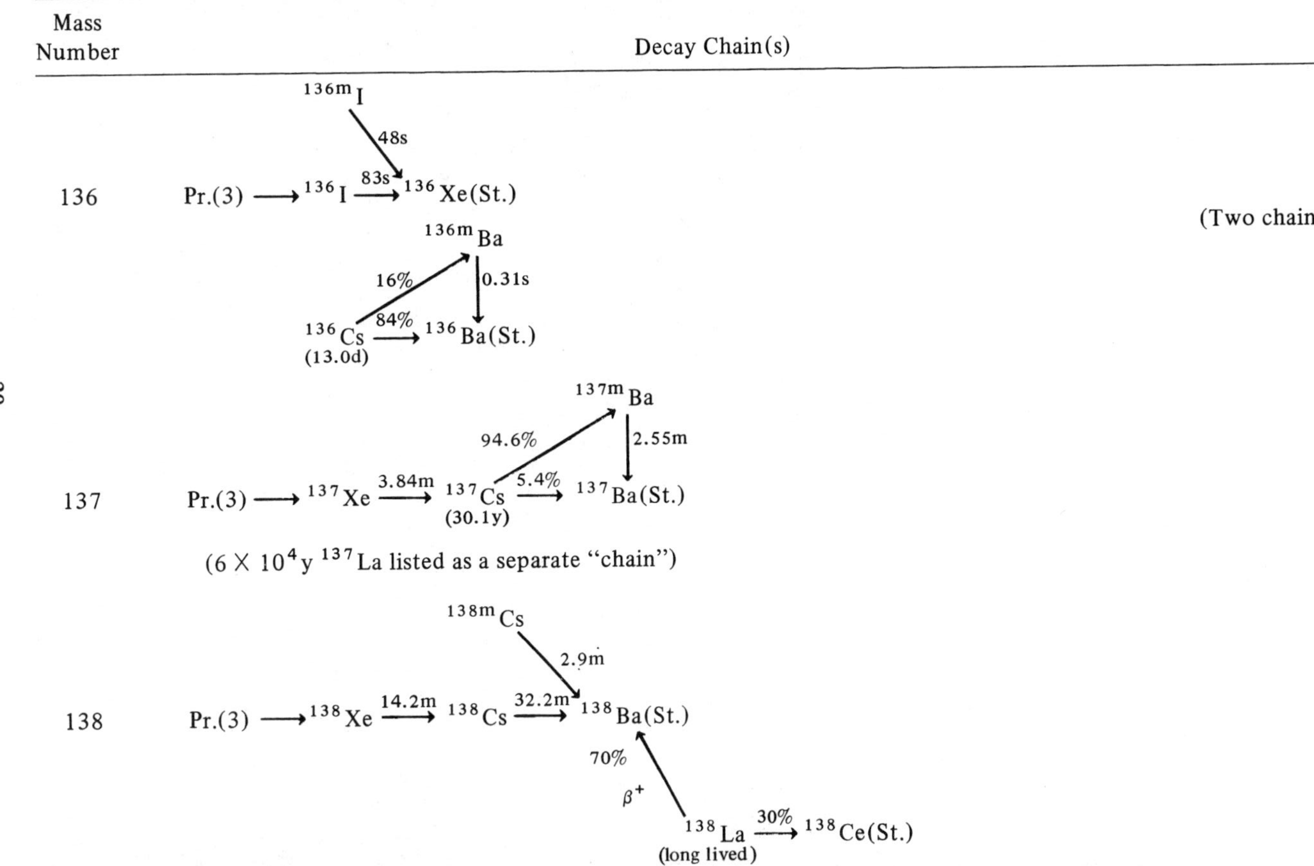

139 $\quad \text{Pr.}(4) \longrightarrow {}^{139}\text{Cs} \xrightarrow{9.3\text{m}} {}^{139}\text{Ba} \xrightarrow{83.3\text{m}} {}^{139}\text{La(St.)}$

140 $\quad \text{Pr.}(3) \longrightarrow {}^{140}\text{Cs} \xrightarrow{64\text{s}} {}^{140}\text{Ba} \xrightarrow{12.8\text{d}} {}^{140}\text{La} \xrightarrow{40.2\text{h}} {}^{140}\text{Ce(St.)}$

141 $\quad \text{Pr.}(4) \longrightarrow {}^{141}\text{Ba} \xrightarrow{18.3\text{m}} {}^{141}\text{La} \xrightarrow{3.87\text{h}} {}^{141}\text{Ce} \xrightarrow{32.5\text{d}} {}^{141}\text{Nd(St.)}$

142 $\quad \text{Pr.}(4) \longrightarrow {}^{142}\text{Ba} \xrightarrow{10.7\text{m}} {}^{142}\text{La} \xrightarrow{92.4\text{m}} {}^{142}\text{Ce (long lived)} \xrightarrow{\alpha} {}^{138}\text{Ba(St.)}$

$${}^{142\text{m}}\text{Pr} \xrightarrow{14.6\text{m}} {}^{142}\text{Pr} \xrightarrow[19.2\text{h}]{} {}^{142}\text{Nd(St.)}$$

(Two chains)

143 $\quad \text{Pr.}(4) \longrightarrow {}^{143}\text{La} \xrightarrow{14.0\text{m}} {}^{143}\text{Ce} \xrightarrow{33.0\text{h}} {}^{143}\text{Pr} \xrightarrow{13.6\text{d}} {}^{143}\text{Nd(St.)}$

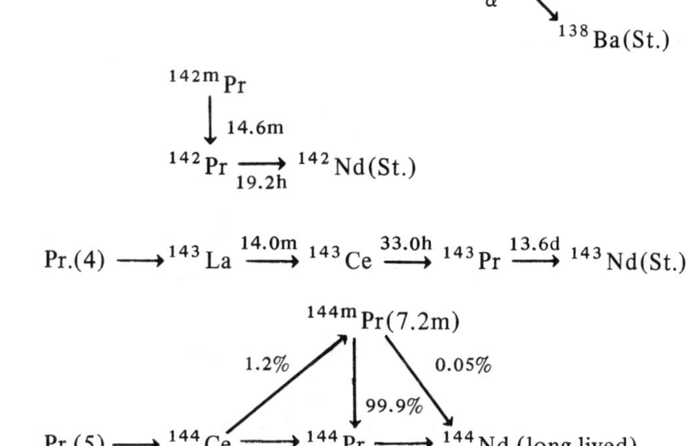

TABLE 1.1 (Continued)

Mass Number	Decay Chain(s)
145	Pr.(5) $\longrightarrow {}^{145}\text{Ce} \xrightarrow{3.3\text{m}} {}^{145}\text{Pr} \xrightarrow{5.98\text{h}} {}^{145}\text{Nd(St.)}$
146	Pr.(4) $\longrightarrow {}^{146}\text{Ce} \xrightarrow{14.2\text{m}} {}^{146}\text{Pr} \xrightarrow{24.2\text{m}} {}^{146}\text{Nd(St.)}$
147	Pr.(5) $\longrightarrow {}^{147}\text{Pr} \xrightarrow{12.0\text{m}} {}^{147}\text{Nd} \xrightarrow{11.0\text{d}} {}^{147}\text{Pm} \xrightarrow{2.62\text{y}} {}^{147}\text{Sm}$ (long lived) $\xrightarrow{\alpha} {}^{143}\text{Nd(St.)}$
148	Pr.(4) $\longrightarrow {}^{148}\text{Pr} \xrightarrow{2.0\text{m}} {}^{148}\text{Nd(St.)}$ ${}^{148m}\text{Pm}$(41.3d) $\xrightarrow{6\%} {}^{148}\text{Pm} \xrightarrow{5.37\text{d}} {}^{148}\text{Sm}$ (long lived) $\xrightarrow{\alpha} {}^{144}\text{Nd(St.)}$ ${}^{148m}\text{Pm}$(41.3d) $\xrightarrow{94\%} {}^{148}\text{Sm}$ (long lived) (Two chains)
149	Pr.(4) $\longrightarrow {}^{149}\text{Pr} \xrightarrow{2.3\text{m}} {}^{149}\text{Nd} \xrightarrow{1.73\text{h}} {}^{149}\text{Pm} \xrightarrow{53.1\text{h}} {}^{149}\text{Sm}$ (long lived) $\xrightarrow{\alpha} {}^{145}\text{Nd(St.)}$

(Two chains)

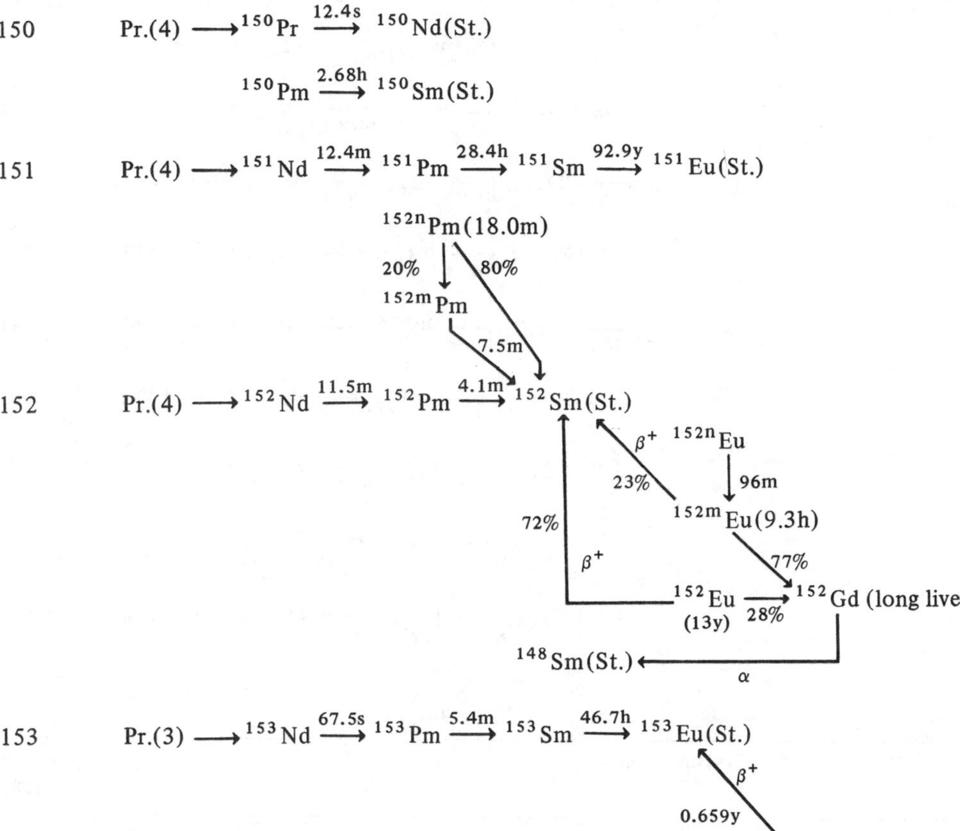
150
Pr.(4) ⟶ ^{150}Pr $\xrightarrow{12.4s}$ ^{150}Nd(St.)
^{150}Pm $\xrightarrow{2.68h}$ ^{150}Sm(St.)
151
Pr.(4) ⟶ ^{151}Nd $\xrightarrow{12.4m}$ ^{151}Pm $\xrightarrow{28.4h}$ ^{151}Sm $\xrightarrow{92.9y}$ ^{151}Eu(St.)
^{152n}Pm(18.0m)
20%
80%
^{152m}Pm
7.5m
152
Pr.(4) ⟶ ^{152}Nd $\xrightarrow{11.5m}$ ^{152}Pm $\xrightarrow{4.1m}$ ^{152}Sm(St.)
β^+
^{152n}Eu
23%
96m
^{152m}Eu(9.3h)
72%
77%
β^+
^{152}Eu
(13y)
28%
^{152}Gd (long lived)
^{148}Sm(St.)
α
153
Pr.(3) ⟶ ^{153}Nd $\xrightarrow{67.5s}$ ^{153}Pm $\xrightarrow{5.4m}$ ^{153}Sm $\xrightarrow{46.7h}$ ^{153}Eu(St.)
β^+
0.659y
^{153}Gd

TABLE 1.1 (Continued)

Mass Number	Decay Chain(s)	
154	$^{154m}Pm(1.8m)$ → (10%) ^{154}Pm; → (90%) $^{154}Sm(St.)$	
	Pr.(3) $\longrightarrow$ $^{154}Nd \xrightarrow{7.73h} {}^{154}Pm \xrightarrow[2.8m]{} {}^{154}Sm(St.)$	(Two chains)
	$^{154}Eu \xrightarrow{8.59y} {}^{154}Gd(St.)$	
155	Pr.(5) $\longrightarrow$ $^{155}Sm \xrightarrow{22.2m} {}^{155}Eu \xrightarrow{4.8y} {}^{155}Gd(St.)$	
156	Pr.(4) $\longrightarrow$ $^{156}Sm \xrightarrow{9.4h} {}^{156}Eu \xrightarrow{15.2d} {}^{156}Gd(St.)$	
157	Pr.(3) $\longrightarrow$ $^{157}Pm \xrightarrow{68s} {}^{157}Sm \xrightarrow{8.0m} {}^{157}Eu \xrightarrow{15.2h} {}^{157}Gd(St.)$	
158	Pr.(3) $\longrightarrow$ $^{158}Sm \xrightarrow{44m} {}^{158}Eu \xrightarrow{45.9m} {}^{158}Gd(St.)$	
159	Pr.(3) $\longrightarrow$ $^{159}Sm \xrightarrow{2.7m} {}^{159}Eu \xrightarrow{18.1m} {}^{159}Gd \xrightarrow{18.6h} {}^{159}Tb(St.)$	
160	Pr.(2) $\longrightarrow$ $^{160}Sm \xrightarrow{5.82m} {}^{160}Eu \xrightarrow{51s} {}^{160}Gd(St.)$	(Two chains)
	$^{160}Tb \xrightarrow{72.3d} {}^{160}Dy(St.)$	
161	Pr.(4) $\longrightarrow$ $^{161}Gd \xrightarrow{3.7m} {}^{161}Tb \xrightarrow{6.92d} {}^{161}Dy(St.)$	

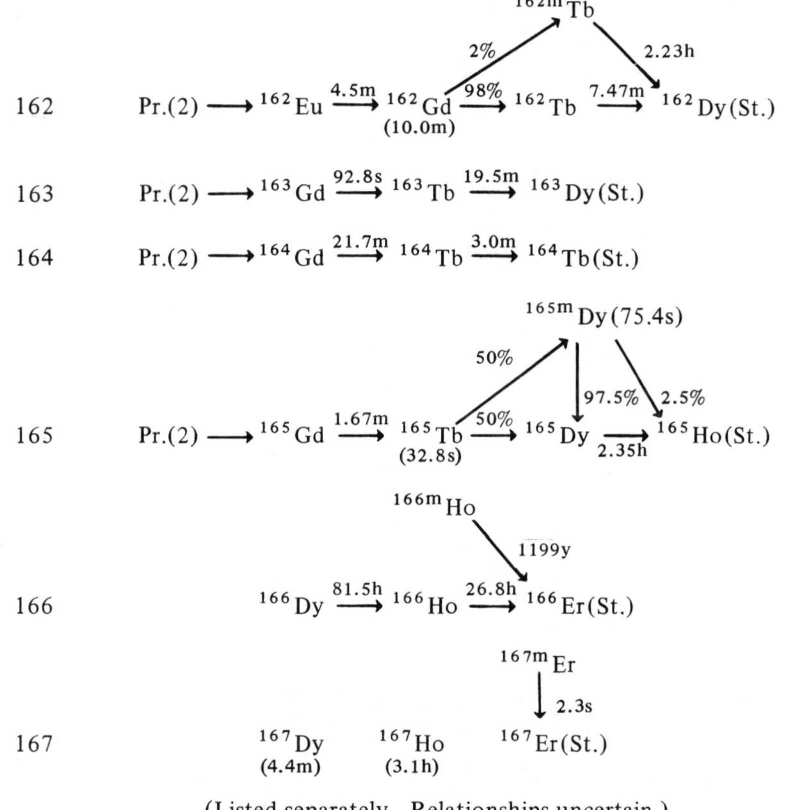

SOURCE: Rose and Burrows, Report BNL-NCS-5045, Reference 1.

lives in excess of one minute.] The existence of these precursors is indicated for each chain by "Pr(n)," where "(n)" gives the number of short-lived species not shown.

(b) Two separate decay chains are produced at certain mass numbers (76, 82, 84, etc.) in fission. Both chains are shown in Table 1.1. Similarly, ^{88}Y(0.292y) is produced directly as a second "chain," as is stable ^{99}Ru, but with no further data given in BNL-NCS-5045.

(c) Some of the very short-lived precursors emit neutrons, particularly at low mass numbers. These "delayed neutron" emitters are not shown in the table. Some of the heavier fission products also have unimportant branching decays to extremely long-lived, low-energy alpha emitters. These decays are shown in essentially all cases, but not in later tables dealing with fission-product properties because of the very low yields and long half-lives (10^{11} to 10^{16}y). These nuclides, for all practical purposes, can be considered stable and make no contribution to the waste-management problem.

1.1.2 Chain Yields

Fission-product yields can be considered in several ways. The yield of a single, particularly accessible isotope (in terms of ease of measurement) can be experimentally determined, but this one datum may ignore the fact that other nuclides earlier or later in the chain may also be directly produced. The occurrence of two chains at the same mass number introduces another complication. "Cumulative" yields for a particular mass number are therefore the numbers usually given, that is, the number of atoms of a particular mass ending up eventually as stable end products of the chain (or chains) per 100 heavy atoms fissioned. (Since each fission produces two new atoms, the summation of all the cumulative yields totals 200 rather than 100.) Cumulative fission yields are given here in Tables 1.2–1.4. The values presented are primarily from the review by Flynn and Glendenin,[2] save for the thermal fission of ^{233}U, ^{235}U, ^{239}Pu and ^{241}Pu, taken from Walker[3] as being a slightly more recent compilation. (Comparable figures in the two reports are actually only slightly different.) A number of other surveys have been published, an incomplete listing being references 4–8.

Comments on Tables 1.2–1.4

(a) Table 1.2 lists fission-product yields for the four major fissile nuclides (^{233}U, ^{235}U, ^{239}Pu, and ^{241}Pu), where cross sections for fission by thermal neutrons are relatively high. The fissiles are also "fissionables," that is, they will fission with relatively low cross sections with fast neutrons, as will all the actinide nuclides. While all of the group are thus fissionables, most are not fissiles, particularly the even-even isotopes (both the atomic mass and the atomic number are even numbers), where cross sections for thermal neutron fission are extremely small. These even-even nuclides are considered in Table 1.3 for the few cases where sufficient data are available.

(b) A "thermal" heading in these yield tables indicates that the fission is caused by slow neutrons. "Fission" shows yields resulting from an unmoderated neutron spectrum as produced in a reactor (that is, fast neutrons). The "14 MeV" heading refers to fission caused by the monoenergetic flux produced by the $^3H(d, n)^4He$ reaction in acclerators. The "8 MeV" data are for yields obtained utilizing the (α, n) reaction on lithium to produce neutrons. Since these last data as well as the 14 MeV numbers for certain targets, are very sparsely available, they are condensed into Table 1.4.

(c) Published yield data are still incomplete to varying degrees for all of the systems that have been studied. Numbers in parentheses in Tables 1.2 and 1.3 have not been experimentally determined, but estimated, primarily on the basis of smooth fission-product yield curves or from theoretical considerations. Flynn and Glendenin[2] estimate that the experimental values they quote are accurate to within 5–20%, the extrapolated numbers being open to error to the extent of 50% either way. Those authors present fission yield versus mass number curves in graphical form for most of the systems covered in Tables 1.2–1.4.

1.1.3 Fission-Product Properties

An identical format was used for the three tables below giving the physical properties of the lighter radioactive nuclides of interest in waste management. Table 1.5 deals with fission products, Table 1.6 with the major radioactive species produced by activation of elements in the irradiated hardware usually associated with reactor fuels, and Table 1.7 with a selected group of isotopes frequently encountered in research and in medical and industrial applications.

(a). The half-lives used in Table 1.5 are from the ENDF/IV-B files as given in Reference 1 and reproduced in Table 1.1. The half-lives shown in Tables 1.6 and 1.7 are generally from the compilations by Heath[9a, 9b] in the CRC *Rubber Handbook*, checked in several ambiguous cases by consultation with Peterson's[10a] listing in *Lange's Handbook of Chemistry*, the "Table of Isotopes,"[11] or Nuclear Wallet Cards.[12]

(b) The "Major Emissions" columns in the tables are meant only as quick indications of the activities to be expected, since in many cases there may be as many as 10–20 other gamma rays emitted, mostly at low intensities, and there may be more than one group of betas. If additional information is needed in some situations, the most readily accessible references are 9a, 9b, 10a, and 11.

TABLE 1.2 Fission Product Yields—Major Fissiles

Mass Number	^{233}U Thermal	^{233}U Fission	^{235}U Thermal	^{235}U Fission	^{235}U 14 MeV	^{239}Pu Thermal	^{239}Pu Fission	^{241}Pu Thermal
72			1.6×10^{-5}		0.0063	0.00011		
73			0.00010		(0.010)			
74			0.00035		(0.016)			
75	(0.0022)		0.0008		(0.028)			
76	(0.0068)		(0.0025)		(0.046)			
77	0.0083		0.20	(0.012)	(0.078)	0.0073		
78	0.020	(0.12)	0.60	(0.025)	(0.13)	0.025		0.0082
79	0.056	(0.195)	0.16	(0.050)	(0.24)	0.050	(0.090)	0.016
80	0.120	(0.35)	0.26	(0.097)	(0.39)	0.12	(0.135)	0.033
81	0.20	(0.55)	0.34	(0.195)	(0.59)	0.18	(0.19)	0.065
82	0.33	(0.82)	0.60	(0.315)	(0.84)	0.22	(0.265)	0.120
83	0.535	(1.2)	1.00	0.615	1.23	0.295	0.366	0.202
84	0.986	(1.75)	1.66	1.07	(1.5)	0.477	0.559	0.360
85	1.33	(2.35)	2.18	1.49	(1.90)	0.558	0.672	0.392
86	1.96	(3.1)	2.80	1.93	(2.35)	0.758	0.882	0.608
87	2.53	(3.95)	3.98	2.66	(2.85)	0.970	1.16	0.750
88	3.59	(4.95)	5.53	3.63	(3.4)	1.37	1.44	0.966
89	4.74	6.30	6.30	5.0	4.16	1.74	1.8	1.20
90	5.82	(6.6)	5.81	5.24	4.5	2.11	2.24	1.55
91	5.95	6.61	6.49	6.1	4.71	2.54	2.58	1.84
92	5.98	(6.7)	6.67	(6.2)	(5.05)	3.06	3.13	2.26
93	7.05	(6.7)	6.41	(6.4)	5.4	3.92	3.91	2.93
94	6.75	(6.6)	6.43	(6.5)	(5.45)	4.45	4.39	3.37

95	6.19	(6.4)	6.53	6.47	4.80	4.98	4.78	3.98
96	5.66	(6.2)	6.30	(6.6)	(5.6)	5.12	5.11	4.39
97	5.36	(5.8)	6.07	6.55	5.29	5.58	5.47	4.73
98	5.10	(5.4)	5.81	6.04	(5.45)	5.81	5.81	5.2
99	5.01	4.75	6.14	5.9	5.25	6.10	5.88	6.2
100	4.36	(3.7)	6.31	6.35	(4.85)	7.00	6.76	6.2
101	3.21	(2.6)	5.07	5.46	(4.5)	6.04	6.88	5.91
102	2.44	(1.55)	4.19	4.65	(4.0)	6.15	6.97	6.29
103	1.8	0.413	3.05	3.5	3.35	5.94	5.85	6.65
104	1.03	(0.23)	1.83	2.35	(2.6)	6.10	6.77	6.77
105	0.53	(0.17)	0.95	1.50	2.09	5.47	(5.77)	6.75
106	0.253	0.16	0.39	0.97	1.76	4.45	4.7	6.05
107	0.130	(0.12)	0.16	(0.50)	(1.55)	3.5	(3.65)	5.3
108	0.070	(0.105)	0.07	(0.17)	(1.4)	2.3	(2.65)	4.0
109	0.047	(0.09)	0.03	0.146	1.17	1.3	1.14	2.5
110	0.029	(0.08)	0.022	(0.078)	(1.21)	0.65	(0.85)	1.2
111	0.023	0.0837	0.018	0.0456	1.11	0.28	0.38	0.55
112	0.015	(0.067)	0.014	0.039	0.93	0.11	0.207	0.20
113	0.014	(0.064)	0.012	0.0342	1.15	0.076	0.133	0.153
114	0.014	(0.062)	0.011	0.0342	(1.03)	0.049	0.099	0.075
115	0.017	0.056	0.0104	0.032	0.97	0.038	0.095	0.040
116	0.014	(0.061)	0.0105	0.0359	(1.00)	0.036	0.064	0.030
117	0.014	0.060	0.0105	(0.032)	(1.01)	0.035	(0.062)	0.026
118	0.0145	0.060	0.0105	(0.032)	(1.02)	0.035	(0.061)	0.025
119	0.015	0.074	0.0105	(0.033)	(1.05)	0.035	(0.061)	0.025
120	0.016	0.083	0.011	(0.034)	(1.08)	0.035	(0.062)	0.025
121	0.018	(0.081)	0.013	(0.036)	1.09	0.038	(0.064)	0.025
122	0.025	0.083	0.013	(0.040)	(1.19)	0.038	(0.071)	0.025
123	0.037	(0.108)	0.016	(0.045)	(1.27)	0.044	(0.082)	0.027

TABLE 1.2 (Continued)

Mass Number	^{233}U Thermal	^{233}U Fission	^{235}U Thermal	^{235}U Fission	^{235}U 14 MeV	^{239}Pu Thermal	^{239}Pu Fission	^{241}Pu Thermal
124	0.060	0.120	0.020	(0.056)	(1.38)	0.055	(0.115)	0.031
125	0.116	(0.165)	0.029	0.073	1.45	0.100	0.19	0.042
126	0.26	0.286	0.053	(0.20)	(1.7)	0.20	(0.33)	0.080
127	0.62	(0.33)	0.127	(0.38)	2.09	0.45	(0.56)	0.17
128	1.00	(0.55)	0.34	(0.71)	(2.5)	0.85	(0.93)	0.37
129	1.70	1.57	0.83	(1.40)	(2.40)	1.50	1.17	0.80
130	2.50	(2.4)	1.7	(2.4)	(3.95)	2.50	(2.7)	1.70
131	3.53	(3.4)	2.80	3.17	4.23	3.73	4.06	3.12
132	4.82	4.36	4.17	4.45	4.70	5.21	5.42	4.64
133	5.99	(5.3)	6.79	6.69	5.6	6.92	6.91	6.72
134	6.14	(5.8)	7.61	7.09	5.9	7.41	7.35	8.08
135	6.21	(6.1)	6.60	6.54	5.7	7.69	7.54	7.06
136	6.88	(6.4)	6.13	5.93	(5.7)	6.47	6.92	7.30
137	6.76	6.28	6.24	6.25	5.9	6.72	6.58	6.50
138	5.84	(6.6)	6.76	6.60	(5.5)	5.74	4.97	6.71
139	6.41	(6.6)	6.53	(6.45)	5.0	5.14	(6.0)	6.3
140	6.39	6.31	6.36	6.21	4.61	5.62	5.59	5.91
141	6.62	6.74	5.87	6.3	3.8	5.27	5.15	4.98
142	6.60	(6.3)	5.96	5.82	4.25	5.02	4.95	4.84
143	5.85	5.0	5.95	5.80	3.81	4.53	4.30	4.52
144	4.62	4.2	5.43	5.15	3.20	3.81	3.68	4.18
145	3.38	(3.1)	3.93	3.85	2.50	3.06	3.05	3.22

146	2.55	(2.3)	2.98	3.00	1.98	2.53	2.53	2.72
147	1.70	1.60	2.26	2.3	1.64	2.16	2.0	2.20
148	1.30	(1.15)	1.68	1.75	(1.16)	1.69	1.69	1.92
149	0.765	0.82	1.08	1.16	(0.85)	1.30	1.36	1.44
150	0.508	(0.73)	0.650	0.832	(0.62)	0.989	1.01	1.17
151	0.314	0.35	0.419	0.438	(0.45)	0.814	0.839	0.882
152	0.213	(0.22)	0.268	0.309	(0.31)	0.619	0.683	0.697
153	0.105	0.12	0.167	0.21	0.22	0.38	0.48	0.522
154	0.0456		0.0743	0.098	(0.153)	0.286	0.324	0.378
155	0.023		0.0321	(0.050)	(0.105)	0.17	(0.21)	0.231
156	0.0176	0.018	0.0132	0.025	0.08	0.120	(0.14)	0.167
157	0.0065	0.0105	0.0061		(0.038)	0.076	(0.09)	0.130
158	(0.0024)		0.0031		(0.023)	0.041		0.090
159	0.00091	0.0018	0.0010	0.0034	0.0127	0.021		0.046
160					(0.0076)	(0.0094)		0.20
161	0.00012	0.00049	0.000082	0.00046	0.0056	0.0039		

SOURCE: Data adopted from References 2 and 3.

TABLE 1.3 Fission Product Yields—Even-Even Nuclides

Mass Number	^{232}Th		^{238}U	
	Fission	14 MeV	Fission	14 MeV
72	0.00033	0.0070		0.003
73	0.00045	0.008		0.0053
74		(0.015)		(0.007)
75		(0.035)		(0.012)
76		(0.075)	(0.0011)	(0.021)
77	0.014	0.13	(0.0040)	0.030
78	(0.035)	0.31	(0.010)	0.041
79	(0.085)	0.95	(0.031)	0.18
80	(0.17)	(0.90)	(0.087)	(0.20)
81	(0.38)	1.21	(0.17)	0.33
82	(0.83)	(1.6)	(0.32)	(0.56)
83	2.06	1.69	0.40	0.75
84	3.78	2.41	0.85	1.26
85	4.01	(3.0)	0.83	1.12
86	6.21	(3.55)	1.38	1.76
87	6.57	(4.2)	(1.93)	(2.0)
88	6.92	(4.9)	(2.36)	(2.4)
89	6.7	5.7	3.3	2.66
90	7.2	(5.8)	3.2	3.35
91	7.45	5.88	3.50	3.66
92	6.6	(5.9)	(4.45)	(4.0)
93	7.24	5.78	4.77	4.4
94	(6.55)	(6.3)	(5.35)	(4.7)
95	5.43	6.7	5.8	5.11
96	(5.35)	(5.4)	(6.0)	(5.2)
97	4.52	2.93	(6.2)	5.27
98	(3.75)	(2.8)	(6.45)	(5.5)
99	2.86	1.92	6.56	5.50
100	(1.35)	(1.5)	(6.55)	(5.7)
101	(0.67)	1.60	(6.35)	5.90
102	(0.33)	0.70	(6.1)	3.65
103	0.146	0.75	5.8	5.15
104	(0.085)	(0.9)	(4.9)	(4.3)
105	0.05	1.06	3.81	3.23
106	0.062	1.07	2.8	2.40
107	(0.057)	(1.0)	1.36	(1.7)
108	(0.054)	(1.05)	(0.68)	(1.4)
109	0.052	1.10	0.26	1.20
110	(0.050)	(1.15)	(0.165)	(1.1)
111	0.082	1.42	0.080	1.00
112	0.090	1.29	0.07	0.81

TABLE 1.3 (Continued)

Mass Number	^{232}Th		^{238}U	
	Fission	14 MeV	Fission	14 MeV
113	0.045	1.23	(0.053)	0.89
114	(0.045)	(1.3)	(0.048)	(0.86)
115	0.065	(1.26)	0.045	0.77
116	(0.045)	(1.24)	(0.042)	(0.82)
117	0.053	(1.19)	(0.041)	(0.80)
118	(0.046)	(1.24)	(0.040)	(0.80)
119	(0.047)	(1.19)	(0.041)	(0.80)
120	(0.049)	(1.14)	(0.042)	(0.81)
121	(0.050)	1.0	(0.043)	0.81
122	(0.052)	(1.02)	(0.045)	(0.86)
123	0.029	(0.95)	(0.048)	(0.89)
124	(0.058)	(0.92)	(0.053)	(0.94)
125	0.026	0.58	0.074	0.84
126	(0.071)	(0.90)	(0.081)	(1.18)
127	0.110	1.21	0.14	1.43
128	(0.205)	(0.92)	(0.34)	(1.40)
129	(0.42)	0.93	0.329	1.30
130	(0.85)	(1.45)	(1.45)	(2.45)
131	1.56	2.44	3.40	4.02
132	2.76	2.68	4.98	4.94
133	3.75	(4.35)	6.00	6.08
134	5.18	(4.75)	7.00	6.50
135	4.66	(5.15)	6.23	5.89
136	5.44	(5.4)	6.28	5.74
137	6.46	(5.6)	5.52	5.08
138	(6.85)	(5.75)	6.10	(5.1)
139	5.92	6.02	(6.10)	(5.02)
140	7.64	5.8	6.15	4.54
141	7.05	5.84	(5.40)	4.84
142	(7.65)	(5.55)	4.92	(4.2)
143	6.72	5.35	4.84	4.26
144	7.52	5.12	4.77	3.4
145	5.52	2.38	3.99	(2.9)
146	4.73	(2.05)	3.60	(2.5)
147	3.09	1.81	2.9	2.03
148	2.08	(0.93)	2.23	(1.8)
149	1.15	0.66	1.80	(1.5)
150	1.04	(0.31)	1.37	(1.2)
151	0.422	0.16	0.90	(0.95)
152	(0.30)	(0.15)	0.57	(0.75)
153	0.198	0.085	0.39	0.42

TABLE 1.3 (Continued)

Mass Number	^{232}Th Fission	^{232}Th 14 MeV	^{238}U Fission	^{238}U 14 MeV
154	(0.036)	(0.057)	0.24	(0.38)
155	(0.009)	(0.034)	(0.15)	(0.25)
156	0.0024	0.036	0.073	0.19
157		0.012	(0.038)	(0.10)
158		(0.0075)	(0.018)	(0.05)
159		0.0044	0.0091	0.026
160		(0.0028)	(0.0038)	(0.015)
161		0.00106	0.0018	0.0089

SOURCE: Flynn and Glendenin, Report ANL-7749. Reference 2.

TABLE 1.4 Fission Product Yields—Miscellaneous Data

Mass Number	8 MeV Neutrons			14 MeV Neutrons	
	^{232}Th	^{235}U	^{238}U	^{233}U	^{239}Pu
77	0.052				
83	2.72				
89	6.7				2.4
91	5.6				2.7
97	5.0				5.3
99	3.1	5.4	6.2	3.5	5.5
103	0.50			2.31	6.25
106	0.53			1.52	4.16
111	0.63			1.22	1.46
115	0.76			1.05	1.03
117	0.37				
131	2.3				
132	1.8			3.98	4.58
137				4.7	5.1
139	9.0				
140					4.35
141				5.0	
144	7.2	3.6	4.1		
147		2.05	2.7		
149		1.25	1.9		
153		0.185	0.41		
154		0.035	0.090		
159		0.0063	0.017		
161		0.0020	0.0043		

SOURCE: Flynn and Glendenin, Report ANL-7749. Reference 2.

TABLE 1.5 Nuclide Properties—Fission Products

Nuclide	$T_{1/2}$	Specific Activity (dis-s/μg)	Ci/g	Major Emissions (MeV)		Q Value (MeV)	W/g	W/Ci	Neutron Cross Sections (Barns)		
				Beta	Gamma				Thermal	Resonant	n-alpha
^{3}H	12.33y	3.57×10^{8}	9650	0.0186	None	0.0186	1.07	0.000110	6×10^{-6}		
^{81}Kr	2.1×10^{5}y	776	0.0210	EC	Br x-rays	0.29	3.61×10^{-5}	0.00172			
^{84}Rb	33d	1.74×10^{9}	47000	1.66 (β^{+}) 0.91	0.511 (42) 0.88 (74)	2.68	748	0.0159			
^{85}Kr	10.7y	1.45×10^{7}	392	0.67	0.514 (0.41)	0.67	1.56	0.00397	1.66	1.80	
^{86}Rb	18.7d	3.00×10^{9}	81000	1.78	1.08 (8.8)	1.78	857	0.0106			
^{87}Rb	4.8×10^{10}y	0.00316	8.57×10^{-8}	0.274	None	0.274	1.39×10^{-10}	0.00162	0.12	2.0	
^{89}Sr	52.0d	1.04×10^{9}	28100	1.463	0.91 (0.01)	1.463	247	0.00867	0.42		
^{90}Sr	28.1y	5.22×10^{6}	141	0.546	None	0.546	0.458	0.00324	0.9		
^{90}Y	64.1h	2.01×10^{10}	5.44×10^{5}	2.27	None	2.27	7320	0.0135	6.5		
^{91}Y	58.6d	9.08×10^{8}	24500	1.545	1.21 (0.3)	1.545	224	0.00916	1.4		
^{93}Zr	9.5×10^{5}y	149	0.00403	0.060	None	0.090	2.16×10^{-6}	0.000533	2.0	33	5×10^{-5} (f)
^{94}Nb	20000y	7020	0.190	0.49	0.702 (100) 0.871 (100)	2.06	0.00232	0.0122	13.6	125	
^{95}Zr	65.5d	7.76×10^{8}	21000	0.396	0.724 (49) 0.756 (49)	1.121	139	0.00664			
^{95m}Nb	86.6h	1.41×10^{10}	3.81×10^{5}	—	0.236 (100)	0.236	534	0.00140			
^{95}Nb	35.1d	1.45×10^{9}	39200	0.160	0.766 (100)	0.925	215	0.00548		7	
^{99}Tc	2.13×10^{5}y	626	0.0169	0.292	None	0.292	2.94×10^{-5}	0.00173	19	340	
^{103}Ru	39.6d	1.18×10^{9}	31900	0.21	0.497 (100)	0.74	140	0.00439			
^{106}Ru	1.01y	1.23×10^{8}	3324	0.039	None	0.0384	0.760	0.000228	0.146	2.6	
^{106}Rh	29.9s	1.31×10^{14}	3.56×10^{9}	3.54	0.512 (21) 0.62 (11)	3.63	7.63×10^{7}	0.0215			
^{107}Pd	6.5×10^{6}y	18.9	0.000513	0.04	None	0.085	2.59×10^{-7}	0.000504			
^{126}I	13d	2.95×10^{9}	79700	1.13 (β^{+}) 1.25	0.388 (34) 0.667 (33)	3.4	1607	0.0202	5960	40600	

TABLE 1.5 (Continued)

Nuclide	$T_{1/2}$	Specific Activity (dis-s/μg)	Ci/g	Major Emissions (MeV)		Q Value (MeV)	W/g	W/Ci	Neutron Cross Sections (Barns)		
				Beta	Gamma				Thermal	Resonant	n-alpha
^{126}Sn	10^5y	1050	0.0284		0.060 0.067 0.092	0.3	5.05×10^{-5}	0.00178			
^{126m}Sb	19m	2.90×10^{12}	7.83×10^7		0.41 0.67						
^{126}Sb	12.4d	3.09×10^9	83400	1.9	0.41 0.69 0.99	3.7	1833	0.0219			
^{129}I	1.6×10^7y	6.40	0.000173	0.150	0.04	0.189	1.94×10^{-7}	0.00112	8.99	12	
^{129m}Xe	8.0d	4.68×10^9	1.26×10^5	—	0.197 (6)	0.236	177	0.00140			
^{129m}Te	33.4d	1.12×10^9	30400	1.60	0.696 (5)	1.58	284	0.00936			
^{129}Te	70m	7.70×10^{11}	2.08×10^7	1.45	0.027 (19) 0.459 (15)	1.48	18200	0.00877			
^{131}I	8.04d	4.59×10^9	1.24×10^5	0.606	0.364 (79) 0.637 (6.8)	0.970	713	0.00575	0.70		
^{131m}Xe	12.0d	3.07×10^9	82900	—	0.164 (2)	0.163	118	0.000966	90		
^{133}Xe	5.29d	6.86×10^9	1.85×10^5	0.346	0.081 (35)	0.427	470	0.00253	190		
^{134}Cs	2.06y	4.78×10^7	1290	0.662	0.57 (14) 0.605 (98)	2.062	15.8	0.0122	140		
^{135}Cs	2.3×10^6y	42.5	0.00115	0.21	None	0.210	1.43×10^{-6}	0.00124	8.7		
^{136}Cs	13.0d	2.73×10^9	74300	0.341	0.818 (100) 1.05 (80) 1.25 (20)	2.54	1110	0.0151	1.3	76	
^{137}Cs	30.0y	3.21×10^6	86.9	0.341	Ba x-rays	1.176	0.607	0.00697	0.11		
^{137m}Ba	2.55m	1.99×10^{13}	5.37×10^8	—	0.662 (94.6)	0.6616	2.10×10^6	0.00392			
^{137}La	60000y	1580	0.0434		Ba x-rays	0.5	1.29×10^{-4}	0.00296		4	
^{140}Ba	12.8d	2.70×10^9	72900	1.02	0.537 (29)	1.05	454	0.00622	1.6	13.6	

^{140}La	40.2h	2.06×10^{10}	5.57×10^{5}	1.36 1.69 (15)	1.596 (97) 0.487 (46) 0.815 (26)	3.769	1250	0.0233	2.7	
^{141}Ce	32.5d	1.05×10^{9}	28400	0.581	0.145 (49)	0.581	98.1	0.00344	29	0.48
^{143}Pr	13.6d	2.48×10^{9}	67000	0.933	None	0.933	371	0.00553	89	190
^{144}Ce	284.9d	1.18×10^{8}	3190	0.31					1	2.6
^{144}Pr	17.3m	2.79×10^{12}	7.53×10^{7}	2.99	1.48 (0.35) 2.19 (0.9)	2.989	1.33×10^{6}	0.0177		
^{147}Nd	11.0d	2.99×10^{9}	80700	0.81	0.091 (27) 0.531 (13)	0.91	436	0.00539		
^{147}Pm	2.62y	3.43×10^{7}	926	0.224	None	0.225	1.24	0.00133	96	1220
^{148m}Pm	41.3d	7.90×10^{8}	21300	0.69	0.551 (95) 0.630 (87) 1.015 (20)	2.59	6.91	0.0154	22000	3600
^{148}Pm	5.37y	1.66×10^{7}	448	2.48	0.551 (50) 1.465 (24)	2.46	312	0.0146	2000	
^{151}Sm	92.9y	9.40×10^{5}	25.4	0.076	0.022 (1.3)	0.076	0.0115	0.00045	15000	3300
^{152}Eu	13y	6.68×10^{6}	181	1.88	1.13 (14) 1.408 (22)	3.7	3.97	0.0219	2300	
^{153}Gd	242d	1.29×10^{8}	1770	EC	0.103 (24)	0.243	5.11	0.00144		
^{154}Eu	8.59y	9.98×10^{6}	269	0.87 1.80 (10)	1.00 (20) 1.278 (37)	1.97	3.16	0.0117	1500	
^{155}Eu	4.8y	1.77×10^{7}	478	0.25	0.087 (72) 0.105 (48)	0.248	0.707	0.00147	4040	
^{156}Eu	15.2d	2.04×10^{9}	55200	2.45	1.97 (5) 2.19 (6) 2.27 (2.4)	2.45	800	0.0145		

SOURCES: Varied, please see text.

Table 1.6 Nuclide Properties–Activation Products

Nuclide	$T_{1/2}$	Specific Activity (dis-s/μg)	Ci/g	Major Emissions (MeV)		Q Value (MeV)	W/g	W/Ci	Neutron Cross Sections (barns)		
				Beta	Gamma				Thermal	Resonant	n-alpha
^{14}C	5730y	1.65×10^5	4.46	0.156	None	0.156	0.00412	0.000925	10^{-6}		
^{45}Ca	165d	6.51×10^8	17600	0.258	0.013	0.258	26.9	0.00153			
^{46}Sc	83.8d	1.25×10^9	33900	0.357	0.899 (100)	2.367	475	0.0140	8		
					1.12 (100)						
^{51}Cr	27.7d	3.42×10^9	92400	EC	0.320 (9)	0.752	412	0.00446			
^{54}Mn	312.5d	2.86×10^8	7720	EC	0.835 (100)	1.379	63.3	0.00817	10		
^{55}Fe	2.7y	8.89×10^7	2400	EC	Mn x-rays	0.232	3.31	0.00138			
^{58}Co	71.3d	1.17×10^9	31600	EC	0.811 (100)	2.309	432	0.0137	1880	6890	
^{59}Fe	45.1d	1.82×10^9	49200	0.475	1.10 (56)						
					1.29 (44)						
^{59}Ni	80000y	2800	0.0756	EC	Co x-rays	1.072	0.000481	0.00635		138	12 (Th)
^{60}Co	5.27y	4.17×10^7	1130	0.315	1.173 (100)	2.819	18.9	0.0167	2	4.3	
					1.332 (100)						
^{63}Ni	92y	2.28×10^6	61.8	0.067		0.067	0.0245	0.000397	23		
^{93}Mo	100y	1.42×10^6	38.4	EC	0.030 (85)						
^{93m}Nb	13.6y	1.04×10^7	282		0.934 (99)						
^{117m}Sn	14.0d	2.95×10^9	79700		0.158	0.317	150	0.00188			
^{119m}Sn	250d	1.62×10^8	4400		0.065 (100)	0.089	2.32	0.000358			
^{121m}Sn	76y	1.44×10^6	38.8	0.35	0.037	0.45	0.104	0.00267			
^{123}Sn	129d	3.04×10^8	8210	1.46	1.08	1.46	71.2	0.00865			
^{124}Sb	60.2d	6.47×10^8	17500	0.621	0.603 (100)	2.916	302	0.0173	6.5		
				2.317	1.69 (51)						
^{125}Sb	2.73y	3.87×10^7	1050	0.619	0.408 (100)	0.764	4.75	0.00453	0.78	0.31	
					0.601 (62)						
^{125m}Te	58d	6.66×10^8	18000		0.110						

SOURCES: Varied, please see text.

TABLE 1.7 Nuclide Properties—Selected Isotopes

Nuclide	$T_{1/2}$	Specific Activity (dis-s/μg)	Ci/g	Major Emissions (MeV)		Q Value (MeV)	W/g	W/Ci	Neutron Cross Sections (barns)		
				Beta	Gamma				Thermal	Resonant	n-alpha
^{24}Na	15.0h	3.22×10^{11}	8.72×10^{6}	1.39	1.37 (100)	5.51	2.85×10^{5}	0.0327			
				4.17	2.75 (100)						
^{32}P	14.3d	1.06×10^{10}	2.86×10^{5}	1.71		1.71	2890	0.0101			
^{35}S	88d	1.57×10^{9}	42500	0.1674		0.1674	42.1	0.000992			
^{40}K	1.28×10^{9}y	0.258	6.97×10^{-6}	1.51 (β^{+})	1.46 (11)	2.855	1.18×10^{-7}	0.0169	30		0.39 (Th)
				1.35							4.4 (n, p)
^{56}Mn	2.56h	8.09×10^{11}	2.19×10^{7}	2.85	0.847 (99)	3.702	4.80×10^{5}	0.0219			
					2.11 (15)						
^{64}Cu	12.9h	1.41×10^{11}	3.80×10^{6}	0.654 (β^{+})	1.35 (0.6)	2.25	50700	0.0133			
				0.573							
^{65}Zn	234.6d	3.05×10^{8}	8260	0.325 (β^{+})	1.12 (50)	1.353	66.1	0.00802			
^{76}As	26.5h	5.76×10^{10}	1.56×10^{6}	2.40	0.559 (100)	2.97	27400	0.0176			
				2.96							
^{82}Br	35.3h	4.01×10^{10}	1.08×10^{6}	0.87 (β^{+})	0.618 (7)	3.092	19900	0.0183			
				2.00	0.666 (1)						
^{160}Tb	73d	4.14×10^{8}	11200	0.95	0.966 (28)	1.72	114	0.0102	525		
					1.178 (17)						
^{170}Tm	128.6d	2.21×10^{8}	5780	0.97	0.084 (3.4)	0.967	34.2	0.00573	92		
^{181}Hf	42.4d	6.29×10^{8}	17000	0.408	0.482 (0.7)	1.023	103	0.00606	40		
^{182}Ta	115d	2.31×10^{8}	6250	0.62	1.19 (16.5)	1.811	67.0	0.0107	8200		
					1.22 (27.9)						
^{191}Os	15.3d	1.65×10^{9}	44700	0.13	0.188	0.310	82.1	0.00184			
^{194}Ir	17.4h	3.44×10^{10}	9.30×10^{5}	1.62	0.328 (33)	2.24	12340	0.0133			
^{197}Pt	18.3h	3.22×10^{10}	8.71×10^{5}	0.67	0.077 (20)	0.75	3870	0.00445			
^{198}Au	2.70d	9.05×10^{9}	2.45×10^{5}	0.961	0.412 (99)	1.374	1990	0.00814	25800		
^{203}Hg	46.6d	5.11×10^{8}	13800	0.210	0.279	0.492	40.3	0.00292			
^{204}Tl	3.8y	1.70×10^{7}	461	0.764		1.103	3.02	0.00654			

SOURCES: Varied, please see text.

The beta energy values shown represent the maximum particle energy (the average is generally around one-third the maximum). "EC" indicates that the decay is by electron capture. If positrons are emitted, 0.511 MeV annihilation gammas are always present. The numbers in parentheses following the maximum beta energy (as is also true in the gamma columns) are *yields*, that is, the average number of emissions seen per 100 disintegrations of the parent nuclide, *not* the percentage of the total betas or gammas emitted. Because of this, the gamma yield of a nuclide such as ^{94}Nb (Table 1.5) or ^{60}Co (Table 1.6) can total 200% or more. The numbers shown in the "Major Emissions" columns were taken primarily from References 9a, 9b, and 11.

(c) "*Q* Values," the total energy in MeV per disintegration event, were essentially all taken from the Heath[9a] compilation.

(d) The neutron-capture cross sections shown in the final columns of the tables are those used in the ORIGEN Isotope Generation and Depletion Code (Chapter 3) as given by Kee[13] and reproduced in Report DLC-38[14] from the Radiation Shielding Information Center at Oak Ridge. "Thermal" refers to capture of slow neutrons at ambient temperatures, that is, with energies of ca. 0.026 ev at 25°C up to 0.092 ev at 800°C.[15] "Resonant" is for capture of neutrons with energies from the thermal value up to roughly 1 MeV. A (Th) following the figure for the (n, α) or other special reactions in the last column indicates that the reaction is brought about by thermal neutrons; (f), by fast neutrons.

(e) The remaining numbers in Tables 1.5-1.7 (Sp. Act., Ci/g, W/g, and W/Ci) were all calculated by means of the equations given in the next section. Because of space limitations and the present ubiquitous availability of pocket calculators, values are not given for g/Ci (possibly of more direct use in waste management, but, in any case, simply the reciprocal of Ci/g), or for heat values expressed in Btu/h, since these are readily obtained from the W/g and W/Ci numbers by multiplication by a factor of 3.415 (one watt equals 3.415 Btu/h). These missing numbers can also be readily calculated through equations 1.3, 1.5, and 1.7 below.

1.1.3.1 Useful Conversion Equations

Useful relationships for practical use can be readily derived from the mathematical expressions for radioactive decay; the equalities between various energy, mass, and time units; and the definitions of commonly used terms as given in Chapter 14. These manipulations yield simple equations where the various unchanging quantities can be combined and expressed as single constants. The equations derived are given below and the numerical values for the constants are summarized in Table 1.8. These will vary in equations 1.1-1.5, depending on the time units in which the half-life of the nuclide is expressed.

TABLE 1.8 Constants for Use in Conversion Equations (1.1)-(1.5)

Given Half-Life Units	K_1	K_2	K_3	K_4	K_5	K_6	K_7
Seconds	4.17×10^{17}	1.13×10^{13}	8.85×10^{-14}	6.66×10^{10}	2.27×10^{11}	0.005927	0.02024
Minutes	6.95×10^{15}	1.88×10^{11}	5.32×10^{-12}	1.11×10^{9}	3.79×10^{9}	0.005927	0.02024
Hours	1.16×10^{14}	3.14×10^{9}	3.18×10^{-10}	1.86×10^{7}	6.35×10^{7}	0.005927	0.02024
Days	4.83×10^{12}	1.31×10^{8}	7.66×10^{-9}	7.74×10^{5}	2.64×10^{6}	0.005927	0.02024
Years	1.32×10^{10}	3.57×10^{5}	2.80×10^{-6}	2120	7240	0.005927	0.02024

Quantity	Abbreviation	Equality	
Specific activity	dis-s/μg*	$\frac{K_1}{(T_{1/2})(A)}$	(1.1)
Curies per gram	Ci/g	$\frac{K_2}{(T_{1/2})(A)}$	(1.2)
Grams per curie	g/Ci	$(K_3)(T_{1/2})(A)$	(1.3)
Watts per gram	W/g	$\frac{(K_4)(Q)}{(T_{1/2})(A)}$	(1.4)
British thermal units per hour per gram	Btu-h/g	$\frac{(K_5)(Q)}{(T_{1/2})(A)}$	(1.5)
Watts per curie	W/Ci	$(K_6)(Q)$	(1.6)
British thermal units per curie	Btu-h/Ci	$(K_7)(Q)$	(1.7)

In the above equations

A = atomic number
$T_{1/2}$ = half-life in time units as given
Q = total energy per disintegration event, expressed in MeV

(Equations and corresponding constants could also be developed for g/W, g/Btu-h, Ci/W, and Ci/Btu-h, but these did not seem to be particularly useful quantities. In any event, they are simple reciprocals of values obtainable through the equations given.)

1.2 HARDWARE-ACTIVATION PRODUCTS

Some fuel cladding and portions of the hardware associated with the assembly are partly dissolved along with the fuel itself, so any activation products generated in these initially inert materials become part of the final high-level waste. Since most of the metals used for these purposes are of the light- or medium-

*Disintegrations per second per microgram, or bequerels per microgram in the SI system of units. Note that in Tables 1.5–1.7, the specific activity is in terms of micrograms, whereas other derived values are per gram.

weight elements, there is some overlap between the resulting activated species and some of the lighter fission products considered in Table 1.5 (^{3}H, ^{89}Sr, ^{91}Y, ^{93}Zr, ^{95}Zr, and ^{99}Tc). The properties of additional frequently seen nuclides are given in Table 1.6.

The selection rules for the listing were again somewhat flexible, but the isotopes shown are basically those given in the ORIGEN code (Chapter 3) as still being present in significant quantities 160 days after discharge of the fuel from the ORIGEN model reactor. Some of the group will still be of importance after a year's cooling, but very few at the end of 10 years, somewhat in contrast to the fission-product case.

1.3 OTHER SELECTED ISOTOPES

Table 1.7 summarizes the properties of a group of nuclides not normally found in quantity in high-level wastes, but used to different levels of frequency in research and in medical and industrial applications. They are thus more likely to be encountered in low- and intermediate-level wastes. The list shown was selected quite arbitrarily, based on the author's impression of the frequency of appearance of a given species in the general literature.

1.4 MISCELLANEOUS CROSS SECTIONS AND OTHER DATA

1.4.1 Thermal and Resonance Integral *n*-Gamma Reaction Cross Sections

These values are presented in Table 1.9 for all of the stable isotopes below ^{209}Bi generating a radioactive product upon capture of a neutron. The data were essentially all taken from *Cross Sections Book* ed 3, BNL 325, issued by the National Neutron Cross Section Center at Brookhaven.[16] (A fourth edition is currently in preparation.) The General Electric Company's "Chart of the Nuclides"[17] and references 9, 10, and 11 were also consulted in some cases where there appeared to be ambiguities.

The resonance integral values (given as I_{γ} in BNL 325) are again the summed capture cross sections for all neutrons having energies between the thermal and 1 MeV levels. BNL 325 gives only a single value for I_{γ} in some cases where neutron activation actually produces one or more excited state nuclides in addition to a ground-state species. In such situations, the portion of the cross section allotted to each product was determined by following the ratio of distribution used in the ORIGEN data base.[14]

TABLE 1.9 Thermal and Resonance Integral *n*-Gamma Cross Sections for the Lighter Elements

Target		Product		Cross Section (Barns)	
Nuclide	Natural Abundance	Nuclide	$T_{1/2}$	Thermal	Resonant
H-1	99.985	H-2	Stable	0.333	0.15
H-2	0.015	H-3	12.33y	0.00053	0.00024
He-4	99.9999	He-5	2×10^{-21} s	–	–
Li-7	92.5	Li-8	0.844s	0.037	0.007
Be-9	100	Be-10	1.6×10^{6} y	0.0092	0.004
B-11	80.0	B-12	0.02s	0.0053	–
C-13	1.11	C-14	5730y	0.0009	0.0013
N-15	0.36	N-16	7.2s	0.000024	–
O-18	0.20	0-19	29s	0.00016	0.00081
F-19	100	F-20	11s	0.0095	0.0176
Ne-22	9.22	Ne-23	37.6s	0.048	–
Na-23	100	Na-24m	0.02s	0.400	0.134
Na-23	100	Na-24	15.02h	0.530	0.177
Mg-26	11.01	Mg-27	9.45m	0.0382	0.025
Al-27	100	Al-28	2.24m	0.230	0.17
Si-30	3.10	Si-31	2.62h	0.107	0.106
P-31	100	P-32	14.28d	0.180	0.08
S-34	4.21	S-35	87.2d	0.240	–
S-36	0.017	S-37	5.03m	0.15	–
Cl-35	75.77	Cl-36	3.01×10^{5} y	43	17
Cl-37	24.23	Cl-38m	0.74s	0.005	–
Cl-37	24.23	Cl-38	37.2m	0.423	0.31
Ar-36	0.34	Ar-37	35d	5	2.5
Ar-38	0.07	Ar-39	265y	0.8	0.4
Ar-40	99.60	Ar-41	1.83h	0.66	0.41
K-41	6.73	K-42	12.36h	1.46	1.42
Ca-40	96.941	Ca-41	1.3×10^{5} y	0.40	0.18
Ca-44	2.086	Ca-45	163d	1.0	0.56
Ca-46	0.004	Ca-47	4.54d	0.7	0.32
Ca-48	0.187	Ca-49	8.8m	1.1	–
Sc-45	100	Sc-46m	18.7s	9.6	4.1
Sc-45	100	Sc-46	83.8d	16.9	7.2
Ti-50	5.2	Ti-51	5.73m	0.179	0.118
V-51	99.75	V-52	3.76m	4.88	2.7
Cr-50	4.35	Cr-51	27.7d	15.9	7.6
Cr-54	2.36	Cr-55	3.55m	0.36	0.18
Mn-55	100	Mn-56	2.58h	13.3	14.0
Fe-54	5.8	Fe-55	2.7y	2.25	1.2
Fe-58	0.3	Fe-59	44.6d	1.15	1.19

TABLE 1.9 (Continued)

Target		Product		Cross Section (Barns)	
Nuclide	Natural Abundance	Nuclide	$T_{1/2}$	Thermal	Resonant
Co-59	100	Co-60m	10.5m	20	40.8
Co-59	100	Co-60	5.67y	17	34.7
Ni-58	68.3	Ni-59	80,000y	4.6	2.2
Ni-62	3.6	Ni-63	100y	14.2	6.8
Ni-64	0.9	Ni-65	2.52h	1.49	1.1
Cu-63	69.2	Cu-64	12.7h	4.5	4.9
Cu-65	30.8	Cu-66	5.1m	2.17	2.4
Zn-64	48.6	Zn-65	243.8d	1.10	2.3
Zn-68	18.8	Zn-69m	13.8h	0.072	0.2
Zn-68	18.8	Zn-69	57m	1.0	3.1
Zn-70	0.6	Zn-71m	4.0h	0.0087	–
Zn-70	0.6	Zn-71	2.4m	0.083	–
Ga-69	60	Ga-70	21.1m	1.68	15.6
Ga-71	40	Ga-72m	0.036s	0.15	1.0
Ga-71	40	Ga-72	14.1h	4.71	30.2
Ge-70	20.5	Ge-71m	0.020s	0.28	0.2
Ge-70	20.5	Ge-71	11.4d	3.15	2.2
Ge-74	36.5	Ge-75m	48.9s	0.143	0.16
Ge-74	36.5	Ge-75	82.8m	0.24	0.27
Ge-76	7.8	Ge-77m	54s	0.092	1.3
Ge-76	7.8	Ge-77	11.3h	0.05	0.7
As-75	100	As-76	26.3h	4.3	60
Se-74	0.9	Se-75	120.4d	51.8	565
Se-76	9.0	Se-77m	17.5s	21	8.7
Se-78	23.5	Se-79m	3.89m	0.33	2.1
Se-78	23.5	Se-79	65000y	0.4	2.6
Se-80	49.8	Se-81m	57.3m	0.080	0.3
Se-80	49.8	Se-81	22.6m	0.53	1.7
Se-82	9.2	Se-83m	70s	0.039	–
Se-82	9.2	Se-83	22.5m	0.0058	–
Br-79	50.69	Br-80m	4.42h	2.6	34.5
Br-79	50.69	Br-80	17.7m	8.5	92
Br-81	49.31	Br-82m	6.1m	2.43	46
Br-81	49.31	Br-82	35.3h	0.26	5
Kr-78	0.35	Kr-79m	55s	0.21	0.2
Kr-78	0.35	Kr-79	34.9h	4.71	5.1
Kr-80	2.25	Kr-81m	13.3s	4.55	37.9
Kr-80	2.25	Kr-81	2.1×10^5 y	9.5	18.2
Kr-82	11.6	Kr-83m	1.86h	20	88.8
Kr-84	57.0	Kr-85m	4.48h	0.090	1.84

TABLE 1.9 (Continued)

Target		Product		Cross Section (Barns)	
Nuclide	Natural Abundance	Nuclide	$T_{1/2}$	Thermal	Resonant
Kr-84	57.0	Kr-85	10.72y	0.042	0.86
Kr-86	17.3	Kr-87	76m	0.060	0.03
Rb-85	72.17	Rb-86m	1.02m	0.050	0.82
Rb-85	72.17	Rb-86	18.65d	0.41	6.68
Rb-87	27.83	Rb-88	17.7m	0.12	2.0
Sr-84	0.56	Sr-85m	67.7m	0.55	7.2
Sr-84	0.56	Sr-85	65.2d	0.26	3.4
Sr-86	9.84	Sr-87m	2.81h	0.84	4.0
Sr-88	82.6	Sr-89	50.5d	0.0058	0.05
Y-89	100	Y-90m	3.19h	1.0	0.001
Y-89	100	Y-90	64h	1.3	0.0008
Zr-92	17.1	Zr-93	1.5×10^6 y	0.26	0.54
Zr-94	17.4	Zr-95	64.0d	0.056	0.30
Zr-96	2.8	Zr-97	16.8h	0.017	5.0
Nb-93	100	Nb-94m	6.26m	–	–
Nb-93	100	Nb-94	20,000y	1.15	8.5
Mo-92	14.8	Mo-93m	6.9h	–	–
Mo-92	14.8	Mo-93	3,500y	0.045	0.52
Mo-98	24.1	Mo-99	66.02h	0.13	6.2
Mo-100	9.6	Mo-101	14.6m	0.20	3.75
Ru-96	5.5	Ru-97	2.89d	0.25	6.6
Ru-102	31.6	Ru-103	39.4d	1.3	4.1
Ru-104	18.7	Ru-105	4.44h	0.47	4.6
Rh-103	100	Rh-104m	4.35m	11	80
Rh-103	100	Rh-104	42s	139	1,020
Pd-102	1.0	Pd-103	17.0d	4.8	–
Pd-106	27.3	Pd-107m	21.0s	0.013	0.2
Pd-106	27.3	Pd-107	6.5×10^6 y	0.29	5.5
Pd-108	26.7	Pd-109m	4.67m	0.2	4
Pd-108	26.7	Pd-109	13.43h	12	246
Pd-110	11.8	Pd-111m	5.5h	0.02	–
Pd-110	11.8	Pd-111	22m	0.20	–
Ag-107	51.83	Ag-108m	130y	3	8
Ag-107	51.83	Ag-108	2.41m	34.2	86
Ag-109	48.17	Ag-110m	252d	4.5	80
Ag-109	48.17	Ag-110	24.3s	89	1,370
Cd-106	1.3	Cd-107	6.5h	1	–
Cd-108	0.89	Cd-109	453d	1.1	–
Cd-110	12.5	Cd-111m	48.7m	0.10	0.4
Cd-112	24.1	Cd-113m	14.6y	2.2	15

TABLE 1.9 (Continued)

Target		Product		Cross Section (Barns)	
Nuclide	Natural Abundance	Nuclide	$T_{1/2}$	Thermal	Resonant
Cd-114	28.7	Cd-115m	49.5d	0.036	2
Cd-114	28.7	Cd-115	53.5h	0.30	18
Cd-116	7.5	Cd-117m	3.4h	0.029	0.074
Cd-116	7.5	Cd-117	2.6h	0.05	0.14
In-113	4.3	In-114n	0.042s	3.1	–
In-113	4.3	In-114m	49.5d	4.4	186
In-113	4.3	In-114	71.9s	3.9	96
In-115	95.7	In-116n	2.2s	92	–
In-115	95.7	In-116m	54.2m	65	2,570
In-115	95.7	In-116	14.2s	45	730
Sn-112	1.01	Sn-113m	20m	0.35	9
Sn-112	1.01	Sn-113	115d	0.8	18
Sn-116	14.7	Sn-117m	14d	0.006	11
Sn-118	24.3	Sn-119m	293d	0.016	7
Sn-120	32.4	Sn-121m	76y	0.001	0.01
Sn-120	32.4	Sn-121	27h	0.14	1.49
Sn-122	4.6	Sn-123m	40m	0.001	0.003
Sn-122	4.6	Sn-123	129d	0.18	0.593
Sn-124	5.6	Sn-125m	9.6m	0.13	6.7
Sn-124	5.6	Sn-125	9.65d	0.004	0.2
Sb-121	57.3	Sb-122m	4.2m	0.055	2
Sb-121	57.3	Sb-122	2.72d	6.2	198
Sb-123	42.7	Sb-124n	21m	0.011	–
Sb-123	42.7	Sb-124m	93m	0.035	0.4
Sb-123	42.7	Sb-124	60.3d	4.28	139.6
Te-120	0.091	Te-121m	150d	0.34	–
Te-120	0.091	Te-121	17d	2.0	–
Te-122	2.5	Te-123m	119.7d	1.1	31
Te-124	4.62	Te-125m	58d	0.04	0.04
Te-126	18.7	Te-127m	109d	0.14	1.3
Te-126	18.7	Te-127	9.4h	0.90	8.7
Te-128	31.7	Te-129m	33.4d	0.015	0.1
Te-128	31.7	Te-129	70m	0.20	1.4
Te-130	34.5	Te-131m	30h	0.02	0.04
Te-130	34.5	Te-131	25.0m	0.27	0.57
I-127	100	I-128	25.0m	6.2	147
Xe-124	0.10	Xe-125m	57s	22	600
Xe-124	0.10	Xe-125	17h	106	3,000
Xe-126	0.090	Xe-127m	72s	0.26	3
Xe-126	0.090	Xe-127	36.4d	3.7	35

TABLE 1.9 (Continued)

Target		Product		Cross Section (Barns)	
Nuclide	Natural Abundance	Nuclide	$T_{1/2}$	Thermal	Resonant
Xe-128	1.91	Xe-129m	8.88d	0.36	0.8
Xe-130	4.1	Xe-131m	11.9d	0.42	0.9
Xe-132	26.9	Xe-133m	2.19d	0.025	0.05
Xe-132	26.9	Xe-133	5.25d	0.36	0.75
Xe-134	10.4	Xe-135m	15.3m	0.003	0.004
Xe-134	10.4	Xe-135	9.09h	0.25	0.32
Xe-136	8.9	Xe-137	3.85m	0.16	–
Cs-133	100	Cs-134m	2.90h	2.5	33
Cs-133	100	Cs-134	2.06y	29	382
Ba-130	0.11	Ba-131m	14.6m	2.5	28
Ba-130	0.11	Ba-131	11.7d	11	122
Ba-132	0.10	Ba-133m	38.9h	0.68	None
Ba-132	0.10	Ba-133	10.7y	8.5	None
Ba-134	2.4	Ba-135m	28.7h	0.16	0.7
Ba-135	6.6	Ba-136m	0.31s	0.014	0.24
Ba-136	7.9	Ba-137m	2.55m	0.010	0.0033
Ba-138	71.7	Ba-139	83.2m	0.35	0.2
La-139	99.911	La-140	40.23h	9.0	12.2
Ce-136	0.19	Ce-137m	34.4h	0.95	9
Ce-136	0.19	Ce-137	9.0h	6.3	61
Ce-138	0.25	Ce-139m	56s	0.015	–
Ce-138	0.25	Ce-139	137.6d	1.1	–
Ce-140	88.48	Ce-141	32.5d	0.57	0.47
Ce-142	11.08	Ce-143	33.0h	0.95	0.73
Pr-141	100	Pr-142m	14.6m	3.9	3.8
Pr-141	100	Pr-142	19.13h	7.6	10.3
Nd-146	17.2	Nd-147	10.99d	1.4	3.2
Nd-148	5.7	Nd-149	17.3h	2.5	19
Nd-150	5.6	Nd-151	12.4m	1.2	14
Sm-150	7.4	Sm-151	93y	102	310
Sm-152	26.7	Sm-153	46.7h	206	3,000
Sm-154	22.6	Sm-155	22.2m	5.5	30
Eu-151	47.9	Eu-152n	96m	4.0	–
Eu-151	47.9	Eu-152m	9.34h	3,300	1,200
Eu-151	47.9	Eu-152	13.4y	5,900	2,100
Eu-153	52.1	Eu-154m	46.1m	–	–
Eu-153	52.1	Eu-154	8.2y	390	1,635
Gd-152	0.20	Gd-153	241.6d	1,100	3,000
Gd-158	24.8	Gd-159	18.6h	2.5	61
Gd-160	21.8	Gd-161	3.7m	0.77	7.0

TABLE 1.9 (Continued)

Target		Product		Cross Section (Barns)	
Nuclide	Natural Abundance	Nuclide	$T_{1/2}$	Thermal	Resonant
Tb-159	100	Tb-160	72.4d	25.5	430
Dy-156	0.057	Dy-157	8.1h	33	960
Dy-158	0.100	Dy-159	144d	43	120
Dy-164	28.1	Dy-165m	1.26m	1,700	237
Dy-164	28.1	Dy-165	2.33h	1,000	140
Ho-165	100	Ho-166m	1,200y	3.5	30
Ho-165	100	Ho-166	26.8h	63.0	670
Er-162	0.14	Er-163	75m	19	480
Er-164	1.56	Er-165	10.36h	13	105
Er-166	33.4	Er-167m	2.28s	15	100
Er-168	27.1	Er-169	9.40d	1.95	36
Er-170	14.9	Er-171	7.52h	5.7	20
Tm-169	100	Tm-170	129d	103	1,720
Yb-168	0.14	Yb-169m	4.6s	–	–
Yb-168	0.14	Yb-169	32.0d	3,470	31,000
Yb-174	31.6	Yb-175m	0.068s	46	23
Yb-174	31.6	Yb-175	4.19d	19	10
Yb-176	12.6	Yb-177m	6.41s	2.4	–
Yb-176	12.6	Yb-177	1.9h	2.4	6
Lu-175	97.4	Lu-176m	3.69h	16.4	620
Lu-176	2.6	Lu-177n	0.16ms	315	–
Lu-176	2.6	Lu-177m	16.1d	7	4
Lu-176	2.6	Lu-177	6.71d	1,780	1,160
Hf-174	0.16	Hf-175	70d	390	465
Hf-176	5.2	Hf-177m	51.4m	–	–
Hf-177	18.6	Hf-178m	31y	1.1	22
Hf-178	27.1	Hf-179m	25.1d	53	1,200
Hf-179	13.7	Hf-180m	5.5h	0.34	5
Hf-180	35.2	Hf-181	42.4d	12.6	43
Ta-181	99.988	Ta-182m	15.9m	0.010	0.3
Ta-181	99.988	Ta-182	115d	21.0	710
W-180	0.13	W-181	121d	3.5	200
W-182	26.3	W-182m	5.2s	–	–
W-184	30.67	W-185m	1.65m	0.002	0.015
W-184	30.67	W-185	75.1d	1.8	14
W-186	28.6	W-187	23.9h	37.8	500
Re-185	37.4	Re-186m	2×10^5 y	0.3	–
Re-185	37.4	Re-186	90.6h	112	1,730
Re-187	62.6	Re-188m	18.6m	73	293
Re-187	62.6	Re-188	16.7h	1.6	7

TABLE 1.9 (Continued)

Target		Product		Cross Section (Barns)	
Nuclide	Natural Abundance	Nuclide	$T_{1/2}$	Thermal	Resonant
Os-184	0.018	Os-185	94d	3,000	–
Os-188	13.3	Os-189m	6.0h	–	–
Os-189	16.1	Os-190m	9.9m	0.26mb	0.003
Os-190	26.4	Os-191m	13.0h	9.1	20
Os-190	26.4	Os-191	15.3d	3.9	9
Os-192	41.0	Os-193	30.5h	2.0	5.4
Ir-191	37.3	Ir-192n	241y	0.38	–
Ir-191	37.3	Ir-192m	1.44m	300	1.5
Ir-191	37.3	Ir-192	74.2d	624	3,750
Ir-193	62.7	Ir-194m	171d	0.035	–
Ir-193	62.7	Ir-194	19.5h	110	1,300
Pt-190	0.013	Pt-191	10.9h	150	–
Pt-192	0.787	Pt-193m	4.3d	2.2	13
Pt-192	0.787	Pt-193	50y	11.8	70
Pt-194	32.9	Pt-195m	4.02d	0.09	0.3
Pt-196	33.8	Pt-197m	94m	0.05	0.5
Pt-196	33.8	Pt-197	18.3h	0.69	7.5
Pt-198	7.2	Pt-199m	14.1s	0.027	0.4
Pt-198	7.2	Pt-199	30.8m	3.7	56
Au-197	100	Au-198m	2.3d	–	–
Au-197	100	Au-198	2.696d	98.8	1,560
Hg-196	0.15	Hg-197m	23.8h	120	59
Hg-196	0.15	Hg-197	64.1h	3,080	413
Hg-198	10.0	Hg-199m	42.6m	0.018	0.7
Hg-202	29.8	Hg-203	46.6d	4.9	4.9
Hg-204	6.9	Hg-205	5.2m	0.43	–
Tl-203	29.5	Tl-204	3.77y	11.0	40
Tl-205	70.5	Tl-206m	3.6m	–	–
Tl-205	70.5	Tl-206	4.20m	0.10	0.7
Pb-204	1.42	Pb-205	1.4×10^7y	0.66	1.7
Pb-206	24.1	Pb-207m	0.796s	–	–
Pb-208	52.4	Pb-209	3.28h	0.0005	–
Bi-209	100	Bi-210m	3.5×10^6y	0.014	–
Bi-209	100	Bi-210	5.0d	0.019	0.19

SOURCES: References 16 and 17. See text.

In several cases, the radioactive product formed by neutron capture itself decays to one or more active species, thus acting as the parent of a short decay chain. No effort was made in Table 1.9 to indicate such situations, but they can be readily evaluated from the various nuclide charts.

1.4.2 Reactor Cross Sections

The effective cross sections in an operating reactor will depend on the neutron energy spectrum. The Radiation Shielding Information Center at Oak Ridge published[18] such cross sections for both Pressurized Water Reactors (PWRs) and Boiling Water Reactors (BWRs); in each case, for all-uranium, all-plutonium, and for mixed U-Pu fuels. Values are given for the stable nuclides and for the most important fission products.

1.4.3 Fast Neutron Reactions

The capture of a fast neutron by a nucleus can produce only a gamma, or particles such as neutrons, protons, or alphas may be emitted. These reactions are primarily important to neutron-activation analysis, and Kenna and Harrison[19] prepared a comprehensive compilation of cross sections for that purpose. Table 1.10 presents selected values from their work of cross sections that could be of interest in evaluating waste compositions. The chief criterion for inclusion in the table was the magnitude of the half-life of the product. (Note that the cross sections are generally small and thus are given in millibarns.)

TABLE 1.10 Fast Neutron Cross Sections

Target		Product		
Nuclide	Natural Abundance	Nuclide	$T_{1/2}$	Cross Section (Millibarns)
		(n, gamma)		
Na-23	100	Na-24	14.9h	0.3
Sc-45	100	Sc-46	85d	1.2
Cu-63	69.09	Cu-64	12.8h	2.6
Br-81	49.46	Br-82	35.9h	3.5
Y-89	100	Y-90	64.2h	2.9
La-139	99.91	La-140	40.2h	1.3
Ce-142	11.07	Ce-143	33h	8.0
Gd-158	24.87	Gd-159	18h	3.4

TABLE 1.10 (Continued)

Target		Product		
Nuclide	Natural Abundance	Nuclide	$T_{1/2}$	Cross Section (Millibarns)
		(n, 2n)		
Na-23	100	Na-22	2.58y	14
Cl-37	24.48	Cl-36	3.1×10^5y	391
Ar-40	99.60	Ar-39	270y	609
V-51	99.76	V-50	4×10^{14}y	660
Cr-52	83.76	Cr-51	27.8d	285
Mn-55	100	Mn-54	0.797y	825
Fe-56	91.66	Fe-55	2.6y	500
Co-59	100	Co-58	71.3d	145
Ni-60	26.23	Ni-59	80,000y	600
Zn-66	27.81	Zn-65	0.671y	518
As-75	100	As-74	12d	1110
Rb-85	72.15	Rb-84	34d	1500
Rb-87	27.85	Rb-86	19d	1210
Sr-86	9.86	Sr-85	64d	592
Y-89	100	Y-88	0.288y	600
Zr-96	2.80	Zr-95	65d	1560
Nb-93	100	Nb-92	10.1d	18.5
Rh-103	100	Rh-102	0.575y	1390
Ag-107	51.35	Ag-106m	8.3d	10.9
Cd-110	12.39	Cd-109	1.288y	1390
Cd-116	7.58	Cd-115m	43d	810
In-115	95.72	In-114m	49d	1540
Sn-124	5.94	Sn-123m	0.373y	1850
Sb-121	57.25	Sb-120m	5.8d	1310
Sb-123	42.75	Sb-122	2.8d	1340
I-127	100	I-126	13.3d	1120
Cs-133	100	Cs-132	6.58d	1550
Ce-142	11.07	Ce-141	33d	1600
Gd-154	2.15	Gd-153	0.647y	2030
W-182	26.41	W-181	0.384y	2290
Pt-194	32.90	Pt-193m	4.5d	2320
Au-197	100	Au-196	5.6d	1722
Ra-226	(100)	Ra-225	14.8d	891
Th-232	(100)	Th-231	1.067d	1440
U-238	99.27	U-237	6.75d	790
Pu-239	—	Pu-238	86.4y	93
		(n, p)		
N-14	99.62	C-14	5700y	100
S-32	95.00	P-32	14.3d	350

TABLE 1.10 (Continued)

Target		Product		
Nuclide	Natural Abundance	Nuclide	$T_{1/2}$	Cross Section (Millibarns)
		(n, p)		
Cl-35	75.52	S-35	87d	122
K-39	93.10	Ar-39	270y	354
Sc-45	100	Ca-45	0.417y	57
Fe-54	5.82	Mn-54	0.797y	373
Co-59	100	Fe-59	45d	80
Ni-58	67.84	Co-58	71.3d	237
Cu-63	69.09	Ni-63	130y	94
Ru-85	72.15	Kr-85	10.3y	32
Y-89	100	Sr-89	50.4d	23
Zr-90	51.46	Y-90	64.2h	240
Zr-91	11.22	Y-91	61d	75
Mo-95	15.72	Nb-95	35d	72
Sb-123	42.75	Sn-123m	0.373y	5.8
I-127	100	Te-127m	0.301y	12
Xe-131	21.18	I-131	8.1d	6.6
Cs-133	100	Xe-133	5.27d	10.5
Ba-136	7.81	Cs-136	12.9d	38
Ce-140	88.48	La-140	40.2h	11
Pr-141	100	Ce-141	33d	4.5
Nd-143	12.17	Pr-143	13.7d	11.5
Sm-144	3.09	Pm-144	0.822y	512
Sm-148	11.24	Pm-148m	41.8d	32
Eu-151	47.82	Sm-151	73y	26
Gd-155	14.73	Eu-155	1.7y	12
Gd-156	20.47	Eu-156	15.4d	6
Dy-161	18.88	Tb-161	6.88d	8
Ir-191	37.30	Os-191	15d	9.2
Pt-196	25.30	Ir-196	9.7d	1.9
Hg-199	16.84	Au-199	3.14d	10.8
Tl-203	29.50	Hg-203	48d	30
Np-237	—	U-237	6.8d	1.3
Pu-239	—	Np-239	2.33d	3.0
		(n, alpha)		
Li-6	7.42	H-3	12.46y	26
Al-27	100	Na-24	14.9h	120
Cl-35	75.52	P-32	14.3d	191
Ti-48	73.98	Ca-45	0.417y	48
V-51	99.76	Sc-48	1.838d	28.6
Fe-54	5.82	Cr-51	27.8d	270

TABLE 1.10 (Continued)

Target		Product		
Nuclide	Natural Abundance	Nuclide	$T_{1/2}$	Cross Section (Millibarns)
		(n, alpha)		
Y-89	100	Rb-86	19d	5.3
Zr-92	17.11	Sr-89	54d	10.1
Nb-93	100	Y-90	64h	9.5
La-139	99.91	Cs-136	12.9d	1.6
Nd-142	27.11	Ce-139	0.384y	2.1
Nd-146	17.22	Ce-143	33h	2.6
Gd-156	20.47	Sm-153	47	3.2
Tb-159	100	Eu-156	15.1d	2.2
Pt-194	32.90	Os-191	15d	1.3

SOURCE: Kenna and Harrison, Reference 19.

TABLE 1.11 Neutron Sources Based on Gamma-*n* Reactions

	Gamma Source		Neutrons Produced	
Nuclide	$T_{1/2}$	γ Energy (MeV)	Energy (MeV)	Yield ($n/10^6$ dis)
		D_2O Target		
Na-24	14.8h	2.76	0.22	7.8
Mn-56	2.56h	2.11	0.22	0.08
Ga-72	14.1h	2.20	0.13	1.9
La-140	40.2h	1.60	0.13	0.2
Ra-226	6.7y	2.62	0.20	2.6
Th-228	1.90y	2.62	0.20	2.6
		Be Target		
Na-24	14.8h	2.76	0.83	3.7
Mn-56	2.56h	2.11	0.1, 0.3	0.8
Ga-72	14.1h	2.20	0.27	1.6
Y-88	107d	1.84	0.16	2.7
In-116	54m	2.11	0.1, 0.3	0.2
Sb-124	60.3d	1.70	0.02	5.1
La-140	40.2h	1.60	0.62	0.06
Ra-226	1622y	1.76	0.12, 0.51	0.8
Ra-228	6.7y	2.62	0.8	0.9

SOURCES: Modified from References 20 and 21.

1.4.4 The Photoneutron Reaction

Certain light elements will emit neutrons when irradiated with gamma rays of sufficient energy. The most practical light element targets for this reaction are heavy water (D_2O) and beryllium metal.

Table 1.11 gives characteristics of some photoneutron sources, compiled from the "Radiological Health Handbook"[20] and the *Compendium on Radiation Shielding*, vol. 1.[21] The yield values in the table are expressed as number of neutrons per 10^6 disintegrations of the gamma source when the latter is placed 1 cm away from 1 gram of the heavy water or Be target.

While some neutrons are undoubtedly generated by gamma action in high-level wastes, they will be quickly absorbed and the photoneutron reaction is thus of minor interest in waste management. The more pertinent (alpha, *n*) reaction (at least for handling concentrated plutonium or americium wastes) is discussed in Section 2.5.

1.5 RULES OF THUMB

Certain rough approximations are useful for quick evaluation of problems encountered in day-to-day handling of radioactivity. Some of these are given below. The list could be extended for certain more specialized situations, as in References 20 and 22.

1. About 200 MeV of energy is released per atom during heavy element fission. Of this quantity, 185 MeV is released instantaneously, the remaining 15 MeV being largely emitted later in the form of radioactive decay.[15]
2. The power produced in a steady-state reactor by the fission of one gram of a heavy isotope is roughly one megawatt, that is, about one gram of fission products is produced per megawatt-day (MWD).
3. In a reactor operating for more than 4 days, the total fission products are about 3 Ci/watt at 1.5 minutes after shutdown; 75 Ci/MWD 2 years after shutdown.
4. The quantity in curies of a short-lived fission product in a reactor that has been operated about four or more times the half-life can be estimated by multiplying the fission yield (see Section 1.1.2) times the reactor power level in watts.
5. The radioactivity of an active isotope is reduced to less than 1% of its original level after seven half-lives and to about 0.1% after 10 half-lives.
6. The average energy of a beta particle spectrum is approximately one-third the maximum energy.

7. The range of a beta particle in air is about 12 feet per MeV.
8. The range of beta particles in any material, expressed as g/cm^2 (thickness of the material in cm divided by the density in g/cm^3) is approximately one-half the maximum energy in MeV.
9. A beta particle must possess at least 70 keV of energy to penetrate the dead layer (0.07 mm) of the skin.
10. The comparable figure for an alpha particle is 7.5 MeV for penetration of the skin barrier.
11. For unshielded sources of gamma rays having energies in the 0.07 to 4 MeV range, the exposure rate at d cm (within 10%) is given by

$$\text{mR/hr} = 5.2 \times 10^6 \times \frac{\text{Ci}\, E}{d^2} \tag{1.8}$$

where Ci is the number of curies of activity in the source and E is the average gamma energy in MeV. The expression assumes an essentially point source and an average of one gamma per disintegration. As a special case:

$$\text{Exposure rate at 1 foot} = (6)(\text{Ci})(E)\ \text{R/hr} \tag{1.9}$$

12. The last two equations are partially derived from the Inverse Square Law, which states that the radiation exposure and dose from an unshielded point source varies inversely as the square of the distance from the source if absorption between the source and the point of measurement can be ignored. When this rule is applied to a small source emitting N neutrons per second, the flux at d cm is

$$\text{n/cm}^2\text{/s} = \frac{N}{4\pi d^2} = \frac{0.080N}{d^2} \tag{1.10}$$

TWO

NUCLIDES OF THE HEAVIER ELEMENTS

2.1 BUILDUP CHAINS

Transuranic element buildup in a reactor follows a fixed pattern because of the restrictions of decay and fission half-lives and of neutron cross sections. There are, of course, numerous side reactions of little importance to waste management because of low yields and/or short half-lives of the products. For all practical purposes, progressive neutron captures on the first-appearing nuclide of a new element will form a longer- or lesser-length series of long-lived alpha emitters until a beta-emitting isotope is produced, as shown in Figure 2.1. (The curium isotope chain is the best example.) The beta decay yields a nuclide of the next highest element, and the capture process then continues. In most cases, the "crossover" beta emitter has a relatively short half-life, so there is no further significant buildup beyond that point of even heavier isotopes of the parent element. Beta-emitting ^{241}Pu is an exception. Its half-life (13.2y) is long enough so that a sufficient number of atoms can accumulate to allow buildup of ^{242}Pu and ^{243}Pu, and thus to maintain the primary production chain. The half-life, however, is short enough to generate a secondary loop in that chain, as shown in Figure 2.2. This loop is of particular importance in waste management since it yields ^{241}Am and ^{242}Cm. The latter is relatively short lived (163d) for a major alpha emitter, and therefore intensely radioactive and a potential problem

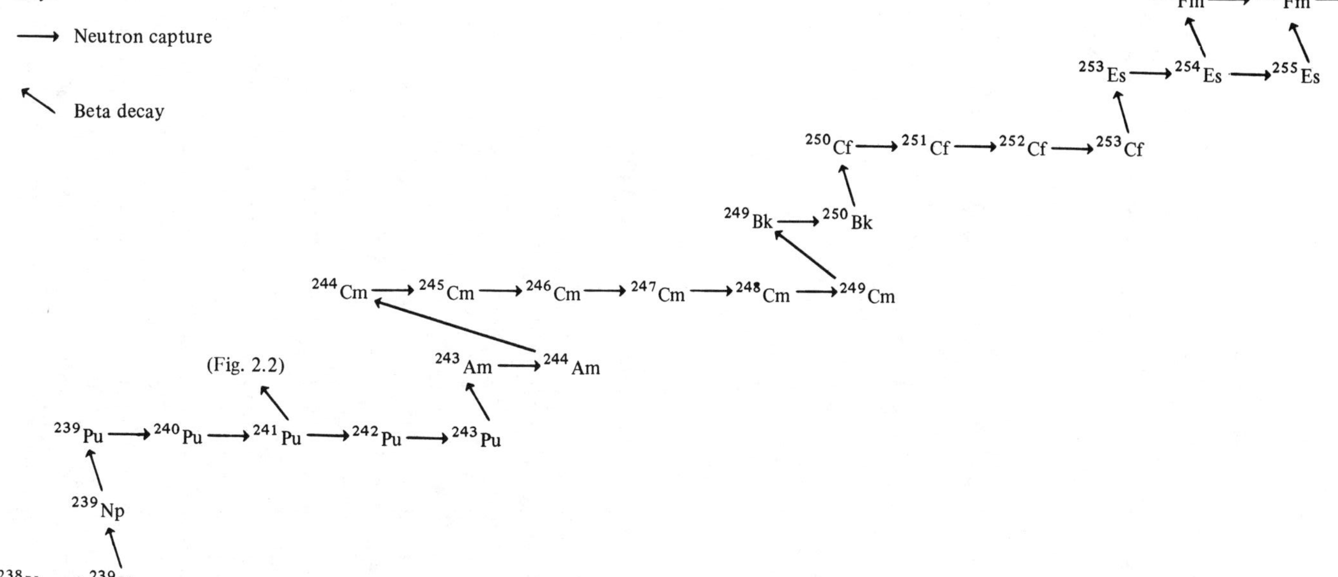

FIGURE 2.1. Transuranic nuclide buildup in reactors (a) the primary chain.

in fresh wastes. It also acts as a significant producer of ^{238}Pu. Americium-241 is longer lived (458y), but because of its mode of generation continues to be produced in the fuel long after reactor discharge and in any plutonium residues in derived wastes, depending on the ^{241}Pu content of the fuel at the time of reactor shutdown.

The isotopes heavier than ^{244}Cm shown in Figure 2.1 are normally only of academic interest to waste-management specialists, although of great importance to the research chemist. Milligram or even gram amounts of ^{252}Cf can be produced in special high-flux reactors with appropriate targets, but yields drop drastically as one goes up the buildup chain, and half-lives typically become much shorter past the curiums. Reactor production of even trace amounts of elements above californium is very difficult.

On the other hand, surprising quantities of two isotopes not shown in Figures 2.1 or 2.2—^{237}Np and ^{238}Pu—are produced in reactor fuels. Sketches of the derivation mechanisms for these nuclides are given in following subsections, along with the production routes of two "nuisance" isotopes (from the point of view of recovering and reusing the corresponding fissile element), ^{232}U and ^{236}Pu.

2.1.1 The Primary Buildup Chain

This is shown in Figure 2.1. In practice, because of sharply declining yields in going up the chain, production of the very heavy elements is based on high-flux irradiation of preseparated targets such as ^{242}Pu, ^{243}Am, or ^{244}Cm. (Use of these or heavier targets eliminates the ^{241}Am-^{242}Cm loop as a side benefit.) The quantity of ^{238}U that would have to be processed would be in the metric

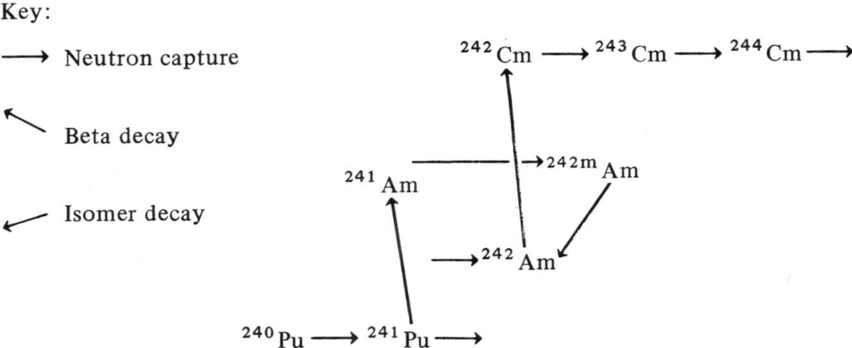

FIGURE 2.2. Transuranic nuclide buildup in reactors (b) the ^{241}Am-^{242}Cm secondary chain.

ton range if one started at the bottom of the chain, with only milligram down to nanogram quantities being produced of the various nuclides at the top.

As an example of buildup, Figure 2.3 shows computer-calculated yield curves for an ^{243}Am target irradiated in a 2×10^{15} n/cm^2/sec thermal neutron flux.[23]

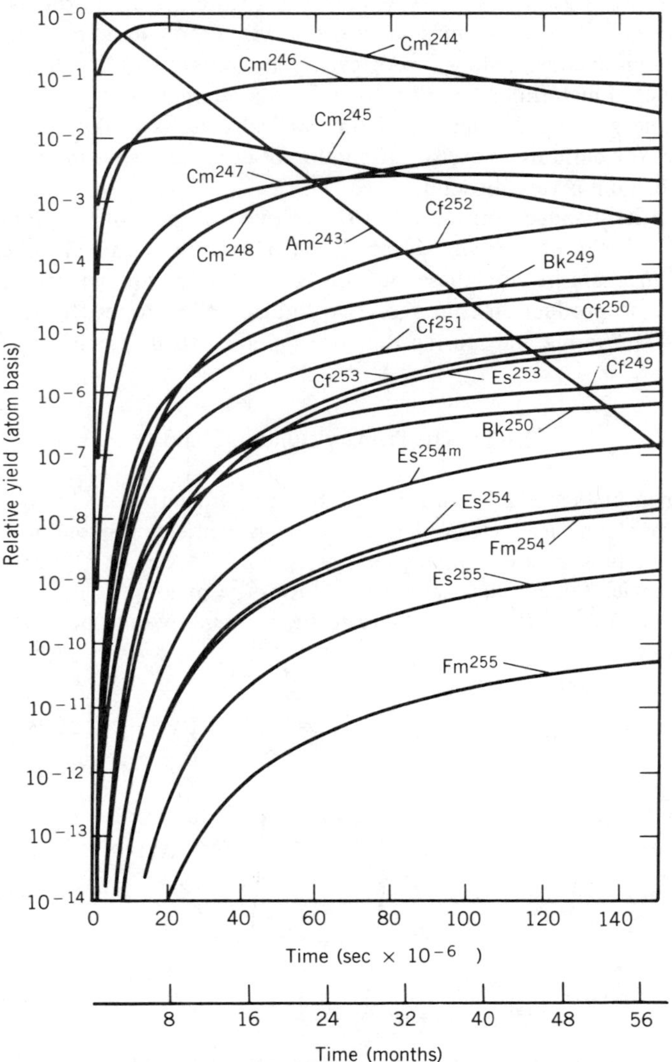

FIGURE 2.3. Example of production of very heavy elements. Yield curves; target: Am^{243}; flux: 7×10^4 neutrons/cm^2/sec.

2.1.2 The ^{241}Am-^{242}Cm Secondary Loop

This is shown in Figure 2.2. It involves a rather unusual (but not unique) situation wherein an excited state (152y ^{242m}Am) has a much longer half-life than the beta-emitting ground state (16.0h ^{242}Am). This latter is the crossover point in the loop to produce 163d ^{242}Cm. This alpha emitter decays to 86y ^{238}Pu, considerably less active, but still a major contributor to the alpha activity in the plutonium fraction of the fuel or in the derived wastes.

2.1.3 Neptunium-237

This isotope has a very long half-life (2.14×10^6 y) and accordingly accumulates during irradiation to the point where, in terms of mass (but not of activity), it becomes the major transuranic nuclide in the fuel next to ^{239}Pu. Production methods are:

1. $^{235}\text{U}\,(n, \gamma)\,^{236}\text{U}\,(n, \gamma)\,^{237}\text{U} \xrightarrow{\text{beta}} {}^{237}\text{Np}$
2. $^{238}\text{U}\,(n, 2n)\,^{237}\text{U} \xrightarrow{\text{beta}} {}^{237}\text{Np}$
3. $^{241}\text{Am} \xrightarrow{\text{alpha}} {}^{237}\text{Np}$

2.1.4 Plutonium-238

Production mechanisms:

1. $^{237}\text{Np}\,(n, \gamma)\,^{238}\text{Np} \xrightarrow{\text{beta}} {}^{238}\text{Pu}$
2. $^{242}\text{Cm} \xrightarrow{\text{alpha}} {}^{238}\text{Pu}$
3. $^{239}\text{Pu}\,(n, 2n)\,^{238}\text{Pu}$

The concentration of ^{237}Np and ^{238}Pu in the fuel and the relative importance of the different production mechanisms varies with the neutron spectrum in the reactor, the irradiation time, the length of the cooling period before reprocessing, and so forth.

2.1.5 Uranium-233 and Uranium-232

Uranium-233 is produced by irradiation of thorium:

$$^{232}\text{Th}\,(n, \gamma)\,^{233}\text{Th} \xrightarrow{\text{beta}} {}^{233}\text{Pa} \xrightarrow{\text{beta}} {}^{233}\text{U}$$

This uranium isotope is essentially the only candidate other than ^{235}U and ^{239}Pu that can be considered for application as reactor fuel. It behaves in a straightforward manner in a reactor, with part of it fissioning and a smaller fraction forming heavier uranium isotopes through neutron capture.

A complication, however, comes about through the reactions

$$^{232}\mathrm{Th}\ (n, 2n)\ ^{231}\mathrm{Th} \xrightarrow{\text{beta}} {}^{231}\mathrm{Pa}\ (n, \gamma)\ ^{232}\mathrm{Pa} \xrightarrow{\text{beta}} {}^{232}\mathrm{U}$$

and

$$^{230}\mathrm{Th}\ (n, \gamma)\ ^{231}\mathrm{Th} \xrightarrow{\text{beta}} {}^{231}\mathrm{Pa}\ (n, \gamma)\ ^{232}\mathrm{Pa} \xrightarrow{\text{beta}} {}^{232}\mathrm{U}$$

The ^{232}U thus produced alpha decays to ^{228}Th and at that point enters the naturally occurring "4n" decay series shown in Section 2.3 below. A series of alpha- and gamma-emitting daughters results, so that reprocessing to recover the ^{233}U is considerably complicated, with a lesser effect on handling of the derived wastes.

2.1.6 Plutonium-236

This is a similar but lesser problem in reprocessing to recover plutonium. The isotope arises by

$$^{237}\mathrm{Np}\ (n, 2n)\ ^{236m}\mathrm{Np} \xrightarrow{\text{beta}} {}^{236}\mathrm{Pu}$$

The 2.85y product alpha decays to ^{232}U and thus generates the same reprocessing problems as discussed for ^{233}U in the last subsection. (Even more difficulty arises because of these contaminants during refabrication of the recovered U or Pu into new fuel.)

2.2 RADIOMETRIC PROPERTIES OF THE VERY HEAVY NUCLIDES

Tables 2.1–2.3 summarize radiometric data for nuclides of the very heavy elements. These are defined here as isotopes of elements Z = 89 (actinium) through Z = 100 (fermium). This classification involves some overlap with the naturally occurring radioactive species considered in the next section. The grouping was chosen because these are the bulk of the "actinide" elements comparable to the "lanthanide" (rare earth) series appearing earlier in the Periodic Table. (The actinides actually extend through element 103.)

Alpha, beta, and EC types of decay are such as to dominate the rate of disappearance of the heavy element isotopes of interest here. The corresponding half-lives are listed as "Normal" in Table 2.1. The values used are from the critical survey made by the nuclear chemistry group at Berkeley Lawrence Laboratory in 1979, and quoted in the Nuclear Wallet Cards.[12] A second new and independent mode of decay in the very heavy nuclides is by spontaneous fission (SF). These heavy species thus have two different half-lives, with the SF mode normally being of much less importance than the normal. Spontaneous fission half-lives given in Table 2.1 were taken from the review by Schirmer and Wächter.[24] Two sets of specific activities can thus be calculated, and are given in the table. Since there are normally many factors of 10 differences in the time spans covered by the two half-lives of a particular isotope, SF decay makes very little contribution to the curie and heat-release totals, and can be ignored. The exception among the isotopes listed here is ^{252}Cf where the value of 541 Ci/g, based only on normal decay, is raised to 557 if disintegration due to spontaneous fission is also included.

The disintegration energy data (Q values) of Table 2.2 were taken from the Heath compilation.[9a] (Since data are not given for ^{234m}Pa, ^{236m}Np, or ^{254m}Es, these nuclides are not included in Table 2.2. Cross-section data for the same three isotopes are also not available and are left out of Table 2.3). Numbers following the decay mode notations in Table 2.2 indicate the percentage of decays by the indicated mechanism.

The capture and fission cross sections in Table 2.3 are from the Brookhaven National Nuclear Center summary.[16] The few fast neutron values available were taken from Kenna and Harrison.[19]

All of the other numbers in the tables were calculated using either the equations given in Section 1.1.3.1 or by Equation 2.1:

$$\text{cm}^3 \text{ helium/day/g} = (3.1 \times 10^{-9})\,(\text{Sp. Act.})\,(\text{fr.}) \qquad (2.1)$$

where the specific activity is per microgram as given in Table 2.1, and "fr." is the fraction of decays generating alpha particles. The volume is at Standard Temperature and Pressure (STP). It should be noted that this equation assumes that the original gram of material remains essentially unchanged from day to day. This obviously will not be the case with a short-lived nuclide unless it is supported by a long-lived parent.

A deliberate omission from the tables is the indication of the energies of the emitted particles or photons. Most of the very heavy nuclides are alpha emitters, and each has a complicated decay scheme[11] that results in the alphas coming off with slightly varying energies. Alphas are not a shielding problem, so small differences in their energies are not of much practical importance in waste man-

Table 2.1 Heavy Nuclide Characteristics—Half-Life Related

Nuclide	"Normal" Decay		Spontaneous Fission		Ci/g
	$T_{1/2}$	Specific Activity (d/s/μg)	$T_{1/2}$	Specific Activity (f/s/μg)	
Actinium					
227	21.77y	2.67×10^{6}			72.2
228	6.13h	8.30×10^{10}			2.25×10^{5}
Thorium					
227	18.72d	1.11×10^{9}			30800
228	1.913y	3.04×10^{7}			821
229	7300y	7900			0.213
230	80000y	717	$>1.5 \times 10^{20}$ y	$<3.8 \times 10^{-13}$	0.0194
231	25.52h	1.97×10^{10}			5.33×10^{5}
232	1.41×10^{10} y	0.00404	$>10^{21}$ y	$<5.7 \times 10^{-14}$	1.09×10^{-7}
233	22.3m	1.33×10^{12}			3.61×10^{7}
234	24.10d	8.56×10^{8}			23200
Protactinium					
231	32800y	1740			0.047
232	1.31d	1.58×10^{10}			4.31×10^{5}
233	27.0d	7.68×10^{8}			20800
234	6.75h	7.35×10^{10}			1.99×10^{5}
234m	1.175m	2.54×10^{13}			6.87×10^{8}

Uranium					
232	72y	7.90×10^{5}	8×10^{13} y	7.11×10^{-7}	21.4
233	1.592×10^{5} y	356	1.23×10^{17} y	4.60×10^{-10}	0.00963
234	2.45×10^{5} y	231	2×10^{16} y	2.82×10^{-9}	0.00623
235	7.038×10^{8} y	0.080	1.9×10^{17} y	2.96×10^{-10}	2.16×10^{-5}
236	2.342×10^{7} y	2.39	2×10^{16} y	2.80×10^{-9}	6.46×10^{-5}
237	6.75d	3.02×10^{9}			81900
238	4.468×10^{9} y	0.0124	7.19×10^{15} y	7.71×10^{-9}	3.36×10^{-7}
239	23.5m	1.28×10^{12}			3.35×10^{7}
Neptunium					
236	1.1×10^{5} y	508			0.0138
236m	22.5h	2.18×10^{10}			5.91×10^{5}
237	2.14×10^{6} y	26.0	$>4 \times 10^{16}$ y	$<1.4 \times 10^{-9}$	7.04×10^{-4}
238	2.117d	9.59×10^{9}			2.60×10^{5}
239	2.35d	9.83×10^{9}			2.33×10^{5}
Plutonium					
236	2.85y	1.96×10^{7}	3.5×10^{5} y	160	531
237	45.4d	4.49×10^{8}			12200
238	87.74y	6.32×10^{5}	4.9×10^{10} y	0.00113	17.1
239	24100y	2290	5.5×10^{15} y	1.00×10^{-8}	0.0620
240	6570y	8370	1.17×10^{11} y	4.70×10^{-4}	0.221
241	14.4y	3.80×10^{6}			103
242	3.76×10^{5} y	145	7.06×10^{10} y	7.72×10^{-4}	0.00392
243	4.956h	9.62×10^{10}			2.61×10^{6}
244	8.1×10^{7} y	0.668	2.5×10^{10} y	0.00216	1.81×10^{-5}
Americium					
241	433y	1.26×10^{5}	2.3×10^{14} y	2.38×10^{-7}	3.42
242	16.01h	2.99×10^{10}			8.11×10^{5}

Table 2.1 (Continued)

Nuclide	"Normal" Decay $T_{1/2}$	"Normal" Decay Specific Activity (d/s/μg)	Spontaneous Fission $T_{1/2}$	Spontaneous Fission Specific Activity (f/s/μg)	Ci/g
Americium (*Continued*)					
242m	152y	3.59×10^{5}	9.5×10^{11} y	5.74×10^{-5}	9.71
243	7370y	7370	3.35×10^{13} y	1.62×10^{-6}	0.199
244	10.1h	4.71×10^{10}			1.26×10^{6}
244m	26m	1.10×10^{12}			2.97×10^{7}
Curium					
242	162.8d	1.23×10^{8}	6.09×10^{6} y	8.96	3330
243	28.5y	1.91×10^{6}			51.5
244	18.11y	2.99×10^{6}	1.346×10^{7} y	4.02	80.8
245	8500y	6340			0.171
246	4700y	11400	1.66×10^{7} y	3.23	0.309
247	1.6×10^{7} y	3.34			9.03×10^{-5}
248	3.5×10^{5} y	152	4.9×10^{6} y	11.6	0.00411
249	65m	4.29×10^{11}			1.16×10^{7}
Berkelium					
249	0.88y	6.02×10^{7}	1.5×10^{9} y	0.0353	1630
250	3.22h	1.44×10^{11}			3.90×10^{6}
Californium					
249	351y	1.51×10^{5}	6×10^{10} y	8.83×10^{-4}	4.08
250	13.1y	4.03×10^{6}	16600y	3180	109

251	900y	58400			1.58
252	2.62y	2.00×10^{7}	85.5y	6.15×10^{5}	541
253	17.8d	1.07×10^{9}			29100
Einsteinium					
253	20.47d	9.33×10^{8}	6.3×10^{5}y	82.8	25300
254	276d	6.88×10^{7}			1870
254m	39.3h	1.16×10^{10}	1.5×10^{5}y	346	3.15×10^{5}
255	38.3d	4.95×10^{8}	3630y	14260	13400
Fermium					
254	3.24h	1.41×10^{11}	246d	7.73×10^{7}	3.82×10^{6}
255	20.1h	2.26×10^{10}	>60y	$<8.63 \times 10^{5}$	6.13×10^{5}

SOURCES: Varied, please see text.

Table 2.2 Heavy Nuclide Characteristics—Energy-Related

Nuclide	Decay Mode	Q Value (MeV)	Watts/Gram	Watts/Curie	Helium ($cm^3/g/d$)
Actinium					
227	Beta-98.62 Alpha-1.38	0.113	0.0485	6.60×10^{-4}	1.18×10^{-4}
228	Beta	2.14	2.65×10^{5}	0.0127	
Thorium					
227	Alpha	6.145	1120	0.0364	3.56
228	Alpha	5.521	26.8	0.0327	0.0976
229	Alpha	5.167	0.00655	0.0306	2.45×10^{-5}
230	Alpha	4.767	5.49×10^{-4}	0.0283	2.30×10^{-6}
231	Beta	0.381	12000	0.00226	
232	Alpha	4.08	2.64×10^{-9}	0.0242	1.30×10^{-11}
233	Beta	1.246	2.66×10^{7}	0.00739	
234	Beta	0.263	36.1	0.00156	
Protactinium					
231	Alpha	5.148	0.00144	0.0305	5.59×10^{-6}
232	Beta	1.34	3410	0.00794	
233	Beta	0.571	70.3	0.00338	
234	Beta	2.23	2.63×10^{5}	0.0132	
Uranium					
232	Alpha	5.414	0.687	0.0321	0.00254
233	Alpha	4.909	2.81×10^{-4}	0.0291	1.14×10^{-6}
234	Alpha	4.856	1.80×10^{-4}	0.0288	7.41×10^{-7}

235	Alpha	4.681	6.00×10^{-7}	0.0277	2.57×10^{-10}
236	Alpha	4.573	1.75×10^{-6}	0.0271	7.67×10^{-9}
237	Beta	0.517	252	0.00306	
238	Alpha	4.268	8.51×10^{-9}	0.0253	3.98×10^{-11}
239	Beta	1.28	2.53×10^{7}	0.00759	
Neptunium					
236	EC-91 Beta-9	0.72	5.9×10^{-5}	0.00427	
237	Alpha	4.956	2.07×10^{-5}	0.0294	8.35×10^{-8}
238	Beta	1.29	1980	0.00765	
239	Beta	0.723	996	0.00429	
Plutonium					
236	Alpha	5.868	18.5	0.00348	0.00629
237	EC-99+ Alpha-0.003	0.23	16.5	0.00136	4.62×10^{-5}
238	Alpha	5.592	0.568	0.0331	0.00203
239	Alpha	5.243	0.00193	0.0311	7.35×10^{-6}
240	Alpha	5.255	0.00707	0.0311	2.67×10^{-5}
241	Beta-99+ Alpha-0.0024	0.021	0.0128	1.24×10^{-4}	2.93×10^{-7}
242	Alpha	4.98	1.16×10^{-4}	0.0295	4.65×10^{-7}
243	Beta	0.59	9110	0.00350	
244	Alpha	4.66	5.00×10^{-7}	0.0276	2.14×10^{-9}
Americium					
241	Alpha	5.640	0.115	0.0344	4.04×10^{-4}
242	Beta-82.7 EC-17.3	0.672	3230	0.00398	

Table 2.2 (Continued)

Nuclide	Decay Mode	Q Value (MeV)	Watts/Gram	Watts/Curie	Helium ($cm^2/g/d$)
Americium (*Continued*)					
242m	IT-99.52	1.50	0.0865	0.00889	5.53×10^{-6}
	Alpha-0.48				
243	Alpha	5.439	0.00644	0.0322	2.37×10^{-5}
244	Beta	0.387	2920	0.00229	
Curium					
242	Alpha	ca. 6.1	120	0.036	0.395
243	Alpha-99.74	6.16	1.89	0.0365	0.00613
	EC-0.26				
244	Alpha	5.902	2.83	0.0350	0.00960
245	Alpha	5.624	0.00573	0.0333	2.09×10^{-5}
246	Alpha	5.476	0.0100	0.0325	3.66×10^{-5}
247	Alpha	ca. 5.3	2.8×10^{-6}	0.032	1.1×10^{-8}
248	Alpha-91.74	5.161	1.26×10^{-4}	0.0306	5.00×10^{-7}
	SF-8.26				
249	Beta	0.9	61700	0.00533	
Berkelium					
249	Beta-99+				
	Alpha-0.0015	0.126	1.22	0.00747	2.90×10^{-6}
250	Beta	1.76	40700	0.0104	

Californium					
249	Alpha	6.295	0.153	0.0373	4.87×10^{-4}
250	Alpha	6.128	3.97	0.0363	0.0129
251	Alpha	5.94	0.0577	0.0352	1.87×10^{-4}
252	Alpha-96.91 SF-3.09	6.217	20.1	0.0368	0.0622
253	Beta-99.69 Alpha-0.31	0.288	49.5	0.00171	0.0106
Einsteinium					
253	Alpha	6.747	1010	0.0400	2.99
254	Alpha	6.623	73.1	0.0393	0.221
255	Beta-92 Alpha-8	ca. 1.04	82.4	0.0062	0.127
Fermium					
254	Alpha-99+ SF-0.059	7.310	1.65×10^{5}	0.0433	452
255	Alpha	7.244	26300	0.0429	72.5

SOURCES: Varied, please see text.

Symbols used: EC, electron capture; IT, internal transfer, SF, spontaneous fission.

Table 2.3 Heavy Nuclide Characteristics—Cross Sections

	Cross Sections (Barns)				
	Capture		Fission		
Nuclide	Thermal	Resonant	Thermal	Resonant	Miscellaneous
Actinium					
227	515		$<2 \times 10^{-6}$		
228					
Thorium					
227	200		1500		
228	123		<0.3		
229	54	1000	30.5	464	
230	23.2	1010	$<1.2 \times 10^{-6}$		0.0046 (*n*, alpha)
231					
232	7.40	85	0.039		1.44 (*n*, 2*n*) 0.0002 (*n*, *p*)
233	1500	400	15		
234	1.8		<0.01		
Protactinium					
231	210	1500	0.010		
232	760		700		
233	21–234m 20–234g		<0.1		
234			<5000		
234m		895	<500		
Uranium					
232	73.1	280	75.2	320	
233	47.7	140	531.1	764	
234	100.2	630	<0.65		0.9 (*n*, 2*n*)
235	98.6	144	582.2	275	0.7 (*n*, 2*n*) 0.0019 (*n*, *p*)
236	5.2	365			
237	411	290	<0.35		
238	2.70	275			0.79 (*n*, 2*n*) 0.0002

Table 2.3 (Continued)

	Cross Sections (Barns)				
	Capture		Fission		
Nuclide	Thermal	Resonant	Thermal	Resonant	Miscellaneous
Uranium (*Continued*)					
					(n, p)
					0.0015
					(n, alpha)
239	22		14		
Neptunium					
236			2500		
236m					
237	169	660	0.019		0.39
					(n, 2n)
					0.0013
					(n, p)
238			2070	880	
239	31–240m				
	14–240g		$<10^{-6}$		
Plutonium					
236			165		
237			2400		
238	547	141	16.5	24	
239	268.8	200	742.5	301	0.093
					(n, 2n)
					0.003
					(n, p)
240	289.5	8013	0.030		
241	368	162	1009	570	
242	18.5	1130	<0.2	5	
243	60		196		
244	1.7	43			
Americium					
241	83.8–242m	202			
	748–242g	1275	3.15	21	
242			2900		
242m	1400	7000	6600	1570	
243	75.2–244m	1709			
	4.1–244g	111	<0.07		
244			2300		
244m			1600		

Table 2.3 (Continued)

	Cross Sections (Barns)				
	Capture		Fission		
Nuclide	Thermal	Resonant	Thermal	Resonant	Miscellaneous
Curium					
242	16	150	<5		
243	225	2345	600	1860	
244	13.9	650	1.2	12.5	
245	345	101	2020	750	
246	1.3	121	0.17	10.0	
247	60	800	90	880	
248	4	275	0.34	13.2	
249	1.6				
Berkelium					
249	1300[a]	1240[a]			
250			960		
Californium					
249	465	765	1660	2114	
250	2030	11600[a]	<350		
251	2850	1600	4300	5900	
252	20.4	43.5	32	110	
253	17.6		1300		
Einsteinium					
253	155-254m	3000			
	<3-254g	4300			
254	<40		2900	2190	
254m	1.3		1840		
255	43				
Fermium					
254	76				
255	26		3400		

SOURCES: References 16 and 19.

[a] Total absorption, that is, capture plus fission.

agement. Heath's very detailed listings of alpha[9a] and gamma[9b] energies are readily available in the *Handbook of Chemistry and Physics*.

2.2.1 Neutron Emission

Wastes containing the heavy elements generate (generally very low) neutron fluxes either from the (alpha, n) reaction or from spontaneous fission. The level of neutrons from the former mechanism depends on many variables, including the presence of very light elements in the waste. Neutron levels from spontaneous fission depend not only on the concentration of heavy nuclides, but also on their nature. The average number of neutrons emitted per fission, "nu-value," increases more or less regularly with increase in the mass of the generating nuclide.

Nu values taken from the Brookhaven Nuclear Data report[16] are given in Table 2.4. The average number of neutrons emitted per spontaneous fission is designated ν_{sp}. The last two columns (presented for comparison only) refer to neutron-induced fission, with ν being the total of both prompt and delayed neutrons, and ν_p the prompt neutrons only. The neutrons per second per microgram figures in column 3 were obtained by multiplying ν_{sp} by the SF activities given in Table 2.1.

(The neutron emission for ^{252}Cf is given as ν. If this value, 3.74, is assumed for ν_{sp}, the resulting emission is 2.3×10^6 n/s/μg.)

2.3 NATURALLY OCCURRING HEAVY ELEMENT RADIOACTIVITIES

The properties of the lighter radioactive nuclides found in nature (^3H, ^{14}C, ^{40}K, etc.) were covered in Chapter 1. The discussion here will be of natural chains of radioactive species arising from uranium and thorium, and of the comparable synthetically produced chain headed by ^{237}Np.

2.3.1 In High-Level Waste

Some modification of the Purex solvent extraction process is used almost universally to recover uranium and plutonium from spent reactor fuels. Recovery of the two elements is never complete, although it is generally in the 98–99% or better range. The uranium and plutonium remaining behind in the aqueous phase (the raffinate) become part of the high-level radioactive waste. Any ^{237}Np produced in the fuel during irradiation tends to fractionate to some degree between the aqueous and solvent streams, but most remains in the raffinate, as do essentially all of the transplutonium elements present, primarily americium and curium. The parents of two of the naturally occurring radioactive chains (^{238}U and ^{235}U) are thus automatically present in the waste, and each chain may be supplemented by decay of heavier reactor-produced precursors, such as ^{238}Pu, ^{239}Pu, and ^{242}Pu. Similarly, the parent (of the third natural chain, ^{232}Th, can arise by decay of the ^{244}Cm-^{240}Pu-^{236}U combination. (This is a smaller contamination in uranium fuel reprocessing, but dominates if and when thorium is used as the target material for production of ^{233}U.) Other heavy precursors, such as ^{241}Am, enter the comparable neptunium series–a

Table 2.4 Nu Values—Average Neutrons Emitted per Fission

Nuclide	Spontaneous Fission			
	ν_{sp}	n/s/μg	ν	ν_p
Th-229			2.14	
Th-232	2.12	$<8.06 \times 10^{-13}$		
U-232				3.13
U-233			2.492	
U-235			2.418	
U-236	1.89	5.30×10^{-9}		
U-238	1.98	1.53×10^{-8}		
Np-236	3.12	–		
Pu-236	2.21	354		
Pu-238	2.24	0.00253		2.90
Pu-239			2.871	
Pu-240	2.17	0.00102		
Pu-241			2.927	
Pu-242	2.10	0.00162		
Pu-244	2.30	0.00497		
Am-241				3.219
Am-242m				3.264
Cm-242	2.48	22.2		
Cm-243				3.43
Cm-244	2.690	10.8		
Cm-245				3.832
Cm-246	2.96	9.56		
Cm-248	3.157	36.6		
Bk-249	3.39	0.120		
Cf-246	3.14	–		
Cf-250	3.50	11100		
Cf-252			3.74	
Cf-254	3.93	–		
Fm-254	3.96	3.07×10^{8}		
Fm-256	3.73	–		
Fm-257	3.77	–		

SOURCE: Neutron emission data from Reference 16.

chain not found in nature, but now known to be produced in reactor fuel during irradiation.

2.3.2 The Four Chains

Kirby gives[25] a detailed history (available in more abbreviated form in most nuclear chemistry texts) of the long and strenuous (and sometimes acrimonious) effort needed to delineate details of the naturally occurring chains since their discovery by Becquerel at about the turn of the century. In retrospect, this was a considerable intellectual and experimental accomplishment. The chemists involved initially had only a bag of mostly short-lived and confusing activities with which to deal, little idea as to which activity was associated with which element and, by today's standards, only primitive measuring instruments available for research.

One result was that the older literature in the field is loaded with an imaginative, but highly confusing, set of aliases for nuclides whose true identities could only be guessed at at the time, but which have now been assigned definite positions in the various decay chains. These older designations (Table 2.5 under "Historical Names") occasionally still appear in current literature, but are now mostly of academic interest, although "ionium" is still a convenient way of referring to ^{230}Th in order to distinguish it from natural thorium (^{232}Th).

It was noted along the line that the mass of any nuclide in the thorium chain (as they were identified), if divided by four, left no arithmetical remainder (the

Table 2.5 Natural and $(4n + 1)$ Series—Half-Lives and Activities

Nuclide	Historical Name	Half-Life	Specific Activity (d/s/μg)	Ci/g
		Thorium $(4n)$ *Series*		
Th-232	Thorium	1.41×10^{10}y	0.00404	1.09×10^{-7}
Ra-228	Mesothorium I	5.75h	8.85×10^{10}	2.40×10^{6}
Ac-228	Mesothorium II	6.13h	8.30×10^{10}	2.25×10^{6}
Th-228	Radiothorium	1.913hy	3.04×10^{7}	821
Ra-224	Thorium X	3.64d	5.92×10^{9}	1.61×10^{5}
Rn-220	Emanation (Thoron)	55.6s	3.41×10^{13}	9.24×10^{8}
Po-216	Thorium A	0.15s	1.29×10^{16}	3.49×10^{11}
Pb-212	Thorium B	10.64h	5.14×10^{10}	1.39×10^{6}
Bi-212	Thorium C	60.60m	5.41×10^{11}	1.46×10^{7}
Po-212	Thorium C′	3.04×10^{-7}s	6.47×10^{21}	1.57×10^{17}
Tl-208	Thorium C″	3.06m	1.09×10^{13}	2.95×10^{8}
Pb-208	Thorium D	Stable		

Table 2.5 (Continued)

Nuclide	Historical Name	Half-Life	Specific Activity (d/s/μg)	Ci/g
		Neptunium (4*n* + 1) *Series*		
Np-237		2.14×10^{6}y	26.0	7.04×10^{-4}
Pa-233		27.0d	7.68×10^{8}	20800
U-233		1.59×10^{5}y	356	0.00963
Th-229		7340y	7860	0.212
Ra-225		14.8d	1.45×10^{9}	39300
Ac-225		10.0d	2.15×10^{9}	58200
Fr-221		4.8m	6.55×10^{12}	1.77×10^{8}
At-217		0.032s	6.00×10^{16}	1.63×10^{12}
Bi-213		46m	7.09×10^{11}	1.92×10^{7}
Po-213		4×10^{-6}s	4.9×10^{20}	1.33×10^{16}
Tl-209		2.2m	1.51×10^{13}	4.09×10^{8}
Pb-209		3.31m	1.00×10^{13}	2.72×10^{8}
Bi-209		$>2 \times 10^{18}$y	$<3.2 \times 10^{-9}$	$<8.5 \times 10^{-16}$
Tl-205		Stable		
		Uranium (4*n* + 2) *Series*		
U-238	Uranium I	4.51×10^{9}y	0.125	3.39×10^{-7}
Th-234	Uranium X_1	24.10d	8.56×10^{8}	23200
Pa-234m	Uranium X_2	1.175m	2.54×10^{13}	6.87×10^{8}
Pa-234	Uranium Z	6.7h	7.35×10^{10}	1.99×10^{5}
U-234	Uranium II	2.47×10^{5}y	231	0.00623
Th-230	Ionium	78000y	735	0.0199
Ra-226	Radium	1602y	3.65×10^{5}	0.986
Rn-222	Emanation (Radon)	3.824d	5.69×10^{9}	1.54×10^{5}
Po-218	Radium A	3.05m	1.05×10^{13}	2.83×10^{8}
Pb-214	Radium B	26.8m	1.21×10^{12}	3.28×10^{7}
At-214	Astatine	2s	9.56×10^{16}	2.59×10^{10}
Bi-214	Radium C	19.8m	1.64×10^{12}	4.44×10^{7}
Po-214	Radium C′	1.64×10^{-4}s	1.19×10^{19}	3.22×10^{14}
Tl-210	Radium C″	1.3m	2.55×10^{13}	6.89×10^{8}
Pb-210	Radium D	22y	2.86×10^{6}	77.3
Bi-210	Radium E	5.01d	4.59×10^{9}	1.25×10^{5}
Po-210	Radium F	138.4d	1.66×10^{8}	45100
Tl-206	Radium E″	4.21m	2.55×10^{13}	2.17×10^{8}
Pb-206	Radium G	Stable		
		Actinium (4*n* + 3) *Series*		
U-235	Actinouranium	7.1×10^{8}y	0.080	2.16×10^{-5}
Th-231	Uranium X	25.52h	1.97×10^{10}	5.33×10^{5}

Table 2.5 (Continued)

Nuclide	Historical Name	Half-Life	Specific Activity (d/s/μg)	Ci/g
Pa-231	Protactinium	32500y	1756	0.0474
Ac-227	Actinium	21.772y	2.67×10^{6}	72.2
Th-227	Radioactinium	18.72d	1.11×10^{9}	30800
Fr-223	Actinium K	22m	1.42×10^{12}	3.83×10^{7}
Ra-223	Actinium X	11.435d	1.89×10^{9}	51400
Rn-219	Emanation (Actinon)	3.96s	4.81×10^{14}	1.30×10^{10}
Po-215	Actinium A	0.00178s	1.09×10^{18}	2.95×10^{13}
Pb-211	Actinium B	36.1m	9.12×10^{11}	2.47×10^{7}
At-215	Astatine	0.00010s	1.94×10^{19}	5.25×10^{14}
Bi-211	Actinium C	2.14m	1.54×10^{13}	4.16×10^{8}
Po-211	Actinium C′	0.55s	3.59×10^{15}	9.74×10^{10}
Tl-207	Actinium C″	4.77m	7.04×10^{12}	1.90×10^{8}
Pb-207	Actinium D	Stable		

SOURCE: Kirby, Reference 26.

"4n" chain, where n is an integer). If the masses in the ^{238}U chain were similarly divided by four, there was always a remainder of two (the $4n + 2$ chain); and for the ^{235}U chain, a remainder of three (the $4n + 3$ chain). Only the $(4n + 1)$ chain was missing in nature. Its discovery had to wait until the ^{237}Np grandparent and ^{233}U parent were produced as part of the intense research activity initiated by the discovery of fission in 1939.

Figures 2.4–2.7, taken from Kirby,[26] diagram present-day understanding of the naturally occurring chains and of the $(4n + 1)$ series, the latter of equal importance in waste management. (In further discussion, the $(4n + 1)$ series will be treated as if it were naturally occurring.) As indicated earlier, the parents in each chain have parents or grandparents of their own, artificially produced, but present in high-level waste raffinates at sometimes significant levels.

2.3.3 Nuclide Properties

These are given in Tables 2.5 and 2.6. Some of the data (for isotopes of actinium through uranium) duplicate information given in earlier tables of this section, but are repeated for convenience. Slightly different half-lives may appear in a few cases since different authorities are quoted. Those here are taken from Kirby[26] so as to agree with the values shown in Figures 2.4–2.7.

Tables 2.5 and 2.6 do not contain columns on the decay mechanisms of the

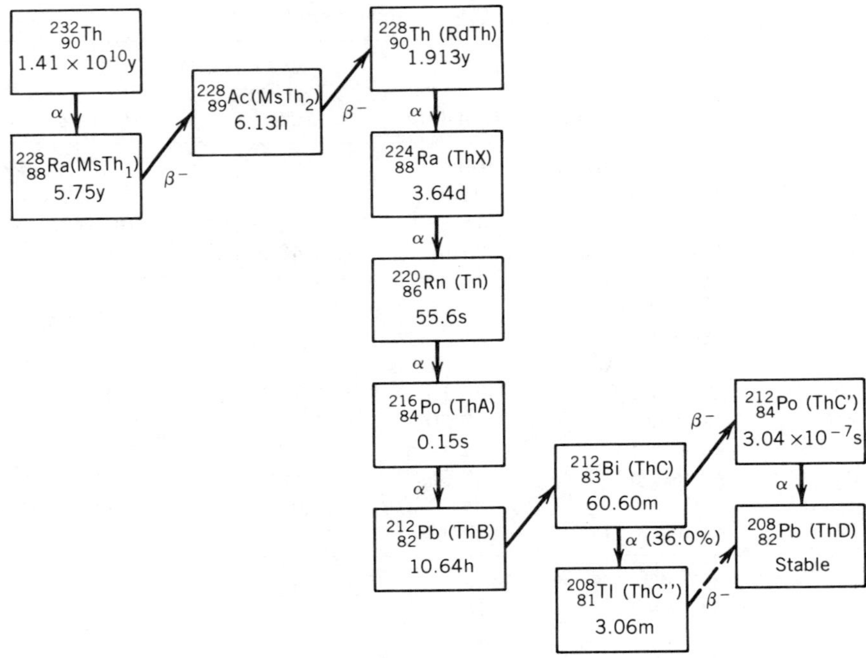

FIGURE 2.4. Thorium series ($4n$). (Reference 26.)

individual nuclides since that information is given in the figures. As with the other heavy isotopes, no effort was made to present particle or photon energies because of the complexity of the decay schemes and because of the ready availability of the data in the compilations by Heath.[9a,9b] Where the decay of a nuclide occurs by two different mechanisms, the Q value used in Table 2.6 was adjusted on a pro rata basis. The Q values themselves were again taken from Heath.[9a] Many of the numbers in Tables 2.5 and 2.6 were calculated from the equations given in Section 1.1.3.1.

2.3.3.1 Nuclear Cross Sections

Thermal and resonant region neutron-capture cross sections for the nuclides in the four chains discussed here are given in Table 2.6 with some duplication of data from earlier tables for chain members at the light and heavy ends. Cross-section values for the intermediate members are few in number because of the brevity of many of the half-lives. The same pattern applies to fast neutron reactions. The only data presented by Kenna and Harrison,[19] not given in pre-

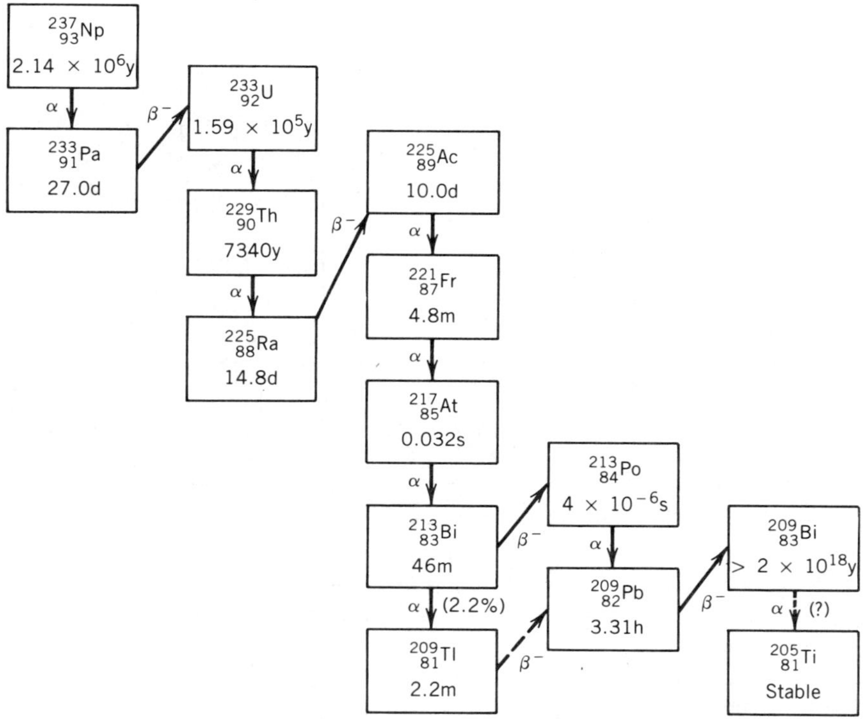

FIGURE 2.5. Neptunium series ($4n + 1$). (Reference 26.)

vious tables, are:

Target	Reaction	Cross Section (Barns)	Product
Pb-207	$(n, 2n)$	2.52	Pb-206
	(n, p)	0.0018	Tl-207
	(fn, γ)	0.005	Pb-208
Ra-226	$(n, 2n)$	0.891	Ra-225

2.3.3.2 *Helium Production*

While also true for the helium production figures given in Table 2.2, it is particularly important in the case of the naturally occurring chains to consider the contribution of the alpha-emitting daughters since the short-lived intermediates allow

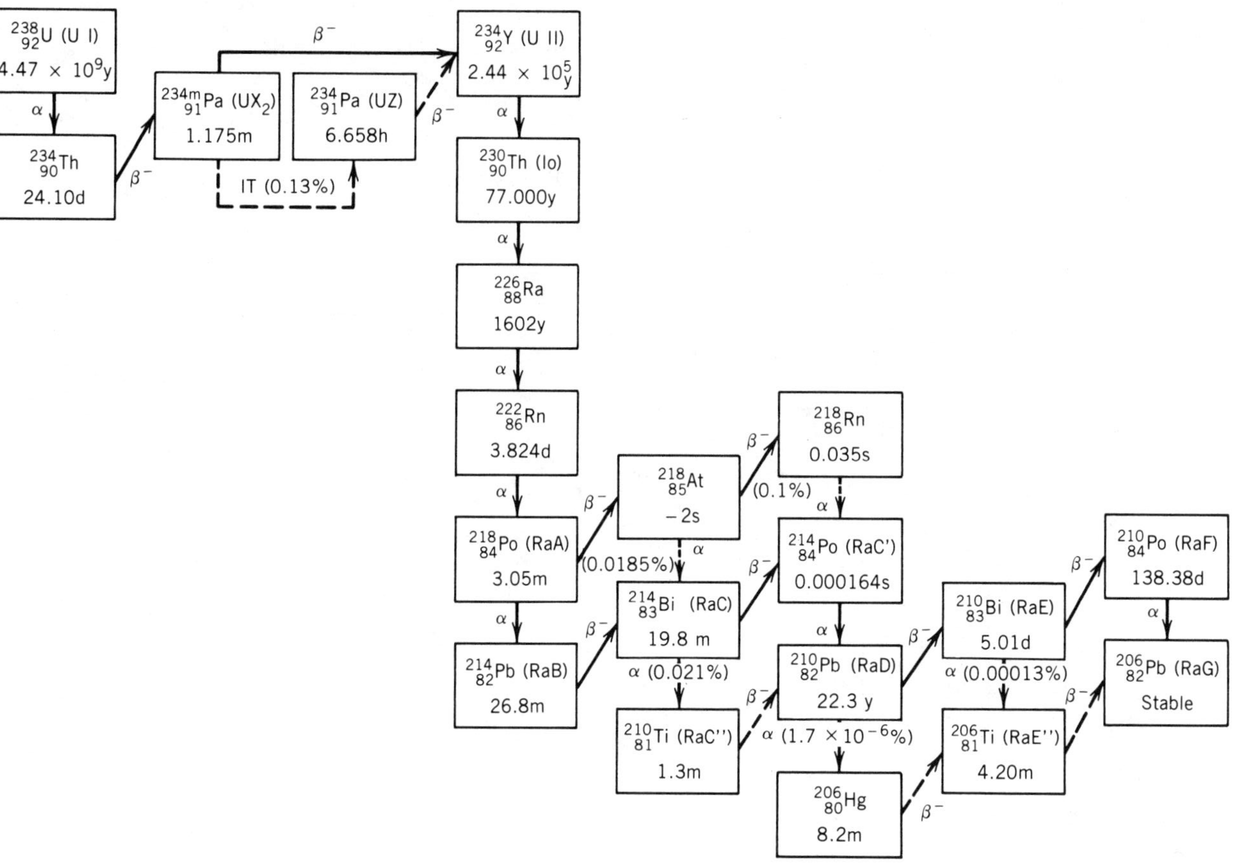

FIGURE 2.6. Uranium-radium series ($4n + 2$). (Reference 26.)

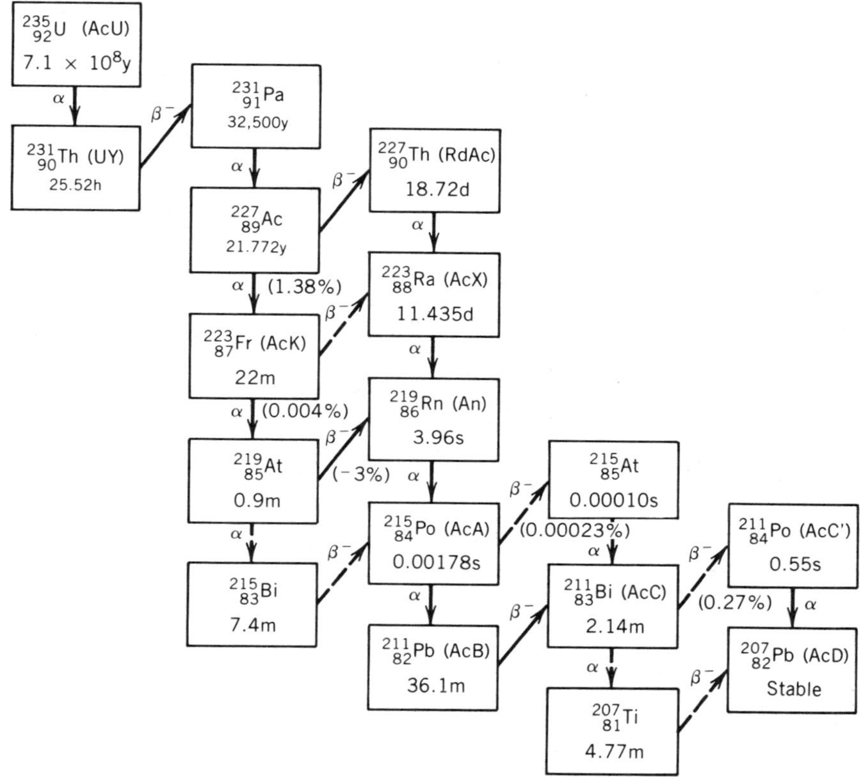

FIGURE 2.7. Uranium-actinium series $(4n + 3)$. (Reference 26.)

rapid establishment of equilibrium in many cases. Such data are presented in Table 2.7 for those nuclides having long enough half-lives to allow isolation of gram or greater amounts of material. The numbers were calculated using Equation 2.1, given earlier.

The case of ^{227}Ac is interesting. It is primarily a beta emitter, so the helium output of a gram of freshly separated actinium is very modest. Its half-life, however, still determines the rate of helium output for the entire chain after equilibrium is reached.

Bismuth-209 is considered as the stable end point of the neptunium series, that is, only seven alpha steps are assumed below the ^{237}Np parent.

2.3.4 Natural Uranium and Thorium

The author could find no literature values for the specific activities and Q values for natural uranium or thorium in equilibrium with all of their respective daugh-

Table 2.6 Natural and (4n + 1) Series—Heat and Cross-Section Data

Nuclide	Q Value (MeV)	W/g	W/Ci	Cross Sections[a] (Barns) Thermal	Resonant
		Thorium (4n) *Series*			
Th-232	4.08	2.64×10^{-9}	0.0242	7.40	85
Ra-228	0.055	780	3.26×10^{-4}		
Ac-228	2.14	2.65×10^{5}	0.0127		
Ra-224	5.787	5490	0.0343	12.0	
Rn-220	6.405	3.49×10^{7}	0.0380	<0.2	
Po-216	6.906	1.42×10^{10}	0.0409		
Pb-212	0.58	4780	0.00344		
Bi-212	3.59	3.10×10^{5}	0.0213		
Po-212					
Tl-208	4.994	8.71×10^{6}	0.0296		
Pb-208	Stable			4.87×10^{-4}	
		Neptunium (4n + 1) *Series*			
Np-237	4.956	2.07×10^{-5}	0.0294	169	660
Pa-233	0.571	70.3	0.00338	21–234m 20–234g	
U-233	4.909	2.81×10^{-4}	0.0291	47.7	140
Th-229	5.167	0.00655	0.0306	54	1000
Ra-225	0.39	90.6	0.00231		
Ac-225	5.931	2040	0.0352		
Fr-221	6.457	6.76×10^{6}	0.0383		
At-217	7.199	6.90×10^{10}	0.0427		
Bi-213	1.44	1.63×10^{5}	0.00853		
Po-213	8.54	6.68×10^{14}	0.0506		
Tl-209	3.98	9.61×10^{6}	0.0236		
Pb-209	0.635	1.02×10^{6}	0.00376		
Bi-209				0.014–210m 0.019–210g	0.19
Tl-205	Stable			0.10	0.7
		Uranium (4n + 2) *Series*			
U-238	4.268	8.51×10^{-9}	0.0253	2.70	275
Th-234	0.263	36.1	0.00156	1.8	
Pa-234m					
Pa-234	2.23	2.63×10^{5}	0.0132		895
U-234	4.856	1.80×10^{-4}	0.0288	100.2	630
Th-230	4.767	5.49×10^{-4}	0.0283	23.2	1010
Ra-226	4.65	0.0272	0.0276	11.5	222

Table 2.6 (Continued)

Nuclide	Q Value (MeV)	W/g	W/Ci	Cross Sections[a] (Barns)	
				Thermal	Resonant
Rn-222	6.68	6090	0.0396	0.72	
Po-218	6.111	1.02×10^{7}	0.0362		
Pb-214	1.04	2.01×10^{5}	0.00616		
At-218	6.39	9.76×10^{8}	0.0379		
Bi-214	3.28	8.59×10^{5}	0.0194		
Po-214	7.835	1.49×10^{13}	0.0464		
Tl-210	5.50	2.24×10^{7}	0.0326		
Pb-210	0.061	0.0280	3.62×10^{-4}		
Bi-210	1.16	853	0.00688	0.054	
Po-210	5.408	144	0.0321	<0.030	
Tl-206	1.524	1.95×10^{6}	0.00903		
Pb-206	Stable			0.0305	0.2
		Actinium ($4n + 3$) *Series*			
U-235	4.681	6.00×10^{-8}	0.0277	98.6	144
Th-231	0.381	12000	0.00226		
Pa-231	5.148	0.00144	0.0305	210	1500
Ac-227	0.113	0.0485	6.60×10^{-4}	515	
Th-227	6.145	1120	0.0364	200	
Fr-223	1.15	2.60×10^{5}	0.00682		
Ra-223	5.977	1814	0.0354	130	
Rn-219	8.16	6.27×10^{8}	0.0484		
Po-215	7.524	1.31×10^{12}	0.0466		
Pb-211	1.37	2.00×10^{5}	0.00812		
At-215	8.16	2.53×10^{13}	0.0484		
Bi-211	6.75	1.66×10^{7}	0.0400		
Po-211	7.592	4.36×10^{9}	0.0450		
Tl-207	1.44	1.62×10^{6}	0.00853		
Pb-207	Stable			0.709	0.4

SOURCES: References 9a and 16 and calculated.

[a]Capture.

ter activities. H. W. Kirby of the Mound Facility, operated by Monsanto Research Corporation, was queried. His reply[27]:

> The best I can offer you is the information that the specific activity of U-238 is 3.36×10^{-7} Ci/g and that there are 8 alphas and 6 betas in the chain, so the specific activity (alpha plus beta) at equilibrium would be 4.7×10^{-6} Ci/g. For U-235, the specific activity is 2.16×10^{-6} Ci/g, and,

Table 2.7 Natural and (4*n* + 1) Series—Helium Production

	Helium Production ($cm^3/g/d$)[a]	
Chain Parent	Parent Only	Plus Daughters At Equilibrium
Th-232	1.30×10^{-11}	7.80×10^{-11}
Th-228	0.0976	0.488
U-232	0.00254	0.0152
Np-237	8.35×10^{-8}	5.85×10^{-7}
U-233	1.14×10^{-6}	6.84×10^{-6}
Th-229	2.44×10^{-5}	1.22×10^{-4}
U-238	4.02×10^{-11}	3.21×10^{-10}
Th-230	2.36×10^{-6}	1.42×10^{-5}
Ra-226	0.00117	0.00585
U-235	2.57×10^{-10}	1.80×10^{-9}
Pa-231	5.54×10^{-6}	3.32×10^{-5}
Ac-227	1.18×10^{-4}	0.0429

SOURCE: Calculated values.

[a] Per gram of chain parent.

with 7 alphas and 4 betas, the specific activity (alpha plus beta) at equilibrium comes to 2.38×10^{-5} Ci/g. Given the 0.72% isotopic abundance of U-235, the natural U specific activity at equilibrium works out to be 4.84×10^{-6} Ci/g. For Th-232, with 6 alphas and 4 betas, the specific activity at equilibrium (alpha plus beta) is 1.09×10^{-6} Ci/g. [Author's interpolation: These numbers work out to be 0.18 d/s per microgram for natural uranium, and 0.040 d/s for pure ^{232}Th ore at equilibrium.]

As for the total disintegration energies, again I know of no reference you can use. The easiest way is to calculate it from the difference in mass excess between the bottom and top of each chain, giving due account for the number of alphas emitted.

Dr. Kirby then presents the pertinent calculations. Using his results gives

Chain	Q Value at Equilibrium (MeV)	W/g
U-238	51.70	1.03×10^{-7}
U-235	46.40	5.90×10^{-7}
Natural U	51.66	1.02×10^{-7}
Th-232	42.66	2.76×10^{-8}

The Q values for a complete chain increase by roughly a factor of 10 as compared to the parents, but these latter have such long half-lives as to make no practical difference in short-time heat release. The picture is, however, very much different when ore bodies over geologic time scales are being considered.

At equilibrium, the number of atoms of each member of a chain are present in quantities directly proportional to their half-lives. Using the longer-lived members of the ^{238}U chain as examples, it can then be calculated that at equilibrium there will be 53.9g ^{234}U per metric tonne of ^{238}U and 0.342g ^{226}Ra per metric tonne of ^{238}U.

The natural isotopic composition of uranium is

^{238}U	99.4746%
^{235}U	0.720%
^{234}U	0.0054%

This composition is almost, but not quite, invariable in nature. It has been discovered that the Oklo uranium ore body in Africa apparently "went critical" millions of years ago and for a long period acted as a buried natural reactor, thus depleting the ^{235}U fraction of the ore. The Oklo phenomenon is of considerable interest in terms of geologic disposal of man-produced radioactive wastes since it has been demonstrated[28] that fission products have migrated for only comparatively short distances from the active core in spite of the long time periods involved.

The isotopic composition of thorium ores varies considerably, depending on the amount of associated uranium and its effect in producing an admixture of ^{230}Th daughter. A survey has shown[29] that the ionium content of thorium ores can vary from almost zero up to as much as 11.6%.

2.4 CALIFORNIUM-252

Compact, reasonably portable isotopic sources of neutrons have found applications in radiation therapy, well-bore logging, reactor instrumentation development, process quality control, field prospecting for ores by neutron activation, and so on. The most compact of these sources is spontaneously fissioning Cf-252, where even quantities in the microgram range have proven useful. Californium-252 is considered briefly in this section, and the somewhat more bulky (alpha, n) neutron sources in the next.

2.4.1 Cf-252—General Properties

The radiometric properties of Cf-252 are given in Tables 2.1-2.3. Chemically, californium behaves primarily as a +3 species, similar to the rare earth elements

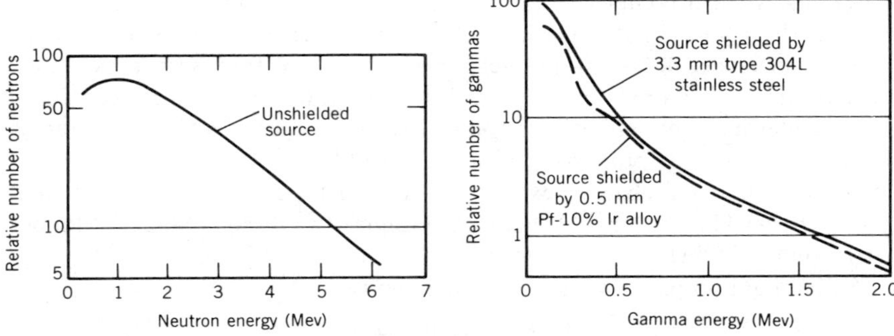

Figure 2.8. Neutron and gamma spectra of Cf-252. (Reference 30.)

and the other +3 actinides. The first portion of Figure 2.8 gives the neutron spectra of an unshielded Cf-252 source, whereas the second graph shows the effect on the gamma spectrum of shielding with appropriate thicknesses of metals commonly used for source encapsulation.

2.4.2 Fission-Product Yields

The three most comprehensive studies of the cumulative chain mass yields from the spontaneous fission of Cf-252 appear to be those of Blachot et al. (1975),[31] Harbour et al. (1971),[32] and Nervik (1960).[33] The data from these three investigations are summarized in Table 2.8.

2.4.3 Exposures and Doses

This subject encroaches somewhat on Chapter 10 on health physics, but is included here because of its pertinence to Section 2.4.1. Table 2.9, based on ICRP Report 21,[34] presents exposure and dose rates for both the neutrons and the photons emitted during Cf-252 decay.

Shielding of neutron sources, and especially Cf-252, involves special problems that will be considered in the Chapter entitled Shielding (Chapter 9).

2.5 (ALPHA, n) NEUTRON SOURCES

Table 2.10 summarizes information relating to various isotopic (alpha, n) sources, cited from a number of different references,[20,21,34–38] many of which consist of similar but shorter compilations. Some of the neutron-yield numbers

Table 2.8 Cumulative Mass Yields from Spontaneous Fission of Cf-252

Mass Number	Yield (%)		
	Blachot et al.	Harbour et al.	Nervik
83			0.021
89			0.32
91	0.41		0.59
92	0.57		
93	1.00		0.83
95	1.28	1.2	1.37
97	1.58	1.8	1.54
99	2.28	2.7	2.57
101		3.7	
102		3.7	
103	5.90	4.8	
105	6.77		5.99
106	7.01	5.8	
109			5.69
111	4.78		5.19
112	5.49		3.65
113			4.23
115	3.12		2.28
117	1.13		
121			0.142
125	0.03		0.011
127	0.12		0.130
129	0.74		0.615
131	1.73	1.8	1.27
132	2.27	2.5	1.75
133		3.9	2.77
134		4.5	
135	3.75	4.9	4.33
136		5.0	
137	4.51	5.2	4.40
138			4.94
139		5.7	5.73
140	6.05	5.8	6.32
141	6.33	5.8	5.9
143	6.29	6.4	5.94
144		5.7	
147	4.80	4.5	4.69
149	3.90		2.65
151	1.77		2.18
153	1.51		1.41
156			0.703
157	0.43		

SOURCE: Reference 31 and References 32 and 33 as quoted in 31.

Table 2.9 Exposure and Dose Rates from 1 gram Cf-252 at 1 meter

Energy Interval (MeV)	Neutrons			Photons	
	$n/cm^2/s$	rad/h	rem/h	$ph/cm^2/h$	rad/h
0.0–0.5	2.2×10^6	13	110	3.7×10^7	17
0.5–1.0	2.9×10^6	35	350	4.5×10^7	61
1.0–1.5				1.4×10^7	30
1.0–2.0	6.1×10^6	91	850		
1.5–2.0				6.1×10^6	16
2.0–3.0	3.7×10^6	59	480		
2.0–2.5				1.8×10^6	5.8
2.5–3.0				8.8×10^5	3.3
3.0–4.0	2.2×10^6	37	290		
3.0–3.5				4.5×10^5	1.9
3.5–4.0				2.4×10^5	1.1
4.0–5.0	1.3×10^6	26	170		
4.0–4.5				1.4×10^5	0.7
4.5–5.0				65,000	0.34
5.0–6.0	4.5×10^5	10	63		
5.0–5.5				39,000	0.23
5.5–6.0				14,000	0.087
6.0–7.0	3.2×10^5	8.0	48		
6.0–6.5				8,000	0.053
7.0–8.0	1.0×10^5	2.5	15		
8.0–10.0	79,000	2.1	12		
10.0–13.0	18,000	0.45	2.7		
0–6.5				1.1×10^8	140
0–13.0	1.9×10^7	280	2,400		

SOURCE: ICRP Publication 21, Reference 34.

are calculated values and, as such, represent upper limits for actual fabricated sources.

A number of factors determines the choice of a source for a particular application: the nature of the neutron spectrum, the neutron intensity generated (the yield), the half-life of the alpha emitter (which determines the useful working life and the physical size of the source), the gamma background, and so on. The spectrum is of utmost importance in most instrument development, calibration and application situations, whereas in waste management, the potential neutron yield from concentrated mixtures of alpha emitters and light elements is of primary interest.

The neutron spectrum varies with the emitter-target combination, the size

Table 2.10 (Alpha, *n*) Isotopic Neutron Sources

Target	Alpha Emitter(s)		Neutrons Produced	
	Nuclide	$T_{1/2}$	Average Energy (MeV)	Yield (n/10^6 α's)
Li	Po-210	138.4d	0.48	1.4
Be	Po-210	138.4d	4.2	69
	Ra-D-E-F	22y	4.5	67
	Rn-222	3.825d		405
	Ra-228	11.435d		610
	Ra-226	1602y	3.9	526
	Ac-227	21.8y	4.8	689
	Th-228	1.91y		540
	Pu-238	87.74y	4.5	79
	Pu-239	24100y	4.5	61
	Am-241	433y	4.5	78
	Cm-242	162.8d		116
	Cm-244	18.11y		66
BeF_4	Ra-226	1602y		68
Be-B-F	Po-210	138.4d		4.1
B-10	Po-210	138.4d		22
Natural B	Ra-226	1602y		189
O-17	Pu-238	87.74y		13
O-18	Pu-238	87.74y		28
Natural O	Ra-226	1602y		189
F	Po-210	138.4d	1.4	5.4
Na	Po-210	138.4d	4.45	1.1

SOURCES: Varied, please see text.

of the source, and the amount of associated moderator material.[38] Frequently, there is only a single energy peak with a maximum in the 1-3 MeV range (^{18}O, ^{17}O, ^{19}F, ^{11}B, etc, targets). The widely used ^{239}Pu-Be source, however, shows a number of peaks in its spectrum, with two major intensities at 3.1 and 4.8 MeV. The ^{210}Po-Be and ^{241}Am-Be spectra are very similar.[38]

Yields in the literature are usually expressed either as n/s/Ci of alpha activity or as n/10^6 alphas, the latter being used in Table 2.10. The high yields for ^{226}Ra, ^{227}Ac and other parents of short-lived alpha chains are for the situation where the indicated nuclides have come to equilibrium with the daughters.

Beryllium as a target consistently produces the best yields, so the ^{9}Be(alpha, n)^{12}C reaction has received considerable study. Anderson and Hertz[37] have developed a thick target yield vs. alpha energy curve for the reaction, shown here as Figure 2.9. The curve can be approximated by the equations

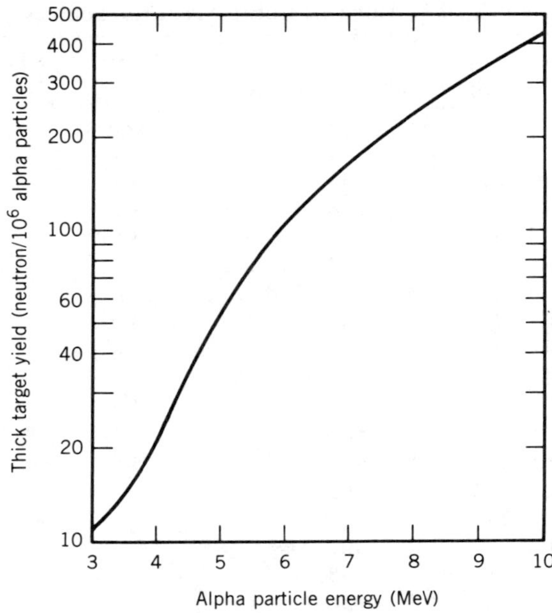

FIGURE 2.9. Thick target yield for ^{9}Be (alpha, n) reaction. (Reference 40.)

$$n/10^6 \text{ alphas} = 0.080^{4.06} \qquad E, 4.1\text{-}5.7 \text{ MeV} \tag{2.2}$$

$$n/10^6 \text{ alphas} = 0.80^{2.75} \qquad E, 5.7\text{-}10 \text{ MeV} \tag{2.3}$$

These equations compare to a similar expression given earlier by Runnals and Boucher[39] for the ^{9}Be(alpha, n) reaction

$$n/10^6 \text{ alphas} = 0.152E^{3.65} \tag{2.4}$$

M. E. Anderson of Mound Laboratory[40] has calculated the threshold energy needed by an alpha particle before it can react with a number of light nuclides ranging from ^{6}Li to ^{40}Ar. This information is of interest in waste management. A few of Anderson's values are

Target	Threshold Energy (MeV)
Li-6	6.62
Li-7	4.38
Be-9	0.00

Target	Threshold Energy (MeV)
B-10	0.00
B-11	0.00
N-15	8.13
O-16	15.17
O-18	0.85
F-19	2.36
Na-23	3.49

If the reaction is energetically possible, the yield will then depend on the target's cross section. Neutrons having maximum energies of up to 12 MeV may be generated by some emitter-source combinations, but these very energetic species appear only in small numbers at the very low-yield end of the spectrum.

The yield, the details of the spectrum, and the maximum obtainable neutron energy depend on the average energy of the alphas of the generating nuclide. No effort has been made in the literature to calculate this average for radioactive chains such as those headed by ^{226}Ra, ^{222}Rn, ^{227}Ac, and so on. For the other most commonly used sources of alphas, the average energies are as follows:

Alpha Source	Average Alpha Energy (MeV)
Po-210	5.30
Pu-238	5.48
Pu-239	5.14
Am-241	5.47
Cm-242	6.10
Cm-244	5.79

THREE

ORIGEN: CALCULATION OF WASTE COMPOSITIONS

3.1 THE COMPUTATIONAL PROBLEM

Bateman,[41] a mathematician in Rutherford's laboratory, was the first to derive equations for calculating the buildup of later members in a radioactive series of reactions. In 1910, his interest was, of course, in the natural decay series discussed in Section 2.3. Many adaptations of his techniques have been published since. Figure 3.1 shows one such variation,[42] chosen because of the relative simplicity of presentation and general applicability.

A four-membered chain is shown, applicable to any series of nuclear events: decay, neutron capture, fission, and so on. Any mixture of reactions can be handled, providing that time units are consistent. N_x expressed the number of atoms of x at time t and N_{A_0} is the original number of atoms in the parent. The nonprimed lamda values are the *destruction* constants for the respective nuclides, and each can be composed of a number of parts as indicated in the figure. (U-235 in a reactor will be undergoing fission, decay, neutron capture, etc.) The single-primed λ_1', while part of the destruction constant for A, for example, is also the *formation* constant for B. Similarly, λ_2' is the formation constant of C, and so forth.

When the equations are being applied to neutron reactions in a reactor

$$\lambda_x' = (f)(\sigma)(10^{-24}) \tag{3.1}$$

$$
\begin{array}{ccccccccc}
 & \lambda_1'' \uparrow & & \lambda_2'' \uparrow & & \lambda_3'' \uparrow & & \lambda_4'' \uparrow & \\
 & A & \xrightarrow{\lambda_1'} & B & \xrightarrow{\lambda_2'} & C & \xrightarrow{\lambda_3'} & D & \xrightarrow{\lambda_4'} \\
 & \lambda_1''' \downarrow & & \lambda_2''' \downarrow & & \lambda_3''' \downarrow & & \lambda_4''' \downarrow &
\end{array}
$$

$$\frac{N_A}{N_{A0}} = e^{-\lambda_1 t}$$

$$\frac{N_B}{N_{A0}} = \frac{\lambda_1'}{\lambda_2 - \lambda_1} e^{-\lambda_1 t} + \frac{\lambda_1'}{\lambda_1 - \lambda_2} e^{-\lambda_2 t}$$

$$\frac{N_C}{N_{A0}} = \frac{\lambda_1'\lambda_2'}{(\lambda_3 - \lambda_1)(\lambda_2 - \lambda_1)} e^{-\lambda_1 t} + \frac{\lambda_1'\lambda_2'}{(\lambda_3 - \lambda_2)(\lambda_1 - \lambda_2)} e^{-\lambda_2 t}$$

$$+ \frac{\lambda_1'\lambda_2'}{(\lambda_1 - \lambda_3)(\lambda_2 - \lambda_3)} e^{-\lambda_3 t}$$

$$\frac{N_D}{N_{A0}} = \frac{\lambda_1'\lambda_2'\lambda_3'}{(\lambda_2 - \lambda_1)(\lambda_3 - \lambda_1)(\lambda_4 - \lambda_1)} e^{-\lambda_1 t} + \frac{\lambda_1'\lambda_2'\lambda_3'}{(\lambda_1 - \lambda_2)(\lambda_3 - \lambda_2)(\lambda_4 - \lambda_2)} e^{-\lambda_2 t}$$

$$+ \frac{\lambda_1'\lambda_2'\lambda_3'}{(\lambda_1 - \lambda_3)(\lambda_2 - \lambda_3)(\lambda_4 - \lambda_3)} e^{-\lambda_3 t}$$

$$+ \frac{\lambda_1'\lambda_2'\lambda_2'}{(\lambda_1 - \lambda_4)(\lambda_2 - \lambda_4)(\lambda_3 - \lambda_4)} e^{-\lambda_4 t}$$

$$\lambda_n = \lambda_n' + \lambda_n'' + \lambda_n''' \qquad (\text{sec}^{-1})$$

N_x = number of atoms of x at time t.

SOURCE: Hanson, USAEC Report ID0-16065, Reference 42.

FIGURE 3.1 Generalized Expressions of the Variations with Time of the Isotopes in a Nuclear Chain.

where f is the flux in n/s/cm^2 and σ is the appropriate cross section in barns: fission, capture, $(n, 2n)$, and so on, for the reaction being considered. Since flux is normally in per second terms, half-lives must also be converted to seconds if they are part of the sum of reaction rates making up the destruction constants. The value of t must also be expressed in seconds.

The computational problem is thus long and tedious, even for the relatively short nuclear chain shown, and obviously prone to human error if calculations are done by hand. Various simplifying assumptions can be made, and shortcuts taken, but the process is still very laborious if one is interested in higher buildup yields in chains such as the one shown in Figure 2.1. This approach, however, was the only choice until the advent of large computers.

At an early stage, the staff at Oak Ridge National Laboratory began the development of computer programs for following the multitude of nuclear reactions occurring simultaneously in an operating reactor, culminating in the publication[43] of a description of ORIGEN (Oak Ridge Isotope Generation and Depletion Code. There is now an ORIGEN-2). The highly versatile tool is

> a collection of programs that: (1) processes a library of nuclear properties to construct a set of first-order linear, ordinary differential equations describing the rates of formation and destruction of the nuclides contained in the library; (2) solves the resulting set of equations, for a given set of initial conditions and irradiation history, to obtain the isotopic composition of the discharged fuel components as a function of postirradiation time; (3) uses these isotopic compositions and the nuclear properties of individual nuclides to construct tables describing the radioactivities, thermal power, potential inhalation and ingestion hazards, and proton and neutron production rates in the discharged fuel. . . . The library of nuclear data that has been compiled for use with the code is sufficiently extensive to treat ^{235}U and ^{239}Pu fuels in both fast and thermal spectra and fission of ^{233}U in thermal spectra.

The assembly of programs is capable of following the fate of nearly a thousand individual nuclides during irradiation and subsequent cooling of the fuel.

The ORIGEN system has been applied by users throughout the reactor physics community to models of all the major reactor types, to safety analyses of specific reactors under design and for fuels based on uranium, plutonium, and mixtures of both elements. An extensive literature[44] has accordingly accumulated for ORIGEN applications, and the code itself is under constant refinement[18,45] as is the basic data library.[13]

If one has access to a large computer, the ORIGEN Code package can be obtained and calculations made to provide good projections of the composition of the high-level waste generated in the fuel of a particular reactor at the time of discharge and during subsequent cooling. There are, however, still situations, particularly for existing wastes, where the large number of unavoidable uncertainties present justify (or necessitate) less precise but more quickly accessible projections. Such situations occur when the waste is derived from a mixture of fuel assemblies from different reactors, each with varying cooling times, incomplete irradiation records, and less than the total 33,000 MWD burnup. (Most of the published ORIGEN data are for fuels with full burnup at time of discharge.) The graphs given in the next four sections are for application to these less-than-optimum situations.

3.2 BUILDUP OF ACTIVE FISSION PRODUCTS

The curves shown in Figures 3.2-3.5 are based on the extensive data tables given in the original ORIGEN paper.[43] The model in this case was a pressurized water reactor (PWR) operating at a power level of 30 megawatts per metric tonne of uranium (MTU) over a period of 1,100 days at a flux level of 2.92×10^{13} n/s/cm^2 to a total burnup of 33,000 (MWD) at discharge. The graphs given here are based on those tables in the report giving the gram-atom concentration per MTU (original charge basis) for approximately 460 active and inert fission-product nuclides, 100 actinides and 200 activation products generated in the hardware associated with the fuel.

Figures 3.2*a* and *b* are for the major longer-lived radioactive fission products. The general criteria for inclusion in the graphs are that the nuclide either represented greater than 0.1% of the total fission-product curie level in the fuel 160 days after shutdown or is present to a concentration of over 0.1 gram-atoms per MTU at discharge after full burnup.

In using the graphs, the power level in megawatts per MTU of the reactor producing the wastes is multiplied by the irradiation days before discharge. This number is the effective burnup and is used to find the appropriate point on each curve to determine the quantities of the individual isotopes.

Figures 3.2-3.5 express the results in gram-atoms of nuclide produced per original tonne of uranium (MTU) fuel charged to the reactor. (The MTU notation is sometimes seen as MTIHM–Metric Tonne Initial Heavy Metal–to cover mixed U-Pu fuels.) The "per gram" conversion equations given in Section 1.1.3.1 can be readily adapted to the gram-atom units:

$$\text{Curies per gram-atom} = \frac{K_2}{T_{1/2}} \tag{3.2}$$

$$\text{Watts per gram-atom} = \frac{(K_4)\,(Q)}{T_{1/2}} \tag{3.3}$$

$$\text{Btu-h per gram-atom} = \frac{(K_5)\,(Q)}{T_{1/2}} \tag{3.4}$$

The K values are taken from Table 1.8 with due regard for the time units in which the half-life is expressed.

3.3 BUILDUP OF INERT FISSION PRODUCTS

Inert nuclides may be produced directly in fission or present in reactor fuels within a short time as end members of decay chains. Their presence adds to the

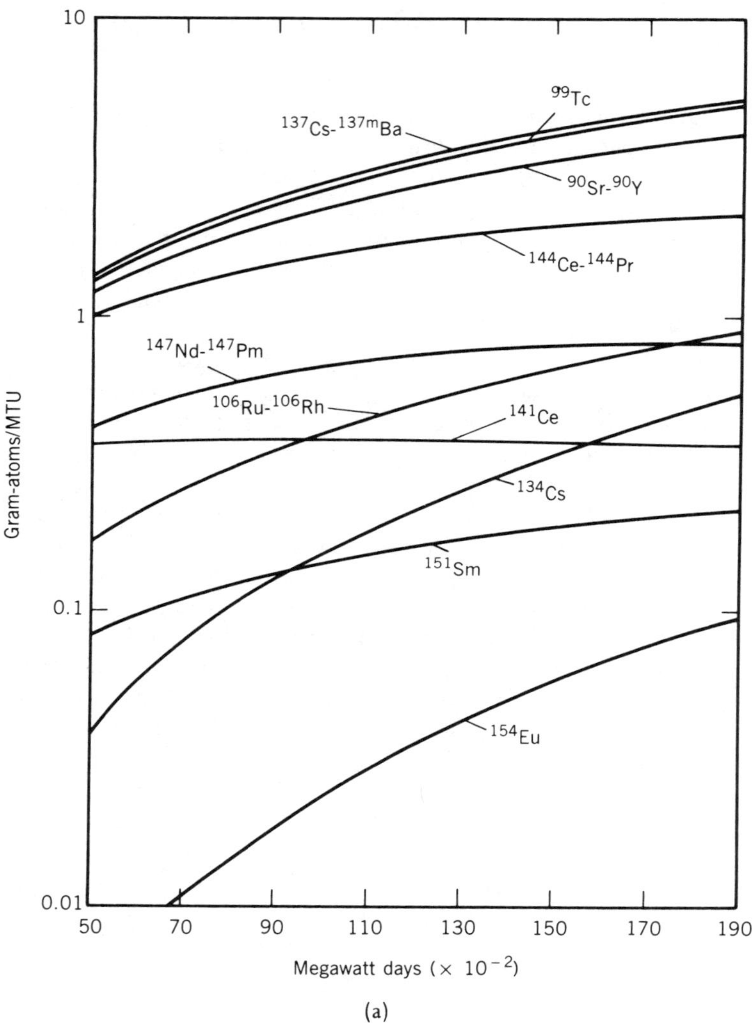

(a)

FIGURE 3.2. Fission product growth in ORIGEN reactor. (Reference 43.)

radioactivity problem only in the sense that they can be activated by neutron capture (^{133}Cs to ^{134}Cs as a significant example). More importantly, however, the accumulation of these inactive species becomes a primary chemical factor in postirradiation handling; complete dissolution of the fuel, reprocessing by the Purex technique, solidification of the wastes for disposal, and decontamination of the exhaust gases from all these steps.

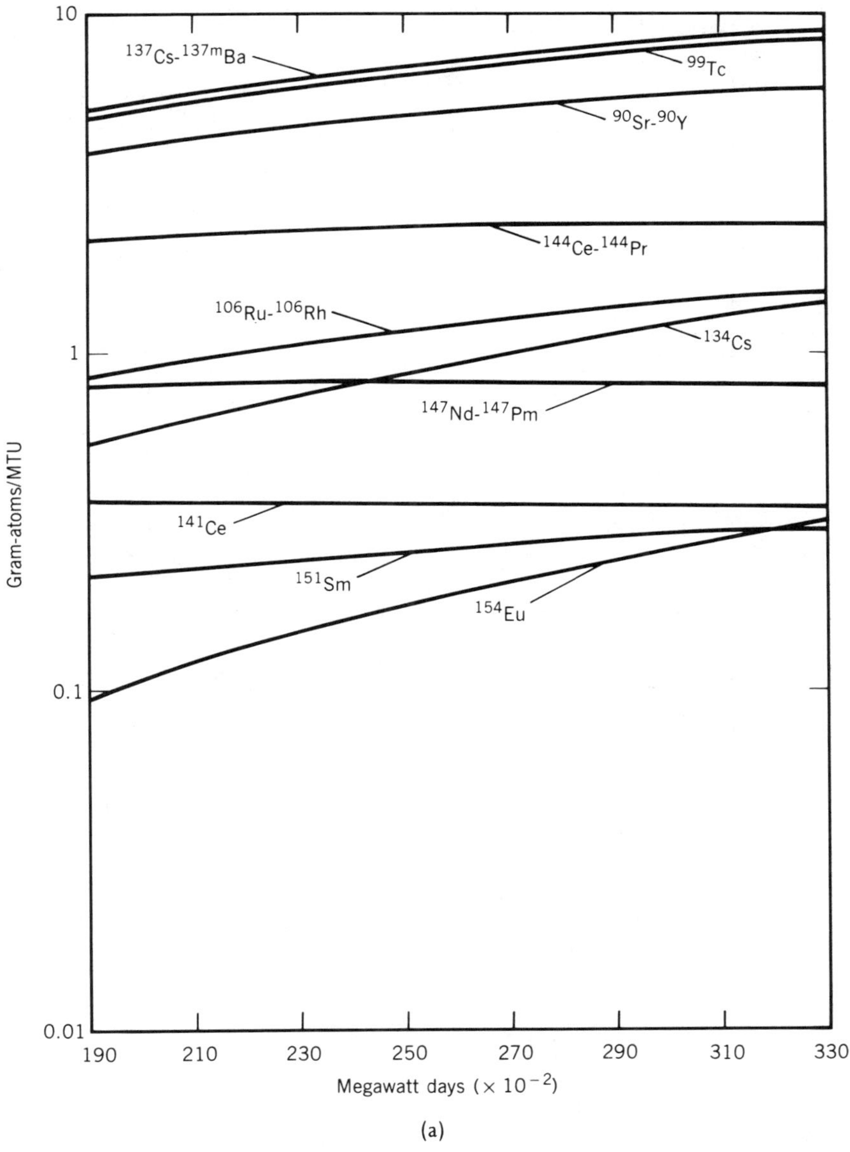

(a)

FIGURE 3.2. *(Continued)*

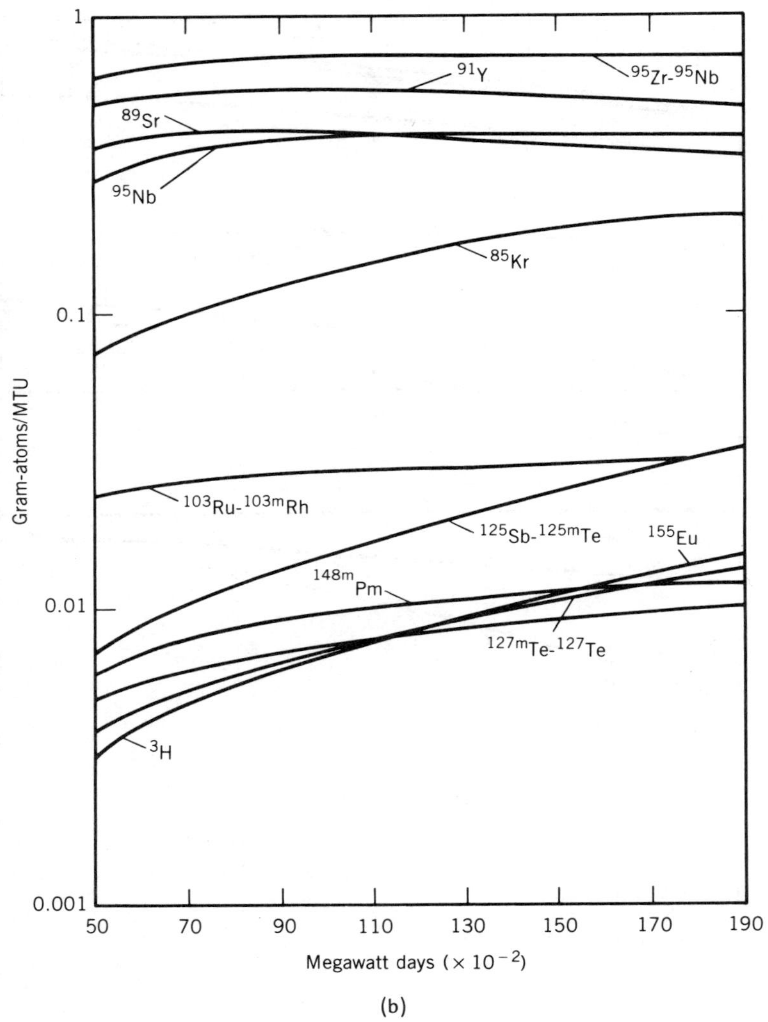

(b)

FIGURE 3.2. *(Continued)*

Table 3.1 shows the buildup of those inert elements contributing roughly 100 grams or more per MTU to the total after 33,000 MWD burnup. The tabular form of presentation is used in this case since many of the curves track each other so closely as to make a graph difficult to read. Note that the results are given in grams rather than gram-atoms per MTU.

The nuclides considered for each element are listed below. The underlined

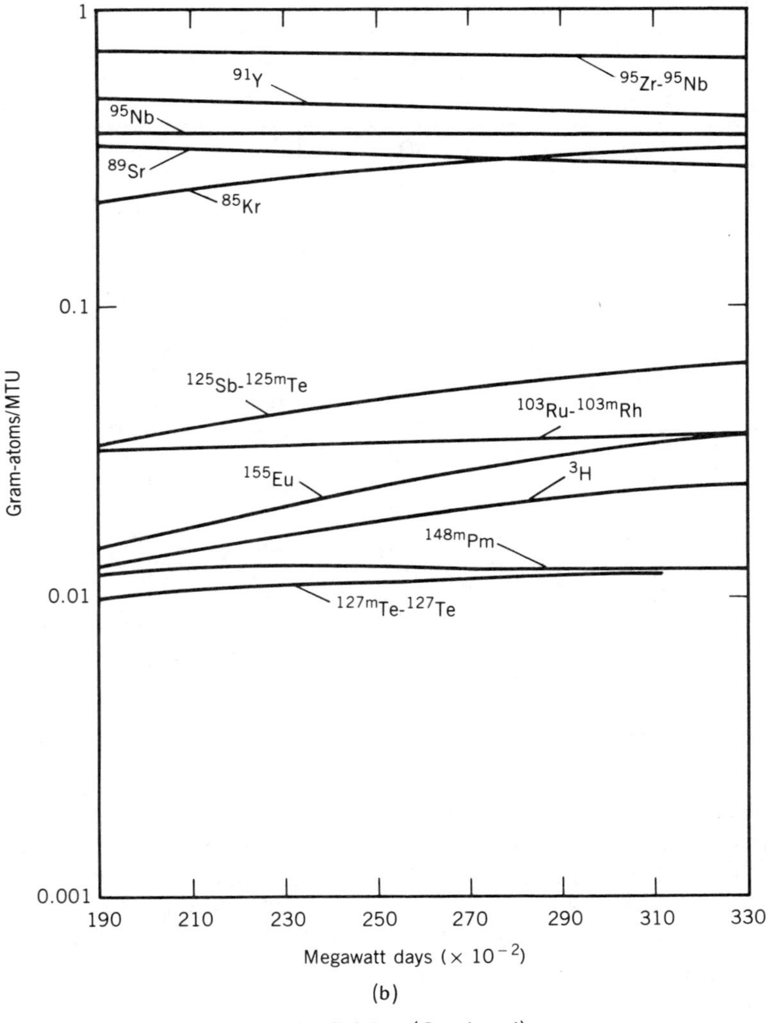

(b)

FIGURE 3.2. *(Continued)*

number following the semicolon in each set is the average atomic weight of the element at the time of discharge after 33,000 MWD burnup. In a few instances, nuclides having extremely long half-lives are treated as inert and are part of the totals.

The nuclides considered are: Kr (83, 84, 86; 84.98), Rb (86; 85), Sr (88; 88), Y (89; 89), Zr (90, 91, 92, 93, 94, 96; 93.36), Mo (95, 96, 97, 98; 96.76), Ru

Table 3.1 Buildup of Inert Fission Products in ORIGEN Reactor[a]

	Grams per MTU at (MWD)					
Element	3,300	6,600	13,200	19,800	26,400	33,000
Nd	312	667	1,413	2,187	2,969	3,767
Zr	371	778	1,523	2,251	2,948	3,618
Mo	186	412	907	1,406	1,900	2,393
Ru	176	362	759	1,182	1,626	2,090
Xe	135	285	592	908	1,233	1,568
Ba	118	240	490	762	1,047	1,352
Cs	149	304	596	865	1,106	1,324
Ce	110	244	507	767	1,026	1,284
La	133	264	523	776	1,026	1,271
Pd	26.6	72.1	211	403	643	926
Sm	39.5	94.5	238	394	554	708
Y	28.2	78.0	180	274	359	436
Te	38.9	79.0	163	247	335	425
Sr	42.9	84.6	160	229	292	350
Rh	26.4	70.3	159	236	300	350
Kr	42.5	82.4	157	224	287	346
I	20.5	44.1	93.4	151	207	265
Eu	4.99	12.4	33.8	62.4	95.3	125
Rb	11.4	22.1	42.2	60.8	78.2	94.4

SOURCE: USAEC Report ORNL-4628, Reference 43.

*Includes a few long-lived species. See text.

(102, 104; 102.14), Rh (103; 103), Pd (104, 105, 107, 108; 105.72), I (127, 129; 128.7), Xe (128, 129, 130, 131, 132; 131.7), Te (130; 130), Cs (133, 135; 133.5), Ba (134, 136, 137, 138; 137.7), La (139, 139), Ce (140; 140), Nd (142, 143, 144, 145, 146, 148, 150; 145.1), Sm (147, 148, 149, 150, 152; 149.6), and Eu (151, 153; 152.95). Some of the elements have stable isotopes not given in this list. In these cases, the fission yields were too small to affect the totals, so their contributions were ignored.

Some of the totals shown in Table 3.1 are augmented after discharge due to decay of major active species (^{137}Cs to ^{137}Ba, etc.). Haug[46] shows data of this kind based on ORIGEN-type calculations made for a liquid water (LWR) reactor model. Over a 10-year cooling period, barium quantities increase by roughly 30%. Changes in the quantities of the other inert fission-product elements are mostly at the lower end of a 0–12% range of increase over the amounts present at discharge.

Conversion factors are available in other sections of the book, but a few are duplicated here for convenience:

One pound = 453.6 grams
One kilogram = 2.2046 pounds
One metric tonne = 1.1023 "short" tons*
Grams/tonne × 0.907 = grams/ton
Grams/tonne × 0.00200 = pounds/ton
Grams/tonne × 0.00220 = pounds/tonne
*A "short" ton is 2,000 lb. A "long" ton is 2,240 lb, 1016.1 kg.

3.4 BUILDUP OF ACTINIDES

Figures 3.3 and 3.4 show the buildup of actinides in the original ORIGEN model LWR. The ^{239}Pu and ^{236}U curves are given separately in Figure 3.4, chiefly as a matter of convenience in graphing, but also because of the importance of the first-named isotope. Its buildup is shown both in terms of gram-atoms and of grams per MTU for the same reason. The concentrations of the major uranium isotopes present in the original fuel are not graphed, but of course also change. In terms of gram-atoms per original MTU:

	Charged	At 33,000 MWD Burnup	Decrease (%)
U-238	4065	3960	-2.58
U-235	140	34.0	-75.7
U-234	1.13	0.519	-54.1

The primary criterion for inclusion of a nuclide in the figures is that the nuclide is present in a concentration of at least 0.001 gram-atoms per MTU before full burnup is attained. Additional information is given in the original report for helium and for heavy nuclides (mostly daughters) ranging from ^{207}Pb to ^{253}Fm. However, the production rates at each end of the mass scale are too low to meet the 0.001 gram-atom criterion given above. The daughter isotopes become more important as they are generated from the heavier parents during extremely long periods of waste storage.

Since there is a direct relationship between the power level of a reactor and the number of fissions occurring in the fuel, the data in Table 3.1 and in Figures 3.1 and 3.2 for fission-product buildups should be reasonably transferable from one thermal reactor to another if the flux characteristics are not too wildly different. As shown in Chapter 2, actinide production is complicated, and the power-level to production-rate relationship is much less direct. As an example, it has been found[47] that application of the ^{239}Pu curve to fuels from the highly thermalized military reactors at Hanford badly underestimates the plutonium actually present. On the other hand, it has been reported[45] that ORIGEN-2

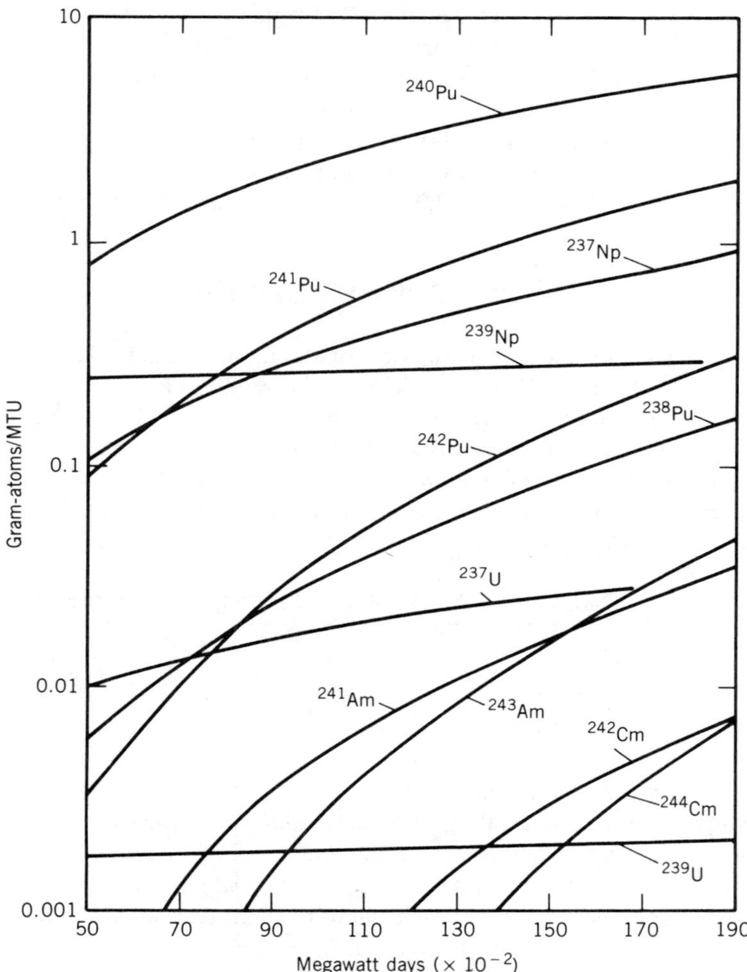

FIGURE 3.3. Actinide nuclide growth in ORIGEN reactor. (Reference 43.)

projects lesser amounts of ^{238}Pu, ^{243}Am, and ^{244}Cm than does ORIGEN-1. The information on actinides given in this subsection must be regarded as approximate and used with some caution.

Since production of the activation products discussed in the next section is uniformly a one-step nuclear process, the reliability of the data for application to different reactor fuels is probably somewhere in between the fission-product and actinide cases.

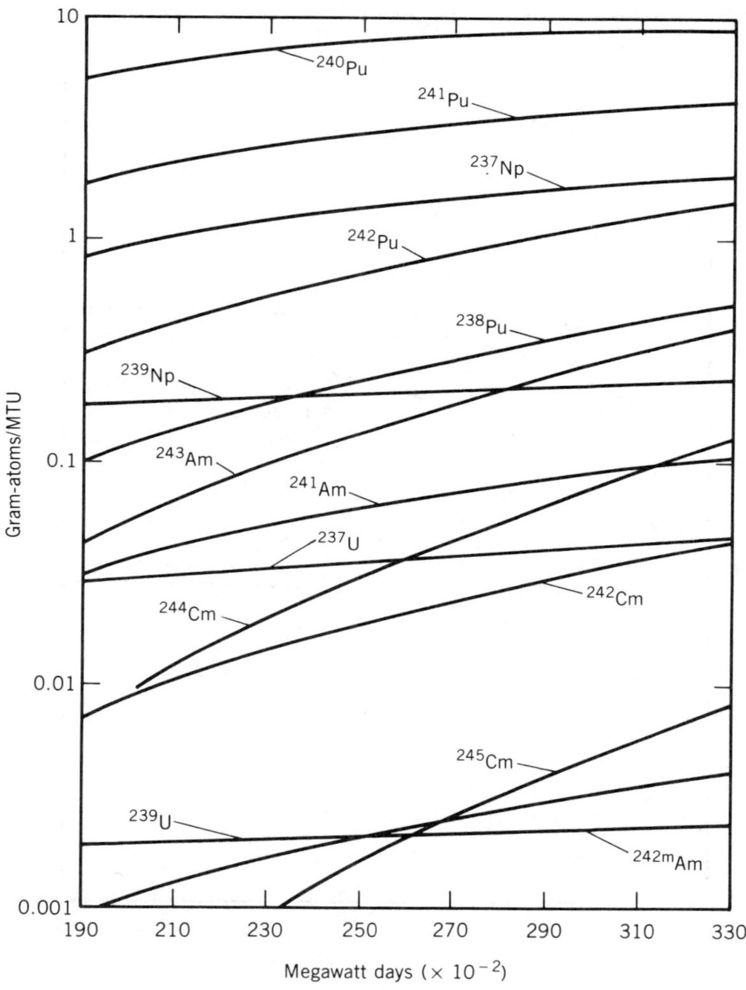

FIGURE 3.3. *(Continued)*

3.5 BUILDUP OF ACTIVATION PRODUCTS

Figure 3.5 shows the buildup of the major radioactivities formed by neutron capture in the cladding, spacers, and other metallic hardware present in the fuel assembly. The quantities of inert elements assumed in the original ORIGEN model to have been initially charged to the reactor (per MTU) for production of these activities are given in Table 3.2. The criterion for inclusion in Figure

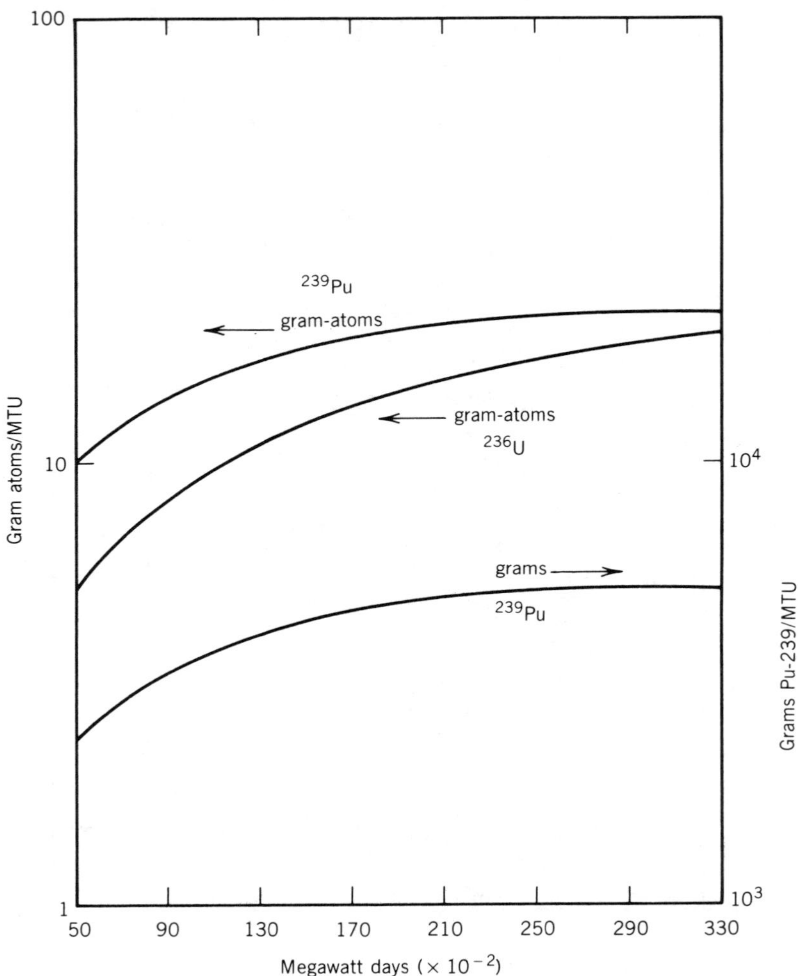

FIGURE 3.4. Pu-239 growth in ORIGEN reactor. (Reference 43.)

3.5 is that the nuclide be 0.1% or greater of the total activation product curie level 160 days after shutdown.

As much as possible of the assembly hardware is mechanically removed before the fuel is dissolved in a reprocessing plant, so all of the activation products do not become part of the high-level waste, but are in the large bulk of material that must be disposed of as solid waste. The report by Croff and his associates[45] describing revisions in the ORIGEN LWR models assumes seven metallic materials in the fuel assemblies: Zircaloy-2, Zircaloy-4, Inconel 718, Inconel X-750,

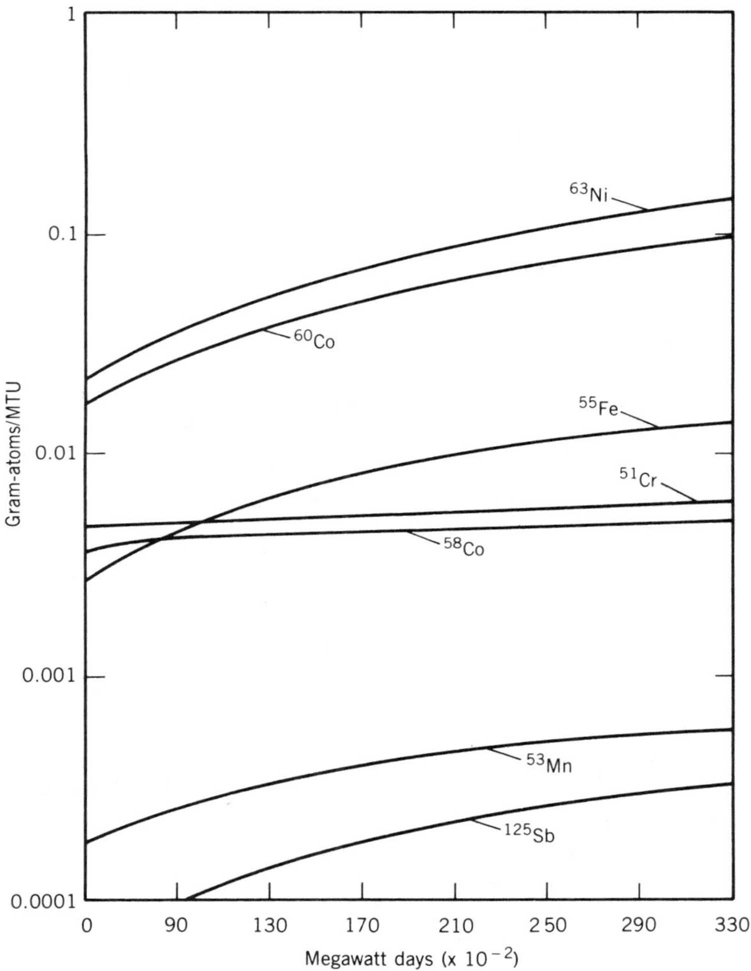

FIGURE 3.5. Activation product growth in ORIGEN reactor. (Reference 43.)

stainless steels 302 and 304, and Microbraze 50. The chemical composition of these alloys is given in the report, and the data are reproduced here as Table 3.3. The report also tabulates the physical distribution of the hardware metals, information of use in planning the solid-waste disposal problem. This information is reproduced here as Table 3.4. (Note that in Table 3.3 the values are given as grams per tonne of alloy, not per MTU. The numbers thus could be labeled as ppm.)

The Croff report also contains two other tables of waste-management interest.

Table 3.2 Initial Charge of Inert Hardware Elements, ORIGEN Reactor

Element	Grams per MTU	Element	Grams per MTU
C	18	Fe	3,734
Al	108	Co	54.9
Si	17.1	Ni	9,570
Ti	180	Zr	248,400
Cr	3,660	Mo	558
Mn	18	Sn	3,760

SOURCE: USAEC Report ORNL-4628, Reference 43.

The first (reproduced as Table 3.5) shows the characteristics of LWR fuel assemblies, which are important if the decision is made to forego reprocessing entirely and to inter complete assemblies at a final disposal site. Table 3.6 is a summary of literature values for the impurity and oxygen levels in LWR oxide fuels. (The oxygen level shown is the stoichiometric amount for UO_2 fuel. The Gd is present as a burnable poison in Boiling Water Reactor fuel rods.)

3.6 ACTIVITY LEVELS DURING COOLING

The data in the previous sections relate to the problem of estimating nuclide quantities in reactor fuels not taken to full burnup. Of equal importance from the waste-management point of view is the fate of these activities between the time of discharge and the time that the high-level wastes resulting after reprocessing are to be converted to solid form for final disposal. In many cases, this is simply a matter of applying half-lives of the individual isotopes to the concentrations at discharge. In others, where there are large amounts of shorter-lived precursors in the fuel at shutdown, the situation is more complicated (the ^{147}Nd-^{147}Pm, ^{125}Sn-^{125}Sb, and ^{241}Pu-^{241}Am pairs are examples).

Tables 3.7–3.9 show the calculated activity levels per initial MTU in the ORIGEN reactor fuel for fission products, actinides, and activation products, respectively, at discharge after full burnup (33,000 MWD) and at intervals thereafter up to 10-years' cooling. The selection rules for inclusion in the tables were: fission products, greater than 100 Ci/MTU after 160-days' cooling; and for the actinides and activation products, greater than one Ci/MTU after the same time interval. Any value less than 0.001 Ci/MTU in the original ORIGEN tables is given here as zero.

Certain of the fission products (the "long-lived beta emitters") and several of the actinide nuclides have such long half-lives that the quantity at discharge

Table 3.3 Assumed Elemental Compositions of LWR Fuel-Assembly Structural Materials

Element	Atomic Number	Structural Material Composition (grams per tonne of metal)						
		Zircaloy-2	Zircaloy-4	Inconel-718	Inconel X-750	Stainless Steel 302	Stainless Steel 304	Microbraze 50
H	1	13	13	0	0	0	0	0
B	5	0.33	0.33	0	0	0	0	50
C	6	120	120	400	399	1,500	800	100
N	7	80	80	1,300	1,300	1,300	1,300	66
O	8	950	950	0	0	0	0	43
Al	13	24	24	5,992	7,982	0	0	100
Si	14	0	0	1,997	2,993	10,000	10,000	511
P	15	0	0	0	0	450	450	103,244
S	16	35	35	70	70	300	300	100
Ti	22	20	20	7,990	24,943	0	0	100
V	23	20	20	0	0	0	0	0
Cr	24	1,000	1,250	189,753	149,660	180,000	190,000	149,709
Mn	25	20	20	1,997	6,984	20,000	20,000	100
Fe	26	1,500	2,250	179,766	67,846	697,740	688,440	471
Co	27	10	10	4,694	6,485	800	800	381
Ni	28	500	20	519,625	721,861	89,200	89,200	744,438
Cu	29	20	20	999	499	0	0	0
Zr	40	979,630	979,110	0	0	0	0	100
Nb	41	0	0	55,458	8,980	0	0	0
Mo	42	0	0	29,961	0	0	0	0
Cd	48	0.25	0.25	0	0	0	0	0
Sn	50	16,000	16,000	0	0	0	0	0
Hf	72	78	78	0	0	0	0	0
W	74	20	20	0	0	0	0	100
U	92	0.2	0.2	0	0	0	0	0
Density, grams/cm^2	–	6.56	6.56	8.19	8.30	8.02	8.02	–

SOURCE: Reproduced from DOE Report ORNL/TM-6051, Reference 45.

Table 3.4 Assumed Fuel-Assembly Structural Material Mass Distribution

	PWR			BWR		
		Mass			Mass	
	Material	kg/MTHM	kg/assembly	Material	kg/MTHM	kg/assembly
Fuel zone						
Cladding	Zircaloy-4	223.0	102.9	Zircaloy-2	279.5	51.2
Fuel channel[a]	—	—	—	Zircaloy-4	227.5	41.7
Grid spacers	Inconel 718 }	12.8	5.9	Zircaloy-4	10.6	1.9
Grid-spacer springs	Inconel 718 }			Inconel X-750	1.8	0.3
Grid-brazing material	Nicrobraze 50	2.6	1.2	—	—	—
Miscellaneous	SS 304[b]	9.9	4.6	—	—	—
Fuel-gas plenum zone						
Cladding	Zircaloy-4	12.0	5.5	Zircaloy-2	25.4	4.7
Fuel channel[a]	—	—	—	Zircaloy-4	20.7	3.8
Plenum spring	SS 302	4.2	1.9	SS 302	6.0	1.1
End fitting zone						
Top end fitting	SS 304	14.8	6.8	SS 304	10.9	2.0
Bottom end fitting	SS 304	12.4	5.7	SS 304	26.1	4.8
Expansion springs	—	—	—	Inconel X-750	2.1	0.4
Total		291.7	134.5		610.6	111.9

SOURCE: Reproduced from DOE Report ORNL/TM-6051, Reference 45.

[a] Assumed to be discarded after one cycle.

[b] Distributed throughout the PWR core in sleeves and so forth.

Table 3.5 Physical Characteristics of LWR Fuel Assemblies

	BWR	PWR
Overall assembly length, m	4.470	4.059
Cross section, cm	13.9 × 13.9	21.4 × 21.4
Fuel-element length, m	4.064	3.851
Active fuel height, m	3.759	3.658
Fuel-element OD, cm	1.252	0.950
Fuel-element array	8 × 8	17 × 17
Fuel elements per assembly	63	264
Assembly total weight, kg	275.7	657.9
Uranium/assembly, kg	183.3	461.4
UO_2/assembly, kg	208.0	523.4
Zircaloy/assembly, kg	99.5[a]	108.4[b]
Hardware/assembly, kg	12.4[c]	26.1[d]
Total metal/assembly, kg	111.9	134.5
Nominal volume/assembly, m^3	0.0864[e]	0.186[e]

SOURCE: Reproduced from DOE Report ORNL/TM-6051, Reference 45.

[a]Includes Zircaloy fuel-element spacers and fuel channel.

[b]Includes Zircaloy control-rod guide thimbles.

[c]Includes stainless steel tie-plates, Inconel springs, and plenum springs.

[d]Includes stainless steel nozzles and Inconel-718 grids.

[e]Based on overall outside dimension.

Table 3.6 Assumed Nonactinide Composition of LWR Oxide Fuel

Element	Concentration (ppm)[a]	Element	Concentration (ppm)[a]
Li	1.0	Mn	1.7
B	1.0	Fe	18.0
C	89.4	Co	1.0
N	25.0	Ni	24.0
O	134,454	Cu	1.0
F	10.7	Zn	40.3
Na	15.0	Mo	10.0
Mg	2.0	Ag	0.1
Al	16.7	Cd	25.0
Si	12.1	In	2.0
P	35.0	Sn	4.0
Cl	5.3	Gd	2.3
Ga	2.0	W	2.0
Ti	1.0	Pb	1.0
V	3.0	Bi	0.4
Cr	4.0		

SOURCE: From DOE Report ORNL/TM-6051, Reference 45.

[a]Parts of element per million parts of heavy metal.

Table 3.7 Fission-Product Decay: Discharge to 10-Years' Cooling

Nuclide	Cooling Time (Ci/MTU)					
	Discharge	30 Days	90 Days	160 Days	1 Year	10 Years
H-3	709	705	699	691	670	403
Kr-85	11,300	11,300	11,100	11,000	10,600	5,960
Sr-89	718,000	481,000	216,000	85,100	5,530	0
Sr-90	77,600	77,500	77,200	76,800	75,700	60,700
Y-90	80,700	77,500	77,200	76,800	75,700	60,700
Y-91	938,000	663,000	327,000	143,000	12,800	0
Zr-95	1,370,000	997,000	526,000	249,000	28,000	0
Nb-95m	28,000	21,100	11,200	5,290	594	0
Nb-95	1,380,000	1,280,000	872,000	473,000	59,500	0
Ru-103	1,220,000	721,000	252,000	74,100	20,500	0
Ru-103m	1,220,000	721,000	252,000	74,100	20,500	0
Ru-106	545,000	515,000	459,000	403,000	273,000	550
Rh-106	740,000	515,000	459,000	403,000	273,000	550
Ag-110m	3,680	3,390	2,880	2,370	1,350	0.17
Ag-110	159,000	441	374	309	176	0.02
Sn-123	8,800	7,520	5,390	3,660	11.7	0
Sb-125	8,700	8,640	8,290	7,890	6,830	0.07
Te-125m	3,110	3,200	3,250	3,190	2,830	281
Te-127m	15,400	13,200	9,000	5,770	1,570	0
Te-127	72,000	13,300	8,900	5,700	1,550	0
Te-129m	57,300	31,300	9,200	2,210	33.8	0
Te-129	337,000	20,000	5,900	1,420	21.7	0
Cs-134	246,000	240,000	227,000	212,000	176,000	8,380
Cs-137	108,000	108,000	107,000	107,000	105,000	85,600
Ba-137m	101,000	101,000	100,000	99,800	98,500	80,000
Ba-140	1,450,000	286,000	11,100	251	0.004	0
La-140	1,500,000	330,000	12,800	289	0.004	0
Ce-141	1,390,000	735,000	204,000	45,600	567	0
Pr-143	1,200,000	294,000	14,100	409	0.013	0
Ce-144	1,110,000	1,030,000	892,000	752,000	456,000	151
Pr-144	1,120,000	1,030,000	892,000	752,000	456,000	151
Pm-147	102,000	106,000	102,000	97,300	83,900	7,750
Pm-148m	38,000	23,700	8,800	2,770	94.0	0
Pm-148	199,000	6,060	709	223	7.55	0
Sm-151	1,250	1,250	1,250	1,250	1,240	1,160
Eu-154	6,990	6,970	6,920	6,860	6,690	4,530
Eu-155	7,480	7,250	6,810	6,330	5,110	163
Eu-156	226,000	56,600	3,540	139	0.011	0
Tb-160	1,280	961	540	275	38.4	0
Total FP	1.38×10^8	1.08×10^7	6,190,000	4,190,000	2,220,000	318,000

SOURCE: USAEC Report ORNL-4628, Reference 43.

shows no perceptible change over the 10-year period. These are:

Product	Ci/MTU
Se-79	0.398
Zr-93	1.89
Tc-99	14.3
Pd-107	0.110
Sn-126	0.546
I-129	0.371
Cs-135	0.286
Pu-240	477
Pu-242	18.1

(The Sn-126 is in equilibrium with its Sb-126m and Sb-126 daughters whose activities follow the parent's 10^5 y half-life.) Ni-59 (Table 3.9) is the only activation product showing no apparent change.

Data similar to those given in this section were calculated at Oak Ridge during the development of the ORIGEN code for the Diablo Canyon reactor (a LWR)

Table 3.8 Actinide Product Decay: Discharge to 10-Years' Cooling

	Cooling Time (Ci/MTU)					
Nuclide	Discharge	30 Days	90 Days	160 Days	1 Year	10 Years
U-237	865,000	39,700	86.3	2.54	2.41	1.57
Np-237	0.333	0.340	0.340	0.340	0.340	0.343
Np-239	1.85×10^7	2,700	18.2	18.2	18.2	18.2
Pu-238	2,720	2,770	2,800	2,820	2,860	2,700
Pu-239	318	323	323	323	323	323
Pu-241	105,000	105,000	104,000	103,000	100,000	65,300
Am-241	85.9	99.7	127	159	250	1,410
Am-242m	9.16	9.16	9.15	9.14	9.12	8.75
Am-242	63,400	9.16	9.15	9.14	9.12	8.75
Am-243	18.1	18.2	18.2	18.2	18.2	18.2
Cm-242	33,400	29,600	22,900	17,000	7,120	7.18
Cm-243	3.71	3.70	3.69	3.68	3.63	2.99
Cm-244	2,440	2,440	2,420	2,400	2,350	1,670
Total Actinide	3.91×10^7	183,000	133,000	126,000	114,000	72,000

SOURCE: USAEC Report ORNL-4628, Reference 43.

Table 3.9 Activation Product Decay: Discharge to 10-Years' Cooling

	Cooling Times (Ci/MTU)					
Nuclide	Discharge	30 Days	90 Days	160 Days	1 Year	10 Years
Sc-46	4.26	3.33	2.03	1.14	0.209	0
Cr-51	29,500	13,900	3,120	545	3.29	0
Mn-54	249	232	202	172	108	0.058
Fe-55	2,000	1,950	1,870	1,780	1,530	139
Fe-59	228	143	56.9	19.4	0.824	0
Co-58	9,320	6,960	3,890	1,970	2,680	0
Co-60	6,440	6,370	6,230	6,080	5,640	1,720
Ni-59	3.85	3.85	3.85	3.85	3.85	3.85
Ni-63	565	565	564	563	561	524
Sr-89	38.1	25.6	11.5	4.52	0.294	0
Y-91	101	71.0	35.0	15.3	1.37	0
Zr-95	28,900	21,000	11,100	5,250	590	0
Nb-95	28,000	26,200	17,900	9,740	1,230	0
Sn-117m	12,100	2,740	140	4.39	0	0
Sn-119m	24.1	22.1	18.7	15.4	8.74	0
Sb-125	44.7	43.8	42.0	40.0	34.6	3.43
Te-125m	16.2	16.5	16.6	16.2	14.3	1.42
Total AP	142,000	80,300	45,200	26,200	10,000	2,400

SOURCE: USAEC Report ORNL-4628, Reference 43.

and for a reference Liquid Metal Fast Breeder Reactor.[48] Haug[46] made similar calculations for "normal" (34,000 MWD) and "high" (45,000 MWD) burnups in LWR's based on uranium fuels, for a LWR with plutonium recycle and for a Fast Breeder reactor with liquid metal cooling. Both the Oak Ridge and Haug reports contain many additional calculations on gamma flux buildups, thermal data, etc.

PART TWO

CHEMICAL DATA

The emphasis in this section is primarily confined to data pertinent to the high-level material resulting from reactor fuel reprocessing. The elements considered are the major fission products (yield of over 1% in the thermal fission of ^{235}U), the actinides from uranium through curium, and the most ubiquitous of the inert elements found in existing wastes as a result of the reprocessing and of any subsequent treatment before tank storage.

The Purex process produces the bulk of the wastes in a nitric acid solution, although other inert anions may be added in smaller amounts during fuel dissolution and the processing. Practically all of the wastes that have been produced in this country to date have then been neutralized with caustic soda so that they could be stored in mild rather than stainless steel tanks. While there is still debate as to the solidification form specified for final disposal in a federal repository, most of the candidate end products require a calcining step somewhere during their preparation, that is, heating of the materials from the tanks to a high temperature to drive off the water and convert the wastes to oxide. Nitrates, hydroxides, and oxides accordingly receive particular consideration in this section.

Table 3.3 gives the chemical composition of the alloys used in reactor assembly construction.

FOUR

GENERAL TABLES

4.1 THE ELEMENTS

As a background reference, Table 4.1 lists the known elements, their atomic numbers (Z), their atomic weights, and the oxidation states in which they can occur. The atomic weights are those recommended by IUPAC in 1973.[49] The weights of the radioactive elements (other than Th and U) are of the longest-lived known isotope.

The oxidation-state data are from various sources.[50–53] Some of the indicated states are of minor importance, and the list is undoubtedly incomplete for some elements. In the case of elements having multiple possible states, the one usually encountered, particularly in nitric acid solution, is underlined. In the case of the actinides and rare earths, oxidation states that have been observed only in solid compounds are given in parentheses.[51]

The rare gases are all shown as having only a zero-oxidation state, although it is now known that Xe, Rn, and Kr have limited chemistries under extreme experimental conditions.

4.2 STANDARD OXIDATION POTENTIALS

These are shown in Table 4.2 for some half-cell reactions (at 25°C) of possible interest in waste management. The data are taken from Latimer.[53] In the con-

TABLE 4.1 The Elements

Atomic Number	Name	Symbol	Atomic Weight	Oxidation States
1	Hydrogen	H	1.0079	1, -1
2	Helium	He	4.0026	0
3	Lithium	Li	6.941	1
4	Beryllium	Be	9.0122	2
5	Boron	B	10.81	3
6	Carbon	C	12.011	2, 4, -4
7	Nitrogen	N	14.0067	1, 2, 3, 4, 5, -1, -2, -3
8	Oxygen	O	15.9994	-2
9	Fluorine	F	18.998	-1
10	Neon	Ne	20.179	0
11	Sodium	Na	22.990	1
12	Magnesium	Mg	24.305	2
13	Aluminum	Al	26.9815	3
14	Silicon	Si	28.0855	4
15	Phosphorus	P	30.9738	5 to -3
16	Sulfur	S	32.06	4, 6, -2
17	Chlorine	Cl	35.453	-1
18	Argon	Ar	39.948	0
19	Potassium	K	39.0983	1
20	Calcium	Ca	40.08	2
21	Scandium	Sc	44.9559	3
22	Titanium	Ti	47.90	2, 3, 4,
23	Vanadium	V	50.9415	2, 3, 4, 5
24	Chromium	Cr	51.996	2, 3, 6
25	Manganese	Mn	54.9380	2, 3, 4, 7
26	Iron	Fe	55.847	2, 3
27	Cobalt	Co	58.9332	2, 3
28	Nickel	Ni	58.70	2, 3
29	Copper	Cu	63.546	1, 2
30	Zinc	Zn	65.38	2
31	Gallium	Ga	69.72	3
32	Germanium	Ge	72.95	2, 4
33	Arsenic	As	74.9216	3, 5, -3
34	Selenium	Se	78.96	4, 6, -2
35	Bromine	Br	79.90	-1
36	Krypton	Kr	83.80	0
37	Rubidium	Rb	85.4678	1
38	Strontium	Sr	87.62	2
39	Yttrium	Y	88.9059	3
40	Zirconium	Zr	91.22	4

TABLE 4.1 (Continued)

Atomic Number	Name	Symbol	Atomic Weight	Oxidation States
41	Niobium	Nb	92.9064	3, 4, 5
42	Molybdenum	Mo	95.94	2, 3, 4, 5, 6
43	Technetium	Tc	(97)	7 to -1
44	Ruthenium	Ru	101.07	2, 3
45	Rhodium	Rh	102.9055	3
46	Palladium	Pd	106.4	2, 4
47	Silver	Ag	107.868	1, 2
48	Cadmium	Cd	112.41	2
49	Indium	In	114.82	3
50	Tin	Sn	118.69	2, 4
51	Antimony	Sb	121.75	3, 5
52	Tellurium	Te	127.60	4, 6, -2
53	Iodine	I	126.9045	-1
54	Xenon	Xe	131.30	0
55	Cesium	Cs	132.9054	1
56	Barium	Ba	137.33	2
57	Lanthanum	La	138.9055	3
58	Cerium	Ce	140.12	3, 4
59	Praseodymium	Pr	140.9077	3, 4
60	Neodymium	Nd	144.24	3
61	Promethium	Pm	(145)	3
62	Samarium	Sm	150.4	2, 3
63	Europium	Eu	151.96	2, 3
64	Gadolinium	Gd	157.25	3
65	Terbium	Tb	158.9254	3, (4)
66	Dysprosium	Dy	162.50	3
67	Holmium	Ho	164.9304	3
68	Erbium	Er	167.26	3
69	Thulium	Tm	168.9342	3
70	Ytterbium	Yb	173.04	2, 3
71	Lutetium	Lu	174.967	3
72	Hafnium	Hf	178.49	4
73	Tantalum	Ta	180.9479	5
74	Tungsten	W	183.85	6
75	Rhenium	Re	186.207	4, 7
76	Osmium	Os	190.2	4, 6, 8
77	Iridium	Ir	192.22	3, 4
78	Platinum	Pt	195.09	2, 4
79	Gold	Au	196.9665	1, 3
80	Mercury	Hg	200.59	1, 2
81	Thallium	Tl	204.37	1, 3

TABLE 4.1 (Continued)

Atomic Number	Name	Symbol	Atomic Weight	Oxidation States
82	Lead	Pb	207.2	2, 4
83	Bismuth	Bi	208.9804	3, 5
84	Polonium	Po	(209)	2, 4
85	Astatine	At	(210)	-1
86	Radon	Rn	(222)	0
87	Francium	Fr	(223)	1
88	Radium	Ra	(226)	2
89	Actinium	Ac	(227)	3
90	Thorium	Th	232.0381	(3), 4
91	Protactinium	Pa	(231)	(3), 4, 5
92	Uranium	U	238.029	3, 4, 5, 6
93	Neptunium	Np	(237)	3, 4, 5, 6
94	Plutonium	Pu	(244)	3, 4, 5, 6
95	Americium	Am	(243)	3, (4), 5, 6
96	Curium	Cm	(247)	3, (4)
97	Berkelium	Bk	(247)	3, 4
98	Californium	Cf	(251)	3
99	Einsteinium	Es	(254)	3
100	Fermium	Fm	(257)	3
101	Mendelevium	Md	(258)	2, 3
102	Nobelium	No	(259)	2, 3
103	Lawrencium	Lr	(260)	3
104	–	–	–	–
105	–	–	–	–
106	–	–	–	–
107	–	–	–	–

SOURCES: Varied, please see text.

vention used by him, the more positive the E° value, the more tendency exists for the reaction to proceed from left to right as shown. Reference 9c lists a number of standard *reduction* potentials where the convention is reversed.

4.3 WEIGHT CONVERSION FACTORS

Data relating to the composition of a waste can be received in several ways: as weight of element present, as weight of oxide, or as weight of a particular salt. Table 4.3 is for convenience in converting these weights to a common basis. As an example and referring to the table, if the weight of $AmCl_3$ is known, but its weight as an element is desired, the given value is divided by 1.438. The number obtained could then be multiplied by 1.132 if the original intent is to convert the weight of the chloride to the weight of dioxide.

A few of the values shown are taken from the table of gravimetric factors in Reference 9. The majority were calculated.

4.4 PROPERTIES OF SELECTED COMPOUNDS

Tables 4.4, 4.5, and 4.6 list some of the properties of most direct interest in waste management for selected nitrates, hydroxides, and oxides, respectively. The data are taken from several sources (not always in agreement, particularly for water solubilities).[9,10,54] The symbol d indicates decomposition; s, soluble; vs, very soluble; i, insoluble; and RT, room temperature. The superscript given with each solubility datum is the temperature in degrees Centigrade, and the value itself is the grams of *anhydrous* compound needed to form a saturated solution at that temperature when dissolved in 100 grams of water.

Data regarding the solubilities of salts of some of the elements are sparse for some elements, technecium being one of them. Keller and Kanellakopulos[55] provide some data at temperatures ranging from 15 to 40°C for a few compounds. At 20°C they give

Compound	g/100 ml Solution
$TlTcO_4$	0.072
$CsTcO_4$	0.412
$AgTcO_4$	0.563
$RbTcO_4$	1.167

Gmelin's handbook[52] gives the composition of saturated barium pertechnetate [$Ba(TcO_4)_2$] solutions as 0.165 *M* at 20°C and 0.361 *M* at 35°C.

4.5 SOLUBILITY CONSTANTS

Solubility data such as those given in the last three tables are only of general value when dealing with complex mixtures such as found in high-level waste. The same is, of course, true of solubility constants, but they allow easier discernment of possible combinations and trends. A selected group of these constants is presented in Table 4.7. The values are taken from the long list in Reference 56, partially reproduced in Reference 10.

TABLE 4.2 Selected Standard Oxidation Potentials

Couple	E° (Volts)	Couple	E° (Volts)
	Acid Solution		
$Zn = Zn^{2+} + 2e^-$	0.763	$Fe^{2+} = Fe^{3+} + e^-$	-0.771
$U^{3+} = U^{4+} + e^-$	0.61	$N_2O_4 + 2H_2O = 2NO_3^- + 4H^+ + 2e^-$	-0.80
$Eu^{2+} = Eu^{3+} + e^-$	0.43	$HNO_2 + H_2O = NO_3^- + 3H^+ + 2e^-$	-0.94
$Cr^{2+} = Cr^{3+} + e^-$	0.41	$PuO_2^+ = PuO_2^{2+} + e^-$	-0.93
$H_3PO_3 + H_2O = H_3PO_4 + 2H^+ + 2e^-$	0.276	$NO + 2H_2O = NO_3^- + 4H^+ + 4e^-$	-0.96
$HCOOH(aq) = CO_2 + 2H^+ + 2e^-$	0.196	$Pu^{3+} = Pu^{4+} + e^-$	-0.97
$Sn = Sn^{2+} + 2e^-$	0.136	$NO + H_2O = HNO_2 + H^+ + e^-$	-1.00
$UO_2^+ = UO_2^{2+} + e^-$	-0.05	$2NO + 2H_2O = N_2O_4 + 4H^+ + 4e^-$	-1.03
$Ti^{3+} + H_2O = TiO^{2+} + 2H^+ + e^-$	-0.1	$Pu^{4+} + 2H_2O = PuO_2^{2+} + 4H^- + 2e^-$	-1.04
$Np^{3+} = Np^{4+} + e^-$	-0.147	$Pu^{4+} + 2H_2O = PuO_2^+ + 4H^+ - e^-$	-1.15
$Sn^{2+} = Sn^{4+} + 2e^-$	-0.15	$NpO_2^+ = NpO_2^{2+} + e^-$	-1.15
$Cu^+ = Cu^{2+} + e^-$	-0.153	$2H_2O = O_2 + 4H^+ + 4e^-$	-1.229
$H_2SO_3 + H_2O = SO_4^{2-} + 4H^+ + 2e^-$	-0.17	$Mn^{2+} + 2H_2O = MnO_2 + 4H^+ + 2e^-$	-1.23
$U^{4+} + 2H_2O = UO_2^{2+} + 4H^+ + 2e^-$	-0.334	$NH_4^+ + H_2O = NH_3OH^+ + 2H^+ + 2e^-$	-1.35

$2I^- = I_2 + 2e^-$	-0.536	$Mn^{2+} + 4H_2O = MnO_4^- + 8H^+ + 5e^-$	-1.51
$MnO_4^{2-} = MnO_4^- - e^-$	-0.564	$Bk^{3+} = Bk^{4+} + e^-$	-1.6
$Ru + 5Cl^- = RuCl_5^{2-} + 3e^-$	-0.60	$Ce^{3+} = Ce^{4+} + e^-$	-1.61
$U^{4+} + 2H_2O = UO_2^+ + 4H^+ + e^-$	-0.62	$Am^{3+} + 2H_2O = AmO_2^{2+} + 4H^+ + 2e^-$	-1.69
$3NH_4^+ = NH_3 + 11H^+ + 8e^-$	-0.69	$Am^{3+} = Am^{4+} + e^-$	-2.18
$Np^{4+} + 2H_2O = NpO_2^+ + 4H^+ + e^-$	-0.75		
	Basic Solution		
$U(OH)_3 + OH^- = U(OH)_4 + e^-$	2.2	$NO_2^- + 2OH^- = NO_3^- + H_2O + 2e^-$	-0.01
$U(OH)_4 + 2Na^+ + 4OH^- = NaUO_4 + 4H_2O + 2e^-$	1.61	$SeO_3^{2-} + 2OH^- = SeO_4^{2-} + H_2O + 2e^-$	-0.05
$Te^{2-} = Te + 2e^-$	1.14	$Mn(OH)_2 + OH^- = Mn(OH)_3 + e^-$	-0.1
$HPO_3^{2-} + 3OH^- = PO_4^{3-} + 2H_2O + 2e^-$	1.12	$Co(OH)_2 + OH^- = Co(OH)_3 + e^-$	-0.17
$Pu(OH)_3 + OH^- = Pu(OH)_4 + e^-$	0.95	$I^- + 6OH^- = IO_3^- + 3H_2O + 6e^-$	-0.26
$SO_3^{2-} + 2OH^- = SO_4^{2-} + H_2O + 2e^-$	0.93	$PuO_2OH + OH^- = PuO_2(OH)_2 + e^-$	-0.26
$Fe + CO_3^{2-} = FeCO_3 + 2e^-$	0.756	$4OH^- = O_2 + 2H_2O + 4e^-$	-0.401
$AsO_2^- + 4OH^- = AsO_4^{3-} + 2H_2O + 2e^-$	0.67	$MnO_2 + 4OH^- = MnO_4^{2-} + 2H_2O + 2e^-$	-0.60
$Fe(OH)_2 + OH^- = Fe(OH)_3 + e^-$	0.56	$RuO_4^{2-} = RuO_4^- + e^-$	-0.60
$Cr(OH)_3 + 5OH^- = CrO_4^{2-} + H_2O + 3e^-$	0.13	$Br^- + 6OH^- = BrO_3^- + 3H_2O + 2e^-$	-0.66
$Mn(OH)_2 + 2OH^- = MnO_2 + 2H^+ + 2e^-$	0.05	$O_2 + 2OH^- = O_3 + H_2O + 2e^-$	-1.24

SOURCE: W. F. Latimer, *The Oxidation States of the Elements and Their Potentials in Aqueous Solutions*, 2nd Edition, copyright 1952, renewed 1980, pp. 340–348. Reprinted by permission of Prentice-Hall, Inc., Englewood Cliffs, NJ. Reference 53.

TABLE 4.3 Weight Conversion Factors

Have Weight of: / Desire Weight of:	Multiply by: → / ← Divide by:	Desire Weight of: / Have Weight of:
Americium-243		
Am	1.438	$AmCl_3$
	1.235	AmF_3
	1.210	$Am(OH)_3$
	1.765	$Am(NO_3)_3$
	1.543	$Am_2(C_2O_4)_3$
	1.099	Am_2O_3
	1.132	AmO_2
Aluminum		
Al	4.942	$AlCl_3$
	2.890	$Al(OH)_3$
	7.894	$Al(NO_3)_3$
	2.186	AlO_2^-
	1.889	Al_2O_3
Ammonia		
NH_3	0.8224	N_2
	2.350	NH_4NO_3
	2.058	NH_4OH
	3.171	N_2O_5
	3.647	NO_3^-
NH_4NO_3	1.350	N_2O_5
Boron		
B	5.719	H_2BO_3
	3.220	B_2O_3
	6.272	BF_3
H_2BO_3	0.5630	B_2O_3

Have Weight of: / Desire Weight of:	Multiply by: → / ← Divide by:	Desire Weight of: / Have Weight of:
Barium		
Ba	1.516	$BaCl_2$
	1.248	$Ba(OH)_2$
	1.903	$Ba(NO_3)_2$
	1.116	BaO
	1.437	$BaCO_3$
	1.699	$BaSO_4$
Beryllium		
Be	8.869	$BeCl_2$
	4.774	$Be(OH)_2$
	14.76	$Be(NO_3)_2$
	2.775	BeO
	11.66	$BeSO_4$
Calcium		
Ca	2.497	$CaCO_3$
	2.769	$CaCl_2$
	1.849	$Ca(OH)_2$
	4.094	$Ca(NO_3)_2$
	1.399	CaO
	2.580	$Ca_3(PO_4)_2$
	3.397	$CaSO_4$
Carbonate ion		
CO_3^{2-}	0.7334	CO_2
	0.9836	HCO_3^-
Cerium		
Ce	1.759	$CeCl_3$
	1.407	CeF_3
	1.364	$Ce(OH)_3$

Element	Factor	Compound
Cerium		
Ce (Continued)	2.327	$Ce(NO_3)_3$
	2.770	$Ce(NO_3)_4$
	1.942	$Ce_2(C_2O_4)_3$
	1.171	Ce_2O_3
	1.228	CeO_2
Cesium		
Cs	1.226	Cs_2CO_3
	1.267	$CsCl$
	1.128	$CsOH$
	1.466	$CsNO_3$
	1.060	Cs_2O
	1.361	Cs_2SO_4
Chromium		
Cr	3.045	$CrCl_3$
	1.654	$Cr(OH)_2$
	4.577	$Cr(NO_3)_3$
	1.308	CrO
	1.462	Cr_2O_3
Cobalt		
Co	2.201	$CoCl_2$
	1.577	$Co(OH)_2$
	3.104	$Co(NO_3)_2$
	1.271	CoO
	1.407	Co_2O_3
Curium-244		
Cm	1.234	CmF_3
	1.209	$Cm(OH)_3$
	1.762	$Cm(NO_3)_3$
	1.541	$Cm_2(C_2O_4)_3$
	1.098	Cm_2O_3
	1.131	CmO_2
Europium		
Eu	1.336	$Eu(OH)_3$
	2.224	$Eu(NO_3)_3$
	2.316	Eu_2O_3
Fluorine		
F	1.053	HF
	2.210	NaF
	1.264	H_2SiF_6
Hydrogen		
H	36.35	HCl
	62.81	HNO_3
	32.57	H_3PO_4
	48.89	H_2SO_4
	8.981	H_2O
$^2H(D)$	9.979	D_2O
$^3H(T)$	10.98	T_2O

TABLE 4.3 (Continued)

Have Weight of: Desire Weight of:	Multiply by: → ← Divide by:	Desire Weight of: Have Weight of:	Have Weight of: Desire Weight of:	Multiply by: → ← Divide by:	Desire Weight of: Have Weight of:
Iodine			Manganese		
I	1.008	HI	Mn	2.183	$HMnO_4$
	1.252	I_2O_4		1.619	$Mn(OH)_2$
	1.315	I_2O_5		3.331	$Mn(NO_3)_2$
Iron				1.583	MnO_2
Fe	2.074	$FeCO_3$		1.291	MnO
	2.270	$FeCl_2$		1.437	Mn_2O_3
	2.904	$FeCl_3$		1.388	Mn_3O_4
	1.609	$Fe(OH)_2$	Molybdenum		
	1.913	$Fe(OH)_3$	Mo	1.688	H_2MoO_4
	3.221	$Fe(NO_3)_2$		1.698	$MoO(OH)_3$
	4.331	$Fe(NO_3)_3$		2.626	$MoO_2(NO_3)_2$
	1.287	FeO		1.336	MoO_2
	1.430	Fe_2O_3		1.500	MoO_3
Lanthanum			Neodymium		
La	1.410	LaF_3	Nd	1.395	NdF_3
	1.367	$La(OH)_3$		1.354	$Nd(OH)_3$
	2.383	$La(NO_3)_3$		2.290	$Nd(NO_3)_3$
	1.950	$La_2(C_2O_4)_3$		1.915	$Nd_2(C_2O_4)_3$
	1.173	La_2O_3		1.166	Nd_2O_3
Lithium			Neptunium-237		
Li	5.324	Li_2CO_3	Np	1.598	$NpCl_4$
	3.452	LiOH		1.287	$Np(OH)_4$
	9.935	$LiNO_3$		1.397	NpO_2NO_3
	2.153	Li_2O		1.135	NpO_2
				1.770	Np_2O_3

Element	Factor	Compound
Nickel		
Ni	2.208	$NiCl_2$
	1.579	$Ni(OH)_2$
	3.112	$Ni(NO_3)_2$
	1.273	NiO
	2.636	$NiSO_4$
Niobium		
Nb	2.908	$NbCl_5$
	2.022	NbF_5
	4.337	$Nb(NO_3)_5$
	1.344	NbO_2
	1.431	Nb_2O_5
Nitrate Ion		
NO_3^-	0.2258	N_2
	1.016	HNO_3
	0.7581	HNO_2
	0.4840	NO
	0.7421	NO_2
	0.8710	N_2O_5
	0.3540	N_2O
Phosphate Ion		
PO_4^{3-}	0.3261	P
	0.8631	H_3PO_3
	1.032	H_3PO_4
	0.7473	P_2O_5

Element	Factor	Compound
Plutonium-239		
Pu	1.327	PuF_4
	1.489	PuF_6
	1.458	$PuCl_3$
	1.385	$Pu(OH)_4$
	2.038	$Pu(NO_3)_4$
	1.653	$PuO_2(NO_3)_2$
	1.148	PuO_2
	1.100	Pu_2O_3
Potassium		
K	1.767	K_2CO_3
	1.907	KCl
	1.435	KOH
	2.586	KNO_3
	2.229	K_2SO_4
	1.205	K_2O
Praseodymium		
Pr	1.755	$PrCl_3$
	1.405	PrF_3
	1.362	$Pr(OH)_3$
	2.320	$Pr(NO_3)_3$
	1.227	PrO_2
	1.170	Pr_2O_3
	2.023	$Pr_2(SO_4)_3$
Promethium-147		
Pm	1.723	$PmCl_3$
	1.388	PmF_3
	1.347	$Pm(OH)_3$

TABLE 4.3 (Continued)

Have Weight of: Desire Weight of:	Multiply by: → ← Divide by:	Desire Weight of: Have Weight of:
Promethium-147		
Pm (Continued)	2.265	$Pm(NO_3)_3$
	1.163	Pm_2O_3
	1.980	$Pm_2(SO_4)_3$
Rubidium		
Rb	1.351	Rb_2CO_3
	1.415	RbCl
	1.199	RbOH
	1.725	$RbNO_3$
	1.094	Rb_2O
Ruthenium		
Ru	2.052	$RuCl_3$
	1.505	$Ru(OH)_3$
	2.840	$Ru(NO_3)_3$
	1.317	RuO_2
	1.633	RuO_4
Samarium		
Sm	1.707	$SmCl_3$
	1.379	SmF_3
	1.339	$Sm(OH)_3$
	2.237	$Sm(NO_3)_3$
	1.160	Sm_2O_3
	1.958	$Sm_2(SO_4)_3$

Have Weight of: Desire Weight of:	Multiply by: → ← Divide by:	Desire Weight of: Have Weight of:
Sodium		
Na	2.862	$NaBO_2$
	2.305	Na_2CO_3
	2.542	NaCl
	1.740	NaOH
	3.697	$NaNO_3$
	1.348	Na_2O
	1.696	Na_2O_2
	2.377	Na_3PO_4
	2.655	Na_2SiO_3
Strontium		
Sr	1.865	$SrCO_3$
	1.809	$SrCl_2$
	1.388	$Sr(OH)_2$
	2.415	$Sr(NO_3)_2$
	1.183	SrO
	2.096	$SrSO_4$
Sulfate Ion		
SO_4^{2-}	0.3338	S
	0.3547	H_2S
	1.021	H_2SO_4
	0.6669	SO_2
	0.8334	SO_3

Technetium-99		
Tc	3.505	$Tc(NO_3)_4$
	1.323	TcO_2
	1.566	Tc_2O_7
	1.828	NH_4TcO_4
	1.879	$NaTcO_4$
	2.340	$Ba(TcO_4)_2$
Thorium		
Th	1.517	$Th(CO_3)_2$
	1.611	$ThCl_4$
	1.327	ThF_4
	1.293	$Th(OH)_4$
	2.069	$Th(NO_3)_4$
	1.759	$Th(C_2O_4)_2$
	1.138	ThO_2
Uranium		
U	1.596	UCl_4
	1.319	UF_4
Uranium		
U (Continued)	1.474	UF_6
	1.655	$UO_2(NO_3)_2$
	2.084	$UO_2(NO_3)_2 \cdot 6H_2O$
	1.134	UO_2
	1.202	UO_3
	1.179	U_3O_8
Yttrium		
Y	2.196	YCl_3
	1.574	$Y(OH)_3$
	3.092	$Y(NO_3)_3$
	1.267	Y_2O_3
	2.621	$Y_2(SO_4)_3$
Zirconium		
Zr	2.555	$ZrCl_4$
	1.746	$Zr(OH)_4$
	3.719	$Zr(NO_3)_4$
	1.351	ZrO_2
	3.106	$Zr(SO_4)_2$

SOURCES: Reference 9 and calculated.

TABLE 4.4 Properties of Selected Nitrates

Compound	Molecular Weight	Melting Point (°C)	Density (g/cm^3)	Solubility in H_2O (g/100g)
$Al(NO_3)_3 \cdot 6H_2O$	375.13	73.5		73.9^{20}, 160^{100}
$Am(NO_3)_3$	429			s
NH_4NO_3	80.04	169.6	1.725	192^{20}, 871^{100}
$Ba(NO_3)_2$	261.35	592	3.24	9.0^{20}, 34.4^{100}
$Ca(NO_3)_2$	164.09	561	2.504	121^{18}, 376^{100}
$Ce(NO_3)_3 \cdot 6H_2O$	434.23	150		vs
$CsNO_3$	194.91	414	3.685	23.0^{20}, 197^{100}
$Cr(NO_3)_3$	238.13			124^{15}, 152^{35}
$Co(NO_3)_2 \cdot 6H_2O$	291.04	55–56	1.87	97.4^{20}, 300^{100}
$Cm(NO_3)_3$	431			s
$Fe(NO_3)_2 \cdot 6H_2O$	287.05	60.5		134^{15}, 266^{60}
$Fe(NO_3)_3 \cdot 6H_2O$	349.95	35	1.684	137.7^{20}, 175^{40}
$La(NO_3)_3 \cdot 6H_2O$	433.02	40		136^{20}, 247^{60}
$LiNO_3$	68.96	261	2.38	70.1^{20}, 175^{60}
$Nd(NO_3)_3 \cdot 6H_2O$	438.35			142^{20}, 211^{60}
$Ni(NO_3)_2 \cdot 6H_2O$	290.81	56.7	2.05	94.2^{20}, 188^{100}
$Pu(NO_3)_4$	492			s
KNO_3	101.11	333	2.019	31.6^{20}, 245^{100}
$RbNO_3$	147.47	310	3.11	52.9^{20}, 374^{100}
$Sm(NO_3)_3 \cdot 6H_2O$	444.46	78–79	2.375	vs
$NaNO_3$	84.99	d 300	2.261	87.6^{20}, 180^{100}
$Sr(NO_3)_2$	211.63	570	2.986	69.5^{20}, 98.4^{90}
$Th(NO_3)_4$	480.06	d 500		191^{20}
$UO_2(NO_3)_2 \cdot 6H_2O$	502.18	60.2	2.807	122^{20}, 474^{100}
$Y(NO_3)_3 \cdot 6H_2O$	383.01		2.68	123^{20}, 200^{60}

SOURCES: Varied, please see text.

Note: All solubilities are as grams of anhydrous salt per 100 grams H_2O.

In a standard solution of an inorganic compound having limited solubility, the product of the concentrations of the ions combining to form the compound is a constant at a given temperature. As shown in Table 4.7, the situation for a saturated solution of $AlPO_4$ would be

$$K_s = [Al^{3+}]\ [PO_4^{3-}] = 6.2 \times 10^{-19}$$

In the case of $Al(OH)_3$, however,

$$K_s = [Al^{3+}]\ [OH^-]^3 = 1.3 \times 10^{-33}$$

TABLE 4.5 Properties of Selected Hydroxides

Compound	Molecular Weight	Melting Point (°C)	Density (g/cm^3)	Solubility in H_2O (g/100g)
$Al(OH)_3$	78.00	300	2.42	0.000104^{18}
$Am(OH)_3$	294			i
$Ba(OH)_2$	171.48	78	4.50	3.89^{20}, 101.4^{100}
$Be(OH)_2$	43.04	d 138	2.01	0.002^{RT}
$Ca(OH)_2$	74.09	580	2.24	0.173^{20}, 0.086^{100}
CsOH	149.91	272.3	3.68	395.5^{18}
$Cr(OH)_2$	86.01			d
$Cr(OH)_3$	103.3			i
$Co(OH)_2$	92.95	d	3.60	0.00032^{RT}
$Cm(OH)_3$	297			i
$Fe(OH)_2$	89.86	d	3.4	0.00015^{18}
$Fe(OH)_3$	106.87		ca 3.7	i
$La(OH)_3$	189.93	d		i
LiOH	23.95	450	2.54	12.8^{20}, 17.5^{100}
$Mo(OH)_3$	146.96	d		0.2^{RT}
$Nd(OH)_3$	195.24			i
$Ni(OH)_2$	99.72	d 230	4.15	0.013^{RT}
$Ni(OH)_3$	109.71	d		i
$Pu(OH)_4$	312			i
KOH	56.11	360.4	2.044	112^{20}, 178^{100}
RbOH	102.49	300	3.203	180^{15}, vs
$Ru(OH)_3$	152.72			v sl s
$Sm(OH)_3$	201.37			i
NaOH	40.00	318.4	2.13	109^{20}, 347^{100}
$Sr(OH)_2$	121.63		3.63	1.77^{20}, 21.8^{100}
$Th(OH)_4$	300.02	d		i
$Y(OH)_3$	139.93	d		i

SOURCES: Varied, please see text.

TABLE 4.6 Properties of Selected Oxides

Compound	Molecular Weight	Melting Point (°C)	Density (g/cm^3)	Solubility in H_2O (g/100g)
Al_2O_3	101.96	2015	3.97	0.000098^{24}
Am_2O_3	524.26		11.77	
AmO_2	275.13		11.68	
BaO	153.34	1918	5.72	3.48^{20}, 90.8^{100}
BeO	25.02	2585	3.02	i
B_2O_3	69.62	450	2.46	1.1^{0}, 15.7^{100}
CaO	56.08	2614	3.3	0.131^{10}, 0.07^{80}

TABLE 4.6 (Continued)

Compound	Molecular Weight	Melting Point (°C)	Density (g/cm^3)	Solubility in H_2O (g/100g)
Ce_2O_3	328.24	1692	6.86	i
CeO_2	172.12	ca. 2600	7.132	i
Cs_2O	281.81	d 400	4.25	vs, d
Cs_2O_3	313.81	400	4.25	s, d
Cr_2O_3	151.99	2266	5.21	i
CrO_3	99.99	196	2.70	61.7^{0}, 67.45^{100}
CoO	74.93	1795	6.45	i
Co_2O_3	165.36	d 895	5.18	i
Cm_2O_3	538			
CmO_2	277			
FeO	71.85	1369	5.7	i
Fe_2O_3	159.19	1565	5.24	i
La_2O_3	325.82	2307	6.51	0.0004^{24}
Li_2O	29.88	1700	2.013	6.67^{0} d, 10.02^{100}
MoO_2	127.94		6.47	i
MoO_3	143.95	795	4.50	0.107^{18}, 2.11^{79}
Nd_2O_3	336.48	2270	7.24	0.00018^{29}, 0.003^{75}
NpO_2	269.00		11.1	i
Np_3O_8	839	d 500		
NiO	74.71	1984	6.67	i
Ni_2O_3	165.38		4.83	i
NbO_2	124.90		5.9	i
NbO	108.91		7.30	i
Nb_2O_5	265.81	1485	4.47	d
PuO_2	276	ca. 1750	11.46	
Pu_2O_3	536		11.47	
K_2O	94.20	d 350	2.32	vs
Rb_2O	186.94	d 400	3.72	s, d
Rb_2O_3	218.94	489	3.53	s, d
RuO_2	133.07	d	6.97	i
RuO_4	165.07	25.5	3.29	2.033^{20}, 2.249^{74}
Sm_2O_3	348.70	2350	8.35	i
SiO_2	60.06	1728	2.32	
Na_2O	61.68	subl. 1275	2.27	d
SrO	103.62	2430	4.7	1.03^{30}, 12.5^{100}
ThO_2	264.04	3220	9.68	i
UO_2	270.03	2878	10.96	i
UO_3	286.03	d	7.27	i
U_3O_8	842.09	d 1300	8.30	i
Y_2O_8	255.81	2410	5.01	0.00018^{29}
ZrO_2	123.22	2715	5.56	

SOURCES: Varied, please see text.

TABLE 4.7 Solubility Constants

Compound	K_s	Compound	K_s
$Ac(OH)_3$	1E-15	$Fe(OH)_3$	4E-38
$Al(OH)_3$	1.3E-33	$FePO_4$	1.3E-22
$AlPO_4$	6.3E-19	LaF_3	7E-17
$Am(OH)_3$	2.7E-20	$La(OH)_3$	2.0E-19
$Am(OH)_4$	1E-56	$La_2(MoO_4)_3$	4E-21
$BaCO_3$	5.1E-9	$La_2(C_2O_4)_3 \cdot 9H_2O$	2.5E-27
BaF_2	1E-6	$LaPO_4$	3.7E-23
$Ba(OH)_2$	5E-3	$Pb(OH)_2$	1.2E-15
$BaMoO_4$	4.0E-8	$PbCO_3$	3.3E-14
$Ba(NbO_3)_2$	3.2E-17	Li_2CO_3	2.5E-2
$Ba(NO_3)_2$	4.5E-2	Li_3PO_4	3.2E-9
BaC_2O_4	1.6E-7	$MgCO_3$	3.5E-8
$Ba_3(PO_4)_2$	3.4E-23	$Mg(OH)_2$	1.8E-11
$BaSO_4$	1.1E-10	$Hg_2(OH)_2$	2.0E-24
$Be(OH)_2$	1.6E-22	$Hg(OH)_2$	3.0E-26
$Bi(OH)_3$	4E-31	$Nd(OH)_3$	3.2E-22
$BiPO_4$	1.3E-23	$NpO_2(OH)_2$	2.5E-22
$Cd(OH)_2$	2.5E-14	$Ni(OH)_2$	2.0E-15
$CaCO_3$	2.8E-9	PuO_2CO_3	1.7E-13
$Ca(OH)_2$	5.5E-6	PuF_3	2.5E-16
$CaMoO_4$	4.2E-8	PuF_4	6.3E-20
$Ca(NbO_3)_2$	8.7E-18	$Pu(OH)_3$	2.0E-20
$Ca_3(PO_4)_2$	2.0E-29	$Pu(OH)_4$	1E-55
$CaSO_4$	9.1E-6	$PuO_2(OH)_2$	2E-25
CeF_3	8E-16	$Pr(OH)_3$	6.8E-22
$Ce(OH)_3$	1.6E-20	$Pm(OH)_3$	1E-21
$Ce(OH)_4$	2E-48	$RaSO_4$	4.2E-11
$Ce_2(C_2O_4)_3 \cdot 9H_2O$	3.2E-26	$Ru(OH)_3$	1E-36
$Cr(OH)_2$	2E-16	$Sm(OH)_3$	8.3E-23
$Cr(OH)_3$	6.2E-31	$Sc(OH)_3$	8.0E-31
$CoCO_3$	1.4E-13	Ag_2CO_3	8.1E-12
$Co(OH)_2$	1.6E-15	$AgCl$	1.8E-10
$Co(OH)_3$	1.6E-44	$AgOH$	2.0E-8
$Co_3(PO_4)_2$	2E-35	$SrCO_3$	1.1E-10
$CuOH$	1E-14	$SrMoO_4$	2E-7
$CuCO_3$	1.4E-10	$Sr(NbO_3)_2$	4.2E-18
$Cu(OH)_2$	2.2E-20	SrC_2O_4	2.8E-9
$Eu(OH)_3$	8.9E-24	$SrSO_4$	3.2E-7
$Gd(OH)_3$	1.8E-23	$Tb(OH)_3$	2.0E-22
$Ga(OH)_3$	7.0E-36	$Te(OH)_4$	3.0E-54
$Hf(OH)_4$	4.0E-26	$Tl(OH)_3$	6,3E-46
$FeCO_3$	3.2E-11	$Th(OH)_4$	4.0E-45
$Fe(OH)_2$	8.0E-16	$Th(C_2O_4)_2$	1E-22
$FeC_2O_4 \cdot 2H_2O$	3.2E-7	$Th_3(PO_4)_4$	2.5E-79

TABLE 4.7 (Continued)

Compound	K_s	Compound	K_s
$Sn(OH)_2$	1.4E-23	$VO(OH)_2$	5.9E-23
$Sn(OH)_4$	1E-56	YF_3	6.6E-11
$Ti(OH)_3$	1E-40	$Y(OH)_3$	8.0E-23
$TiO(OH)_2$	1E-29	$Y_2(C_2O_4)_3$	5.3E-29
UO_2CO_3	1.8E-12	$Zn(OH)_2$	1.2E-17
$UO_2(OH)_2$	1.1E-22	$ZrO(OH)_2$	6.3E-49
$UO_2C_2O_4 \cdot 3H_2O$	2E-4	$Zr_3(PO_4)_4$	1E-132
$(UO_2)_3(PO_4)_2$	2.0E-47		

SOURCES: References 56 and 10.

In other words, if more than one ion of the same kind enters into the composition of the molecule, the concentration of that ion is raised to a power equal to the number needed to form the compound. All concentrations are expressed in terms of gram-weights of ion per liter. The data given in Table 4.7 are for the room temperature range (18–25°C). The "*E-x*" convention used in presenting the numerical values of K_s can be deduced from the example given above. The numbers before *E* are used to multiply 10 taken to the indicated power. All of the exponents in Table 4.7 are negative, and, of course, a high negative exponent indicates a highly insoluble compound.

4.6 DATA FOR PROCESS CHEMICALS

4.6.1 Common Acids and Bases

Properties of the more commonly used acids and bases are given in Table 4.8. The numbers given under the "Commercial Product" heading are in most cases for the "concentrated" form, but other grades are generally available.[54] The halogen acids (and NH_4OH) are solutions of gases, so specific gravities of the pure compounds are not given. Some of the other materials are commonly purchased as solids, although NaOH and KOH in particular are available in various concentrations in solution. A 10 wt% value for citric acid and 8 wt% value for oxalic acid were chosen as being reasonable levels for their use in solution as decontaminating agents.

4.6.2 Extractants and Solvents

Certain extractants and solvents have had considerable application in the nuclear industry, either in fuel-rod reprocessing (diethyl ether, hexone, TBP), in labora-

TABLE 4.8 Common Acids and Bases

Name	Pure Compound		Commercial Product[a]		
	Molecular Weight	Specific Gravity	Weight %	Molarity	Specific Gravity
Acids					
Acetic	60.05	1.049			
Dilute			36	6.3	1.04
Glacial			99-99,5	17.6	1.06
Citric(s)	192.14	1.665	(10)	0.54	1.04
Fluosilicic	144.12		61	5.9	1.4
Formic	46.03	1.220	85-90	23.5	1.20
HBr(g)	80.92	–	(40)	6.8	1.38
HCl(g)	36.46	–	36-37	12.1	1.19
HF(g)	20.01	–	51	30.6	1.20
Lactic	90.08	1.206	85	11.4	1.21
Nitric	63.01	1.503	68-70	15.8	1.41
Oxalic(s)	90.04	1.90	(8)	0.93	1.04
Perchloric	100.46	1.764	71.6	11.8	1.66
Phosphoric	98.00	1.834	85	14.7	1.7
Sulfuric	98.08	1.834	96-98	18.0	1.84
Bases					
NH_4OH	35.0		58.6	15.1	0.90
KOH(s)	56.11	2.044	(45)	11.6	1.45
NaOH(s)	40.00	2.130	(40)	14.3	1.43

SOURCES: Varied, plus calculated values.
[a] Please see text.

tory analyses (TTA, EDTA), in decontamination (EDTA), or as diluents for TBP in the Purex process (benzene, *n*-butyl ether, CCl_4 *n*-dodecane, toluene). Kerosine and "Varsol" have also been used for this last purpose. The first material is defined as "a distilled hydrocarbon from petroleum or shale oil having a boiling range from about 100 to 300°C." "Varsol" is the trademark (Humble Oil and Refining Company) for straight petroleum aliphatic solvents "to conform to CS 3-40, the U.S. Department of Commerce commercial standard . . . and to have minimum Tag closed cup flash points of 100°F."[54]

Di-butyl phosphate is the first hydrolysis product of TBP, and chiefly of importance because of its adverse effects on product recovery and decontamination in the Purex process. Formaldehyde has been used (chiefly in Europe) to destroy nitrate ions before calcination of high-level wastes.

Data for Table 4.8 are taken from References 9 and 10 and from commercial literature.

TABLE 4.9 Extractants, Solvents, and Other Organics

Compound	Molecular Weight	Specific Gravity (RT)	Boiling Point (°C)	Air Tolerances (ppm)
Diethyl ether	74.12	0.7138	34.5	400
Methyl isobutyl ketone (hexone)	100	0.8042	115.8	100
Tri-*n*-butyl phosphate (TBP)	358.9	0.978	177–178 (627 mm)	1
Acid-di-*n*-butyl phosphate (HDBP)	302.9			1
Thenoyltrifluoro acetone (TTA)	222.		ca. 43 (MP)	
Ethylene diamine tetra acetic acid (EDTA)	292		240 (dec) (MP)	
Benzene	78.12	0.8765	80.1	10[a]
n-Butyl ether	130.32	0.7689	142	
Carbon tetrachloride	156	1.585	76.7	
n-Dodecane	170.34	0.7487	216.3	
Toluene	92.15	0.8669	110.6	200[a]
Formaldehyde (g)	30.05	1.08 (37% solution)	98 (37% solution)	

SOURCES: Varied, please see text.
[a]Eight-hour weighted average.

4.6.3 Ion Exchangers

Since there are a number of manufacturers, and each usually offers an extensive line of products for various applications, a very large number of ion-exchange resins are available commercially. Only a limited number of these are considered here—namely, those known to have been investigated or used for radioactive waste treatment. It should be noted that while there are dozens of resins on the market, many of them are essentially the same, and substitution of one manufacturer's product for that of another is often possible if one knows the equivalent trade designations. An extensive comparison of this type is given in Table 4 of Reference 57.

Each resin will be briefly discussed and the manufacturer listed. Properties not covered in the text are given in Table 4.10. The data presented are, for the most part, taken from manufacturers' literature.

TABLE 4.10 Ion-Exchange Resins

Trade Name	Total Exchange Capacity meq/ml	Total Exchange Capacity meq/g	Shipping Density lb/ft^3
Duolite ARC-9359	1.2(H)		20
Chelex-100	0.7		
Duolite CS-100	1.0		38–45
Amberlite IRA-401	1.0		43 (690 g/l)
Dowex-1	1.33	3.5	44(Cl)
Dowex-2	1.33	3.5	44(Cl)
Dowex-50	1.7(H)	5.0(H)	50(H)
	1.9(Na)	4.8(Na)	53(Na)

SOURCES: Manufacturers' literature.

Note: Notations in parentheses indicate ionic form.

Duolite ARC-359 (Diamond Shamrock[58]). This is a strong acid-type resin, but has a macroporous phenolic matrix rather than the more usual polystyrene type and methylene sulfonic acid functional groups, rather than the nuclear sulfonic acid groups used in other strong acid cation resins. The material has a particularly useful selectivity for cesium over sodium. It is accordingly receiving consideration for the removal of ^{137}Cs from the alkaline, high-salt supernatants of the wastes stored at Savannah River[59] and at West Valley.[60] The resin has a pH operating range of 0–13, good radiation stability, and a maximum operating temperature at 40°C. (ARC-359 is now designated ARC-9359 by the manufacturer. It is the nuclear grade of Duolite C-3R).

Chelex-100 (Bio-Rad Laboratories[57]). The Savannah River report describes the use of this resin for removal of ^{90}Sr from alkaline supernatants. Chelex-100 is a chelating resin with a styrene-divinyl benzene matrix and iminodiacetic acid functional groups. It is versatile in that variations in pH and in complexing ions present allow a number of different separations to be made. Its selectivity for divalent over monovalent ions is approximately 5,000 to one. The resin is classified as a weak acid cation exchanger and shows very low metal absorption below pH 2, a sharp increase between 2 and 4, and a maximum above 4.

Duolite CS-100 (Diamond Shamrock). Oak Ridge has developed processes for removal of ^{137}Cs and ^{90}Sr from alkaline low-level wastes by use of this resin. The matrix in this case is a phenol-formaldehyde condensate; the functional groups are carboxylic and phenolic. It is classified as a granular weak acid cation exchanger. The operating pH range is 6–14, and the resin can be used to an operating temperature of 80°C.

Amberlite IRA-401 (Rohm and Haas[61]). In the late 1960s, Panesko[62] devised a process for separating rhodium, palladium, and technetium from Purex wastes using Amberlite IRA-401. This is a Type I strong base resin utilizing a styrene matrix cross-linked with divinylbenzene. The Type I in the description indicates a trimethylamine functional group in such resins: Type II, dimethyl-β-hydroxylethylamine. The Type I products are more stongly basic than Type II, but are more difficult to regenerate. Type II has a higher chemical stability, but is more sensitive to oxidants.[63] The strong base anion and strong acid cation resins are produced in large volumes and are available with different degrees of cross-linking (divinylbenzene content). The OH^- form of the 8% cross-linked strong base type can be used up to about 50°C, the Cl^- and other forms to 150°C. They should not stand in high nitric acid solution for any length of time since they become unstable and may react explosively. The operating pH range otherwise is 0-14.

Dowex-1 (Dow Chemical U.S.A.[64]). This resin has also been used for Tc removal from alkaline Purex wastes.[65] It is also a Type I strongly basic anion resin, and the statements in the previous paragraph all apply. Dowex-2 is Dow's Type II strong base anion resin.

Dowex-50 (Dow Chemical U.S.A.). This strong acid cation resin, as its equivalents from other manufacturers, does not appear to have had much direct application in high-level radioactive waste management, although used extensively in reactor water-treatment systems and in nuclear research. It has the polystyrene matrix with nuclear sulfonic acid functional groups and is available in a number of different levels of cross-linking. On a laboratory scale, these strong acid cation resins have been used extensively for rare earth and actinide element separations. The operating pH range is 0-14 and the maximum thermal stability 150°C.

Nuclear reactors require exceptionally pure water at various points in their operation, particularly for cooling water where trace contaminants could become neutron-activated or cause corrosion. Most of the resin manufacturers have developed suites of resins for reactor use, but these will not be considered here. The materials are usually carefully purified versions of the company's everyday products. Similarly, highly purified and carefully graded resins can be obtained[56] for analytical and biochemical use.

The literature is replete with laboratory-scale schemes for absorbing ^{137}Cs and/or ^{90}Sr on various column materials, but the few ion-exchange resins already mentioned and certain zeolites appear to be the prime candidates for large-scale application. The zeolites are naturally occurring or synthetically prepared inorganic minerals or compounds having absorbent or ion-exchange properties. The one of most current interest for Cs absorption is Linde AW-500 (Union Carbide Corporation[66]). This material is a synthetic crystalline metal alumino-silicate produced in the form of $\frac{1}{8}$- or $\frac{1}{16}$-inch pellets. The shipping density is 45.4 lb/ft^3, the pellet density, 72.7 lb/ft^3.

PART THREE

RADIOACTIVE WASTES

The sources of radioactive wastes produced in the Light Water Reactor Fuel Cycle[67] are listed in Table 5.1. Contaminated materials also originate from medical and industrial applications, research activities, etc., but these materials fall into the low-level or intermediate-level categories of wastes, discussed in Chapter 6. This chapter presents only data relevant to the characteristics and potential disposal forms of the high-level liquid waste (HLLW) items of Table 5.1.

The Nuclear Regulatory Commission defined[68] high-level waste as

> (1) irradiated reactor fuel, (2) liquid wastes resulting from the first-cycle solvent extraction system, or equivalent, and the concentrated wastes from subsequent extraction cycles, or equivalent, in a facility for reprocessing irradiated reactor fuel, and (3) solids into which such liquid wastes have been converted.

(Tables 3.3-3.6 apply to the unprocessed irradiated fuels if they are treated as waste. It has been estimated[69] that there are more than 8000 tonnes of such fuels currently in storage in the country.)

TABLE 5.1 Classification of Primary Wastes from the Post-Fission LWR Fuel Cycle

Waste Category	Fuel Cycles	General Description
Gaseous	All	Predominantly two types: (1) large volumes of ventilation air, potentially containing particulate activity, and (2) smaller volumes of vessel vent and process off-gas, potentially containing volatile radioisotopes in addition to particulate activity.
Compactable trash and combustible wastes	All	Miscellaneous wastes include paper, cloth, plastic, rubber, and filters. Wide range of activity levels dependent on source of waste.
Concentrated liquids, wet wastes, and particulate solids	All	Miscellaneous wastes including evaporator bottoms, filter sludges, and resins. Wide range of activity levels dependent on source of waste.
Failed equipment and noncompactable, noncombustible wastes	All	Miscellaneous metal or glass wastes including massive process vessels. Wide range of activity levels dependent on source of waste.
Spent UO_2 fuel	Once-through	Irradiated PWR and BWR fuel assemblies containing fission products and actinides in ceramic UO_2 pellets sealed in zircaloy tubes.
High-level liquid waste	Uranium recycle and uranium–plutonium recycle	Concentrated solution containing over 99% of the fission products and actinides, except U and Pu, in the spent fuel. Contains about 0.5% of the U and Pu in the spent fuel.
Hulls and assembly hardware (fuel residue)	Uranium recycle and uranium–plutonium recycle	Residue remaining after UO_2 is dissolved out of spent fuel. Includes short segment of zircaloy tubing (hulls) and stainless steel assembly hardware. Activity levels are next highest to HLLW.
PuO_2	Uranium recycle	Purified PuO_2 powder.

SOURCE: Report DOE/EIS-0046, Reference 67.

FIVE

HIGH-LEVEL LIQUID WASTES

5.1 TYPES AND COMPOSITIONS OF HLLW

There are approximately 83 million liters (22 million gallons) of high-level liquid waste (HLLW) at the Savannah River Plant in Aiken, South Carolina; 190 million liters (50 million gallons) at the Hanford Reservation, Richland, Washington; and 1,500 cubic meters (52,000 cubic feet) of solid HLW at the Idaho National Engineering Laboratory, Idaho Falls. All of these wastes were produced during reprocessing of nuclear reactor fuels used for the production of defense nuclear materials, and thus are classified as "defense wastes."[70]

The wastes generated at Hanford and Savannah River were neutralized with caustic soda as they were obtained and subsequently stored in underground tanks. Since water was also evaporated off in order to reduce the volume, these wastes, if they have not been further manipulated, physically consist of a mixture of alkaline sludge, a crystalline phase (mostly $NaNO_3$) and a concentrated, high-salt supernatant. At Hanford, a program was started in the 1960s to extract the ^{137}Cs and ^{90}Sr from the supernates. As of 1978, about 80% of the former and 65% of the latter had been recovered, most of which was still in solution and stored in tanks having cooling coils, but some 10% of the total had been converted to solid form. The cesium is fixed as CsCl and stored in stainless steel containers, the strontium as SrF_2 in nonferrous nickel. The containers are stored in water pools similar to those for storing spent reactor fuel elements.[71]

The method of handling the HLLW at Idaho follows a different pattern.

There the wastes are calcined in a fluidized bed reactor and stored as powder, with about 2,000 tonnes having been produced since the program began in 1963. At one stage, it was thought that the powder might be the final disposal form, but there is now general agreement that the calcine must undergo additional stabilization before it can be placed in a repository.

"Commercial" wastes are defined as those originating in a privately owned reprocessing plant. The relatively small quantity (560,000 gallons, less than 1% of the defense wastes) in existence in the U.S. was produced during the 1966–1972 period at the Western New York Nuclear Service Center at West Valley, New York, and is stored at that site. As with the defense wastes, the bulk of the material was neutralized as produced, and now exists as a supernate covering an alkaline sludge. There is also a smaller quantity (12,000 gallons) of an acidic waste from a single reprocessing run on an experimental thorium fuel. This is stored separately in a stainless steel tank.[72]

Production of ^{239}Pu with a low content of other plutonium isotopes for use in nuclear weapons requires relatively short irradiation times. This, combined with the fact that the defense wastes, on the average, have been undergoing decay for more than 20 years, means that the average activity level is comparatively low, perhaps of the order of a Ci per gallon. The West Valley wastes are also aging, and were partially derived from low-irradiation fuel furnished by Hanford, so their activity level is also modest–about 50 Ci/gal. On the other hand, the data of Table 3.7 can be used to calculate that the wastes from a LWR fuel having full 33,000 MWD exposure and processed 160 days after discharge would have an activity level of around 40,000 Ci/gal. (This assumes 378 liters of waste produced per MTU of fuel processed).

Most of Chapter 3 on application of the ORIGEN code, particularly the figures and Tables 3.7–3.9, is concerned with the question of HLLW composition. In addition, Table 5.2 gives the chemical makeup of wastes from LWR fuel reprocessing, including the chemicals added during the operation. Table 5.3 provides the chemical information for an average Savannah River waste (average because there can be wide differences between storage tanks). Table 5.4 gives the activity levels of *fresh* Savannah River waste. Table 5.5 presents chemical data for wastes resulting from the processing of fuels from a High Temperature Gas Reactor and a Liquid Metal Fast Breeder Reactor as well as for a LWR. Table 5.6 gives LWR data in another format as a matter of possible convenience. Information on the West Valley wastes is not given here but is available in a series of tables in several publications.[70, 72]

It should be noted that data such as discussed in the last paragraph are almost completely based on calculations of the ORIGEN type plus estimates of the average quantities of process chemicals needed and the average volume of wastes resulting per MTU of fuel processed. It is quite certain that no HLLW sample has undergone complete chemical and radiometric analysis, although the concen-

TABLE 5.2 Typical HLLW Composition

Concentration (moles/l @ 378 l/MTU)			Kilograms Nonvolatile Oxide/MTU	
Constituent	Possible Range	Reference for This Document[a]	Constituent	Reference for This Document
Inerts (reprocessing chemicals)				
H^+	1–7	2.0	–	–
Na	0–3.0	0.01	Na_2O	0.12
Fe	0.05–1.4	0.054	Fe_2O_3	1.6
Cr	0.01–0.04	0.0096	Cr_2O_3	0.28
Ni	0.005–0.02	0.0034	NiO	0.19
PO_4^{3-}	0.025–0.30	0.042	P_2O_5	1.1
SO_4^{2-}	0–0.90	0	SO_4^{2-}	0
NO_3^-	2.7–20	3.6	–	–
F^-	0–0.25	0	F^-	0
Gd[b]	0–0.2	0.150	Gd_2O_3	10.0
B[b]	0–5.0	0	B_2O_3	0
Cd[b]	0–1.5	0	CdO	0
			Subtotal	13.0
Fission products				
Rb	c	0.0095	Rb_2O	0.34
Sr	c	0.017	SrO	0.68
Y	c	0.0095	Y_2O_3	0.41
Zr	c	0.074	ZrO_2	3.5
Mo	c	0.071	MoO_3	3.9
Tc	c	0.017	Tc_2O_7	1.0
Ru	c	0.044	RuO_2	2.2
Rh	c	0.011	Rh_2O_3	0.53
Pd	c	0.030	PdO	1.4
Ag	c	0.0015	Ag_2O	0.067
Cd	c	0.0016	CdO	0.078
Sn	c	0.0009	SnO_2	0.052
Sb	c	0.002	Sb_2O_3	0.013
Te	c	0.0078	TeO_2	0.47
Cs	c	0.039	Cs_2O	2.2
Ba	c	0.023	BaO	1.3
La	c	0.018	La_2O_3	1.1
Ce	c	0.034	CeO_2	2.2
Pr	c	0.016	Pr_6O_{11}	1.1
Nd	c	0.055	Nd_2O_3	3.5
Pm	c	0.0005	Pm_2O_3	0.035
Sm	c	0.012	Sm_2O_3	0.80
Eu	c	0.002	Eu_2O_3	0.13
Gd	c	0.001	Gd_2O_3	0.076
			Subtotal	27.0

TABLE 5.2 (Continued)

Concentration (moles/l @ 378 l/MTU)			Kilograms Nonvolatile Oxide/MTU	
Constituent	Possible Range	Reference for This Document[a]	Constituent	Reference for This Document
Actinides				
U	0.011–0.22	0.053	U_3O_8	5.7
Np	[c]	0.003	NpO_2	0.31
Pu	0.0001–0.006	0.002	PuO_2	0.17
Am	[c]	0.009	Am_2O_3	0.96
Cm	[c]	0.003	Cm_2O_3	0.26
			Subtotal	7.4
			Total	47.0

SOURCE: Report ERDA-76-43, Vol. 2. Reference 73.

[a]The reference HLLW is representative of a large state-of-the-art commercial reprocessing plant such as BNFP.

[b]Potential soluble poisons that may be used during fuel dissolution.

[c]Depends upon burnup of the fuel being reprocessed, should not vary over a factor of two from the reference waste composition (burnup = 25,000 MWD/MTU).

TABLE 5.3 Average Chemical Composition of Fresh Savannah River Plant High-Level Waste

Constituent	Concentration (molar)
$NaNO_3$	3.3
$NaNO_2$	<0.2
$NaAl(OH)_4$	0.5
NaOH	1
Na_2CO_3	0.1
Na_2SO_4	0.3
$Fe(OH)_3$	0.07
MnO_2	0.02
$Hg(OH)_2$	0.002
Other solids	0.13[a]

SOURCE: Report ERDA-77-42, Vol. 1. Reference 74.

[a]Assuming an average molecular weight of 60.

TABLE 5.4 Average Radionuclide Composition of Fresh[a] Savannah River Plant High-Level Waste

Radionuclide	Activity (Ci/gal)	Radionuclide	Activity (Ci/gal)
^{144}Ce–^{144}Pr	68	^{241}Am	1×10^{-3}
^{95}Zr	60	^{99}Tc	5×10^{-4}
^{91}Y	47	^{239}Pu	3×10^{-4}
^{89}Sr	36	^{154}Eu	1×10^{-4}
^{95}Nb	15	^{93}Zr	1×10^{-4}
^{141}Ce	12	^{240}Pu	6×10^{-5}
^{147}Pm	12	^{135}Cs	4×10^{-5}
^{103}Ru	10	^{126}Sn–^{126}Sb	1×10^{-5}
^{106}Ru–^{106}Rh	4	^{79}Se	1×10^{-5}
^{90}Sr	3	^{233}U	2×10^{-6}
^{137}Cs	3	^{129}I	1×10^{-6}
^{129}Te	2	^{238}U	6×10^{-7}
^{127}Te	2	^{107}Pd	5×10^{-7}
^{134}Ce	1	^{237}Np	4×10^{-7}
^{151}Sm	8×10^{-2}	^{152}Eu	2×10^{-7}
^{238}Pu	1×10^{-2}	^{242}Pu	6×10^{-8}
^{241}Pu	2×10^{-3}	^{158}Tb	6×10^{-8}
^{244}Cm	1×10^{-3}	^{235}U	3×10^{-8}

SOURCE: Report ERDA 77-42, Vol. 1. Reference 74.
[a]After reprocessing fuel that has been cooled six months after discharge from reactor.

trations of key elements and isotopes may be occasionally checked to be certain that the predicted and actual values are not too wildly different.

5.2 SOLIDIFICATION OF HLLW

In 1970, the Atomic Energy Commission (now the Nuclear Regulatory Commission) issued regulations relating to the final disposition of high-level liquid wastes. These rules appeared in 10CFR50, Appendix F (Title 10, Appendix F of Part 50 of the *Code of Federal Regulations*). The portions of most interest here read:

> A fuel reprocessing plant's inventory of high-level wastes will be limited to that produced in the prior 5 years. · · · High-level liquid radioactive wastes shall be converted to a dry solid as required to comply with this inventory limitation, and placed in a sealed container prior to transfer to a Federal repository in a shipping cask meeting the requirements of 10CFR Part 71. The dry solid shall be chemically, thermally and radiolytically stable to

TABLE 5.5 Wastes From Various Reactor Types

Material[b]	Grams/MT from Reactor Type[a]		
	LWR[c]	HTGR[d]	LMFBR[e]
Reprocessing chemicals			
Hydrogen	400	3,800	1,300
Iron	1,100	1,500	26,200
Nickel	100	400	3,300
Chromium	200	300	6,900
Silicon	–	200	–
Lithium	–	200	–
Boron	–	1,000	–
Molybdenum	–	40	–
Aluminum	–	6,400	–
Copper	–	40	–
Borate	–	–	98,000
Nitrate	65,800	435,000	244,000
Phosphate	900	–	–
Sulfate	–	1,100	–
Fluoride	–	1,900	–
Subtotal	68,500	452,000	380,000
Fuel product losses[f,g]			
Uranium	4,800	250	4,300
Thorium	–	4,200	–
Plutonium	40	1,000	500
Subtotal	4,840	5,450	4,800
Transuranic elements[g]			
Neptunium	480	1,400	260
Americium	140	30	1,250
Curium	40	10	50
Subtotal	660	1,440	1,560
Other actinides[g]	<0.001	20	<0.001
Total fission products[h]	28,800	79,400	33,000
Total	103,000	538,000	419,000

SOURCE: Reference 75, with permission. Original based on Reference 76.

[a]Water content is not shown; all quantities are rounded.

[b]Most constituents are present in soluble, ionic form.

[c]U-235 enriched PWR, using 378 liters of aqueous waste per metric ton, 33,000 MWD/MT exposure. (Integrated reactor power is expressed in megawatt-days [MWD] per unit of fuel in metric tons [MT].)

[d]Combined waste from separate reprocessing of "fresh" fuel and fertile particles, using 3,785 liters of aqueous waste per metric ton, 94,200 MWD/MT exposure.

[e]Mixed core and blanket, with boron as soluble poison, 10% of cladding dissolved, 1,249 liters per metric ton, 37,100 MWD/MT average exposure.

[f]0.5% product loss to waste.

[g]At time of reprocessing.

[h]Volatile fission products (tritium, noble gases, iodine, and bromine) excluded.

TABLE 5.6 HLLW Composition as Oxides[a]

Constituent	Wt %	Atom %	Constituent	Wt %	Atom %
Rb_2O	0.937	0.489	Nb_2O_3	–	–
Cs_2O	6.804	2.638	MoO_3	10.798	4.099
SrO	2.476	1.305	Tc_2O_7	3.123	1.101
BaO	3.745	1.334	RuO_2	7.043	2.892
Y_2O_3	1.323	0.640	Bi_2O_3	–	–
La_2O_3	3.351	1.124	RhO_2	1.182	0.479
CeO_2	7.646	2.427	PdO	3.692	1.648
Pr_2O_3	3.166	1.049	CdO	0.219	0.093
Nd_2O_3	10.837	3.519	Ag_2O	0.156	0.073
Pm_2O_3	0.282	0.090	Sb_2O_3	0.049	0.018
Sm_2O_3	2.257	0.707	SnO_2	0.185	0.067
Eu_2O_3	0.506	0.157	U_3O_8	14.684	2.858
Gd_2O_3	0.282	0.085	NpO_2	1.216	0.247
Dy_2O_3	0.005	0.001	PuO_2	0.165	0.033
SeO_2	0.151	0.074	Am_2O_3	0.384	0.079
TeO_2	1.702	0.583	Cm_2O_3	0.107	0.022
ZrO_2	11.629	5.157			
Sum of atom % = 35.091			Atom % of oxygen = 64.909		

SOURCE: Reprinted by permission of the publisher from *Scientific Basis for Nuclear Waste Management*, S. V. Topp, ed., pp. 297–311. Copyright 1982 by Elsevier Publishing Co. Inc. Reference 77g.

[a]Composition of LWR waste: 3% U-235 fuel, burnup 30,000 MWD/THM; reprocessed 150 days after discharge from reactor; 0.5% U and 0.5% Pu not separated.

> the extent that the equilibrium pressure in the sealed container will not exceed the safe operating pressure for that container during the period from canning to a minimum of 90 days after receipt (transfer of physical custody) at the Federal repository. All of these high-level radioactive wastes shall be transferred to a Federal repository no later than 10 years following separation of fission products from the irradiated fuel.

Since the West Valley reprocessing plant was already licensed, the wastes in storage at that site were specifically excluded from the above procedure, which is just as well since there is still no Federal repository to accept the materials. Plans are now being made[78] to revamp the West Valley facilities, and to proceed with a waste solidification program. In the overall national waste-management plan, the next step will be to solidify the Savannah River defense wastes, with the materials at Hanford and Idaho awaiting solidification until more experience is gained from the West Valley and Savannah River Plant projects.

A considerable amount of development work on solidification techniques had been carried out both in this country and abroad before issuance of these AEC-NRC regulations, primarily on calcines and glasses. There has been criticism of both of these as final disposal forms, chiefly on the question of their ability to retain active species if there was a water intrusion into the repository, particularly at higher temperatures. Recent years have seen much activity in searching for the ultimate stable form. Table 5.7 summarizes the potential disposal products having received at least some investigation. The table is thought to be fairly complete for work in this country, although overseas development is somewhat slighted. At least one specific reference is given in each case where detailed information can be obtained. There are also a number[71, 79a, 80–83] of more general surveys comparing the properties and relative difficulty of preparation of the different forms.

Some of the forms given in Table 5.7, such as aqueous silicate and normal concrete, are probably more suitable for application to disposal of intermediate- or low-level wastes. Concrete, in particular, has a long history of use for such purposes.

5.2.1 Comparison of Forms

Since the Savannah River materials are the first defense wastes scheduled for solidification, much study of the various solidification possibilities has been carried out by Savannah River Laboratory. Table 5.8 lists the criteria established for one such study concerned with solidifying Savannah River Laboratory waste sludges,[80] although the same criteria would apply to any HLLW solidification form. Each of 11 candidate forms was rated on a scale of 1 to 5 for each of the criteria, these being divided between desirable features in the product and desirable features in processing. Since some of the criteria are more important than others (leachability vs. development status, as an example), a weighting factor was applied to each criterion. The various forms were then given a "weighted sum" and an overall score calculated, the results appearing in Table 5.9. The decision was then made that for Savannah River Plant sludges, high-silica glass, supercalcine, SYNROC, and coated ceramics appeared to be most promising for further investigation as possible improvements on the front-runner, borosilicate glass.

Table 5.10 gives a somewhat similar comparison, but no attempt at numerical quantification. It is estimated in the table that at least 15 years more development work would be required before some of the forms could be applied on a full-scale basis.

5.2.2 Product Descriptions

Capsule descriptions are given in this subsection of the less-developed solid forms listed in Table 5.7, followed by somewhat longer discussions of calcines and borosilicate and phosphate glasses since these are the only products that have been manufactured to date with full-level HLLW and thus the only ones immedi-

TABLE 5.7 Solid Disposal Forms for HLLW

Solid Form	Developer	References
Clays		
Aqueous silicate	ARHCO, PNL	73, 84
Lean clay		
Rich clay		
Concretes		
Normal	SRL, Penn State U., ORNL, BNL	85–87
Hot-pressed	Penn State U., RHO	88
FUETAP	ORNL	89
Calcines		
Pot	ORNL, PNL	90
Spray	PNL	91
Fluidized bed	ICPP	92
Pelletized	ICPP	93
Stabilized	ICPP	92
Rotary kiln	Marcoule	73
Stirred bed	Mol	73
Glasses		
Phosphate	BNL, PNL, Julich, USSR	90, 94, 77b
Borosilicate	PNL, SRL, Overseas	95–97
High silica	Catholic U., NPD Nuclear Systems	80, 98
Ceramics		
Supercalcine	Penn State U., RHO, PNL	99, 100
Glass ceramic	ICPP	101
Clay ceramic	RHO	102
SYNROC	Australia U., LLL, ANL, N. C. State U.	103, 104
Tailored	Penn State U., RHO	99
Titanates	Sandia Lab., RHO	105
Multibarrier		
Glass in metal	PNL, ANL, Eurochemic	106, 107
Ceramic in metal	PNL, ICPP	106, 108
Cermets	ORNL	79b, 109
Coated ceramic	Battelle-Columbus, PNL	107, 110
Coated ceramic via Sol-gel	ORNL	111

SOURCES: References 71, 79a, 81 and varied.

Note: Abbreviations used: ANL, Argonne National Laboratory; ARHCO, Atlantic Richfield Hanford Company; BNL, Brookhaven National Laboratory; ICPP, Idaho Chemical Processing Plant; LLL, Lawrence Livermore Laboratory; ORNL Oak Ridge National Laboratory; PNL, Pacific-Northwest Laboratory; RHO, Rockwell Hanford Operation; SRL, Savannah River Laboratory.

TABLE 5.8 Savannah River Waste Form Selection Criteria

Related to Final Product

Development status.

Waste loading—weight fraction and density of waste in the solid form; sensitivity to variations in waste composition and loading.

Leachability—a measure of release of radionuclides to the environment when the waste form contacts water. Leachability typically is measured for ^{90}Sr, ^{137}Cs, and ^{239}Pu. Leachability under possible repository conditions (temperature and pressure; brine or ground water) is of principal interest. However, data are most often available for leaching in distilled water at ambient conditions.

Long-term stability—chemical and mechanical durability under conceivable repository conditions, including the long-term effects of heat, radiation, and transmutation.

Thermal conductivity.

Thermal stability—including emission of gases at high temperatures.

Transportation safety—fire resistance; dispersability and impact resistance; compressive and tensile strength.

Related to Formation Process

Complexity—number of operations; use of high temperatures and/or high pressures; remotability. Capital and operating costs are not considered here.

State of development—including applicable large-scale experience.

Quality assurance—ease of determining acceptable products during production; ease of sampling and analysis.

Yield and recycle—emphasizing ease of reworking off-specification products.

Process safety—unusual hazards.

SOURCES: Stone et al., Report DP-1545, Reference 80.

ately available for industrial application. Because of the importance of the leachability question, that topic will also be considered separately.

Aqueous Silicate. This process has been specifically developed for immobilization of the relatively low-activity salt cakes and residual liquors of the Hanford defense wastes into alumino-silicate mineral form. The wastes are mixed with powdered clays such as kaolin or bentonite to form a cancrinite-type mineral at 30–100°C temperatures. If excess clay is used to act as a binder for the cancrinite crystals, the procedure is known as the Rich Clay Process. The original mixture has a peanut-butter consistency, but with aging hardens to resemble adobe. In the Lean Clay Process, only the stoichiometric amount of clay is used,

TABLE 5.9 Savannah River Ratings of Waste Forms

	Product Factors								Process Factors						
	Development Status	Waste Loading	Leachability	Long-Term Stability	Thermal Conductivity	Thermal Stability	Transportation Safety	Weighted Sum	Complexity	State of Development	Quality Assurance	Yield and Recycle	Process Safety	Weighted Sum	Overall Score
Borosilicate glass	5	3	3	3	2	5	4	66	4	5	3	3	4	69	135
High-silica glass	2	2	5	3	2	5	4	68	3	2	3	2	4	51	119
FUETAP concrete	3	1	1	2	2	2	2	33	5	4	3	2	4	69	102
Hot-Pressed concrete	2	3	1	4	2	3	3	49	3	1	3	2	3	45	94
Supercalcine ceramic	2	5	4	5	2	5	5	81	2	1	1	1	3	30	111
SYNROC ceramic	3	5	4	5	2	5	5	83	2	1	1	1	2	27	110
Cermet (urea process)	2	3	3	1	2	5	5	52	1	2	5	1	2	36	88
Glass marbles in metal matrix	4	2	3	3	3	4	2	57	3	4	4	3	3	60	117
Ceramic pellets in metal matrix	3	4	4	3	4	4	3	67	2	2	2	1	3	36	103
Coated ceramic	3	4	5	5	2	5	5	86	2	2	2	1	2	33	119
Coated ceramic via Sol-Gel	1	4	5	5	2	5	5	82	2	3	2	1	4	42	124
Factor weights	2	2	5	5	1	2	2		6	3	3	3	3		

SOURCE: Stone et al., Report DP-1545, Reference 80.

TABLE 5.10 High-Level Nuclear Waste Immobilization Forms—Properties Comparison

Waste form	Devel. status	Process complexity	Process flexibility	Waste loading	Dispersion impact resistance	Long-term stability	Fire resistance	Leachability 100 °C	Leachability 350 °C
Calcine	Available	Low	Excellent	High	Very low	High	Poor	Poor	Poor
Rich clay	Available	Low	Excellent	Low	Low	?	Poor	Medium	Poor
Normal concrete	Available	Medium	Excellent	Medium	Medium	Medium	Poor	Medium	Poor
Hot pressed concrete	5 years	High	Excellent	Medium	High	Medium	Medium	Good	Poor
Pelletized calcine	5 years	High	Excellent	Medium	Medium	Medium	Medium	Good	Poor
Glass	Available	High	Excellent	Medium	High	High	Excellent	Excellent	Poor
Clay ceramic	5 years	High	Poor	Medium	High	Medium	Medium	Good	Poor
Supercalcine	15 years	Very high	Poor	High	Very high	High?	Best	Best	Poor
SYNROC	15 years	High	Poor	Very low	Very high	High?	Best	Best	Good
Glass ceramic	18 years	Very high	Poor	Medium	High	High	Excellent	Excellent	Poor
Pellet in metal matrix	5 years	Very high	Good	Low	Very high	High	Poor	Excellent	Poor
Coated supercalcine in Metal matrix	15 years	Highest	Poor	Medium	Very high	High?	Excellent	Best	Poor
Cermet	10 years	Highest	Poor	Medium	High	High	Excellent	Excellent	Poor

Most attractive   Intermediate   Least attractive

thus reducing the final volume. The product must be washed to remove excess salts. Binders such as Portland cement or an organic polymer may be added to increase mechanical strength. If the salt-clay mixture is molded into bricks, allowed to air dry, then fired at 800°C, the product is nepheline bricks—the Clay Calcination Process. Leach rates for the three products are roughly

Product	g/cm^2-day
Rich clay	10^{-2} to 10^{-3}
Lean clay	1-4 × 10^{-4}
Clay calcination	2-6 × 10^{-4}

Normal Concrete. There has been extensive industrial experience with cement solidification systems, and these have been used widely throughout the world to solidify or encapsulate all types of radioactive but non-high-level wastes. While serious consideration was given[86] to concrete as a disposal form for Savannah River defense wastes, the disadvantages of a bulky product coupled with residual water and nitrate content susceptible to radiolytic decomposition were adverse factors. A careful study of the economics involved[79a] also indicated, rather surprisingly, that there would be no financial advantage to concrete over borosilicate glass as a final disposal form. Concrete might still be considered for immobilization in place of older, lower-level wastes where removal of the material from a tank would not justify the risks and costs involved.

Oak Ridge has developed[87] a "shale-fracture" method for disposal of intermediate-level waste. A waste plus cement plus an absorptive materials grout is pumped under pressure into underground shale beds. The grout distributes itself from the charging well into pregenerated horizontal fractures in the bed where it hardens to a rocklike consistency. Careful study over a decade's application reveals no problem with the technique.

Hot-Pressed Concrete. This process is designed to improve concrete properties by elimination of void spaces and of nonchemically combined water. Dense concrete monoliths are formed by hot-pressing at 150–200°C and 25–50,000 psi. Waste loading would be about 25 w/o. Estimated leachabilities: 10^{-5} g/cm^2-day for ^{137}Cs, 10^{-7} for ^{90}Sr, and 10^{-10} for ^{239}Pu. The product would be stable to 760°C, have a thermal conductivity in the range of 0.6 to 1.0 Btu/hr-ft-°F, a density of 175 lb/ft^3, and a compressive strength of around 20,000 psi.[80]

FUETAP. This acronym is for Formed Under Elevated Temperature And Pressure and is similar to hot-pressed concrete save that milder temperatures and pressures would be required. The major difference is that the wastes are first dried at 105°C to a powder containing less than 0.8% water. This powder is then blended with cement and a minimum amount of water, then cast directly into the disposal container. The product is cured for a day at 150°C and 100 psi, then heated to 250°C for 48 hours to drive off excess water. Waste loading (oxide basis) about 12 w/o; thermal conductivity, 0.3 Btu/hr-ft-°F; tensile strength, 200–400 psi; compressive strength, 2000–4000 psi; leachability in distilled water, about 10^{-6} g/cm^2-day for ^{137}Cs, 10^{-4} for ^{90}Sr, and less than 10^{-8} for ^{239}Pu.

Special Concretes. These are not listed in Table 5.7, but the researchers at Penn State have suggested[88] that ceramic forms such as supercalcine could be cast into concrete as a form of multibarrier protection.

Supercalcine. Glass is thermodynamically unstable, so the intent of all of the proposed ceramic-type disposal forms is to incorporate the waste elements into a crystalline lattice in order to improve stability. Supercalcine is an alumino-silicon product of this type. Tailored additions of Al_2O_3, SiO_2, CaO, and so on, are added to the liquid waste, the mixture is spray dried at 600–800°C, the resulting calcine is then outgassed at 1000°C, then isostatically compressed at 1000–1300°C and 50,000 psi to full density. Up to 95 w/o of waste as oxides can be incorporated, and the product is stable to better than 1000°C. The expected thermal conductivity would be about 1 Btu/hr-ft-°F and leachabilities would be some 10 times less than glass. Table 5.11 shows the phases in which the waste elements are fixed in supercalcine.

TABLE 5.11 Supercalcine Fixation Phases (Primary Containment Phases for Principal HLW Radionuclides in Supercalcine)

Constituent	Fixation Phase	Structure Type
Cs, Rb	$(Cs, Rb)AlSi_2O_6$	Pollucite
Sr, Na, Mo	$(Ca, Sr)_2[NaAlSiO_4]_6(MoO_4)_2$	Sodalite
Sr, Ba, Mo	$(Ca, Sr, Ba)MoO_4$	Scheelite
Sr, RE[a] $[PO_4]$	$REPO_4$	Monazite
	$(Ca, Sr)_2RE_8[SiO_4]_6O_2$	Apatite
Ce, U, Zr	$(Ce, U, Zr)O_2 + x$	Fluorite
	$(Zr, Ce)O_2$	
Fe, Ni, Cr	$(Fe, Ni)(Fe, Cr)_2O_4$	Spinel
	$(Fe, Cr)_2O_3$	Corundum
Ru	RuO_2	Rutile

SOURCE: Report NUREG/CR-0895, Reference 71.

[a]RE = rare earths, particularly La, Pr, Nd, Sm, and Gd (and probably Am, Cm).

Glass Ceramics. The Idaho Processing Plant has several investigative programs to develop methods for further stabilizing their inventory of stored calcines. In the glass ceramics approach, the calcine is mixed with glass frit and then sintered or hot pressed at comparatively low temperature. Nucleating agents can be added to produce an assemblage of small crystals. The product has good thermal shock resistance, low leachability, and high temperature stability.

Clay Ceramics. The process for making clay ceramics is similar to the Clay Calcination described under Aqueous Silicate. The wastes are mixed with bentonite or kaolin. The mixture is extruded and cut into pellets that are fired at 1200°C. The leachability characteristics of the final form are only fair.

SYNROC. Whereas supercalcine is designed to produce alumino-silicon crystalline forms, SYNROC is titanate based. Table 5.12 shows the minerals formed and the waste elements associated with each phase. The required additives are mixed with the HLLW to form a slurry that is calcined at 800°C. Powdered Ni metal is added to reduce iron to the plus-two state, and the mixture is then cold-pressed at 50,000 psi, sintered at 1000°C, cooled, canned, and finally isostatically pressed at 1200°C and 50,000 psi. Waste loading can be up to 70 w/o, and the expected thermal conductivity about 1 Btu/hr-ft-°F. Leach rates at 95°C may be several factors of 10 less than borosilicate glass, depending on the element examined, a differential that becomes even greater at higher temperatures.[110]

Tailored Ceramics. These are similar to supercalcine, but with the additives being varied to obtain maximum stability and retention of particular waste elements of special interest.

Titanates. Sandia Laboratory has developed a series of hydrous inorganic ion-exchangers (titanates, niobates, zirconates). HLLW is passed through columns of the materials as an activity-concentration device. The contents of the loaded columns are then mixed with about 30 w/o glass frit and sintered at 900–1100°C to give low leachability products.

Glass Marbles in Metal Matrix. The multibarrier concept, as the name implies, would provide additional stability and resistance to possible leaching by surrounding the solidified waste by an inert material inside the disposal container. In the cases where metal would be used as the encapsulating agent, a considerable improvement in thermal conductivity can also be obtained, an important consideration with fresh wastes where container centerline temperatures must be kept within reason.

TABLE 5.12 SYNROC Mineral Phases

	Typical Composition of SYNROC and Its Constituent Minerals (weight percent)				Distribution of Principal High-Level Waste Elements among Constituent Minerals of SYNROC[a]			
	Hollandite	Zirconolite	Perovskite	Bulk SYNROCK	Hollandite	Zirconolite	Perovskite	Metal[b]
TiO_2	71.0	50.3	57.8	60.3	Cs^+	U^{4+}	Na^+	Ru
ZrO_2	0.2	30.5	0.2	10.8	Rb^+	Th^{4+}	Sr^{2+}	Tc
Al_2O_3	12.9	2.5	1.2	6.3	K^+	Pu^{4+}	Pu^{3+}	Mo
CaO	0.4	16.8	40.6	16.2	Ba^{2+}	Cm^{4+}	Am^{3+}	Ni
BaO	16.0	–	–	6.4	Fe^{2+}	Am^{4+}	Cm^{3+}	Pd
Total	100.5	100.1	99.8	100.1	Cr^{3+}	Np^{4+}	Np^{3+}	Rh
					Ni^{2+}	ACT^{3+}	RE^{3+}	Te
					Mo^{4+}	RE^{3+}	ACT^{4+}	S
						Sr^{2+}		Fe

SOURCE: Reproduced with permission of the Scientific Society of America. Reference 104.

[a] ACT = actinides. RE = rare earths.

[b] When prepared by heat treatment under the preferred nonoxidizing conditions, SYNROC may contain a few percent of dispersed metal particles.

TABLE 5.13 Eurochemic's Vitromet Multibarrier—Some Mechanical Properties of Glass Beads, a Matrix Alloy, and a Metral Matrix Composite

Material	Tensile Strength (kg/mm^2)	Compression Strength (kg/mm^2)	Charpy Impact (kg/cm^2)	Elasticity (Modulus) (kg/cm^2)	Expansion Coefficient ($^\circ C^{-1}$)
Vitromet[a]	0.4	230	0.43	—	$28.8 \cdot 10^{-6}$
Phosphate glass	—	~8	—	665	$1.1 \cdot 10^{-5}$ (<400°C) $4.0 \cdot 10^{-5}$ (>400°C)
Lead-1% Sb	2.4	300	4.3	1600	$28.8 \cdot 10^{-6}$

SOURCE: Reference 113 as quoted in Reference 106.

[a]Vitromet consists of phosphate glass beads dispersed in a lead-1 wt % antimony alloy.

Eurochemic's Vitromet consists of waste-containing glass beads encapsulated in a matrix of 99% Pb–1% Sb alloy. Some of its properties are given in Table 5.13. The comparable development in this country has been towards the formation of waste borosilicate glass in a ceramic melter at 1150°C. The glass, as it is drained from the melter, is formed into 6-mm glass marbles. These are loaded into a container (about 50% of the final volume) and the interstices and an annular outside sheath filled with a molten 90% Pb–10% Sn alloy at 300°C. The container is then cooled upwards from the bottom. The glass density is about 2.37 g/cm^3, the alloy, 10.7 g/cm^3. The thermal conductivity of the final product is estimated at 4.5 Btu/hr-ft-°F, and the overall waste oxide loading at 4.4 w/o.

Ceramic Pellets in Metal Matrix. This concept is similar to the one above, save that the waste would be converted to supercalcine or to a sintered ceramic rather than to glass. Some work at ICPP has also been done on metal encapsulation of ordinary calcine.

Stone describes[80] a process utilizing supercalcine as the waste form. The wastes are converted to supercalcine as previously described and the product formed into 6-mm diameter pellets. After sintering at 1200°C, the pellets are loaded into a container and the voids filled with 88% Al–12% Si alloy at 600°C. The lightweight alloy allows waste loadings of up to 56 w/o (oxide basis) with 50% alloy–50% supercalcine loading. The expected thermal conductivity would be around 30 Btu/hr-ft-°F, the density of the sintered pellet, 3.4 g/cm^3, and of the alloy, 2.6 g/cm^3.

Cermets. In a cermet, the waste is dispersed as a tailored ceramic in the form of 1-μm diameter particles in a continuous metal phase. Some reducible waste elements may alloy with the metal.

In the Oak Ridge process, metals as nitrates, plus ceramic forming additives, are mixed with HLLW to form a slurry that is then heated with molten urea at 150°C. The urea chemically destroys the water and dissolves any solids. The mixture is then calcined at 800°C. The calcine is cooled, binder is added, and the material formed into pellets. These are conveyed to a 1200°C continuous-feed furnance where: (1) the binder is removed by heat, (2) CO gas is added to reduce the metal oxides to metal, and (3) the pellets are sintered. The finished product is loaded into containers and the voids filled with sand. The cermets themselves are about 25% ceramic phase–75% metal phase and have a density of 6–8 g/cm^3. The thermal conductivity is projected as 8 Btu/hr-ft-°F. This is reduced by a factor of 10 in the pellet plus sand-filled canisters, but still adequate for defense wastes. Leach rates are of the order of 10^{-7} to 10^{-8} g/cm^2-day. The metal phase is 70% Fe, 20% Ni, 5% Cu, and 5% miscellaneous.

Coated Ceramics. Supercalcine or SYNROC in the form of 1–4 mm diameter pellets is first coated with a 40-μm layer of pyrolytic carbon, then with 60 μm of alpha Al_2O_3, the latter providing mechanical strength and oxidation protection. The pellets are loaded into canisters and the void again loaded with sand. Short-term leach tests on the coated pellets have shown no loss of activity. The thermal conductivity is low, so a copper coating might be needed for full-activity commercial wastes. The pellets are sintered at 1300°C, the coatings applied at 1100°C.

Coated Ceramic via Sol-Gel. This process is largely in the conceptual stage. The wastes plus ceramic-forming additives would be peptized to form a sol of micron or smaller-diameter particles. This would then be pumped through an orifice and the liquid stream broken up in a controlled manner to form individual droplets that gel to rigid spheres and, after sintering, are coated with pyrolytic carbon and Al_2O_3 as before. Sand would again be used to fill the void after the spheres were loaded into a container.

Table 5.14 gives the bulk properties of some of these multibarrier waste forms as prepared on a small scale. The symbol PyC is for pyrolytic carbon.

5.2.3 Calcines

Figure 5.1 sketches the primary processes now being used for production of HLLW calcines. The oldest of the techniques is pot calcination, so-called because the entire procedure takes place in the "pot" or final disposal canister. The liquid waste is added batch-wise to the container, which is enclosed in a zoned furnace. As the water boils off and oxides form, a crust of calcine forms on the sides and bottom. The process is continued until the container is about 90% filled. The batch nature of this process limits its use for any large-scale application, and quality control is more difficult than in other methods.

TABLE 5.14 Bulk Properties of Multibarrier Waste Forms

Inner Core	Coating	Matrix	Encapsulation Method	Bulk Density (g/cm^3)	Thermal Conductivity[a] (W/m-K)	Maximum Use Temperature (°C)
ICM-11 Waste glass	None	None	—	3.4	0.84	550
ICM-11 Waste glass marble	None	Pb-10Sn	Vacuum cast 400°C	6.2	8.3	250
Supercalcine SPC-2[b]	None	None	—	4.88	0.91	1200
Supercalcine SPC-4[c]	None	None	—	4.03	N.D.[d]	1200
Supercalcine SPC-2[c]	None	None	—	2.35	N.D.	1200
Supercalcine SPC-4	None	Al-12Si	Vacuum cast 650°C	3.5	51.0	550
Supercalcine SPC-4	Glass (~1 mm)	None	—	3.66	N.D.	600
Supercalcine SPC-4	Glass (~1 mm)	Al-12Si	Vacuum cast 650°C	3.4	45.0	550
Supercalcine SPC-2	PyC (40 μm) Al_2O_3 (60 μm)	None	—	2.5	N.D.	1200
Supercalcine SPC-2	PyC (40 μm) Al_2O_3 (60 μm)	Cu	Gravity sintered 900°C, 8 hr	3.48	24.0	1000

SOURCE: Rusin, Report PNL 2668-2, Reference 107.

[a] Determined by comparative method at 200°C.

[b] Hot-pressed.

[c] Disc pelletized.

[d] Not determined.

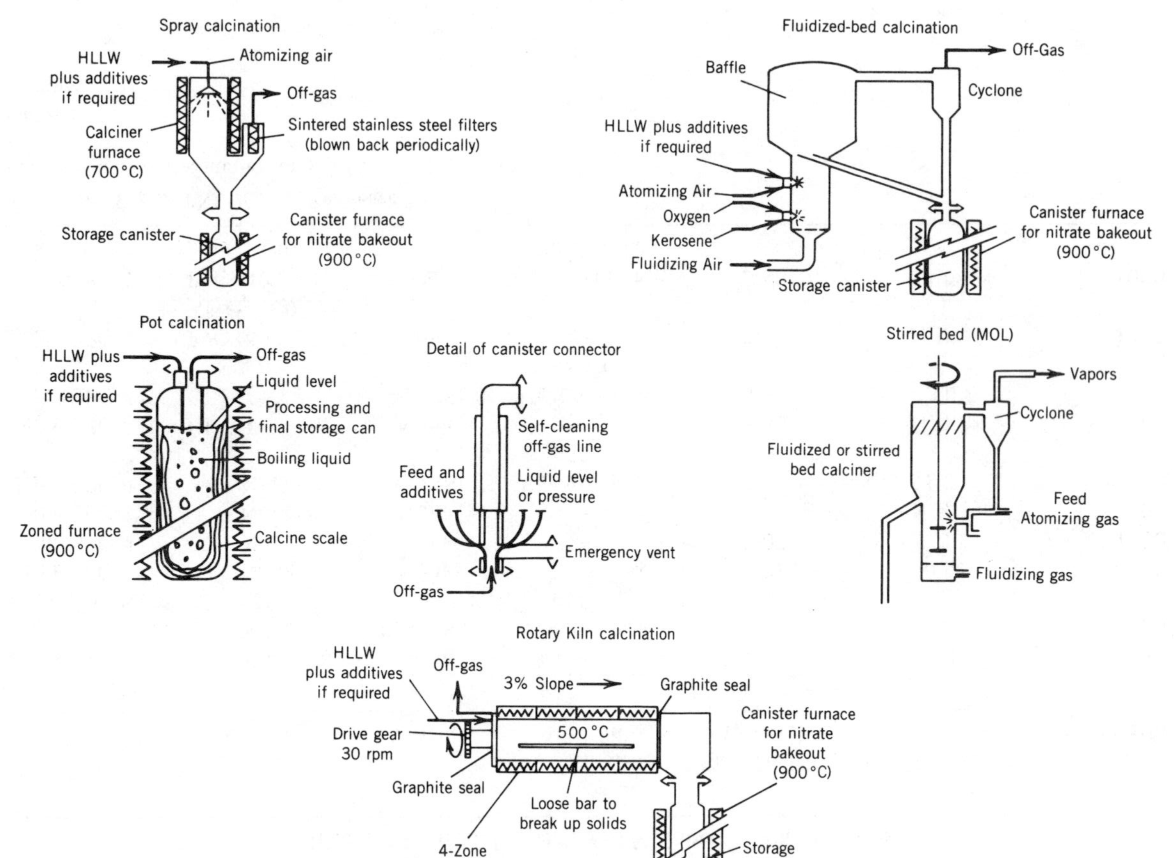

FIGURE 5.1. Current waste calcination processes. (Reference 71, based on Reference 73.)

The two major techniques in this country are Fluidized Bed Calcination at ICCP and Spray Calcination at PNL. In the latter process, the HLLW is introduced into the top of an externally heated reaction chamber through a spray orifice along with a jet of atomizing air. Water is driven off of the falling droplets, the waste is largely converted to oxides and is collected in the disposal container in the form of a fine powder. The reaction furnace is operated to produce a 700°C wall temperature, although the calcine temperature itself is typically in the 350–550°C range. The technique accepts wastes of almost any concentration and has been demonstrated on a large scale with full-activity wastes.

This latter is also true of the Fluidized Bed Calcination Process, which has been used routinely at ICCP since 1963 for calcining HLLW. The wastes are atomized into a bed of inert oxides kept in suspension by air jets from below and heated internally to 500–600°C. Evaporation occurs on the surfaces of the original bed particles and results in a product consisting of granular bed material and powdered calcine, both of which are removed from the reactor on a continuous basis. Heating of the bed was originally done by means of an exterior furnace, but this resulted in high losses of Ru and Cs. Heating is now accomplished by kerosene combustion in the bed, and element volatility is no longer a major problem.

The rotary kiln calciner has been largely developed by the French. The equipment consists of an externally heated (500°C) rotating cylinder operating at a slight angle from the horizontal so that HLLW introduced at the upper end is dried and almost completely denitrated before it exits at the lower end. A loose bar in the barrel keeps the calcine free-flowing and prevents wall deposits.

As shown in Figure 5.1, the Stirred Bed reactor at Mol in Belgium is a variation of the fluidized bed approach, the materials being stirred as well as being fluidized by air jets. The stirring approach allows better control of particle size in the finished product. The Eurochemic wastes are high in aluminum content. Addition of phosphate to the HLLW feed produces metal phosphates and a substantial amount of aluminum phosphate that acts as a secondary containment (the LOTES Process). Leachability characteristics are considerably improved, and the product is considered a simple form of supercalcine.

Calcines are typically fine powders and thus relatively dispersible, a serious consideration in the event of a transportation accident. The untreated materials are also easily soluble in water. These two characteristics have largely eliminated calcines from consideration as a final disposal form for HLLW. They are, however, highly reactive chemically and excellent raw materials for further processing.

A high sodium nitrate content, typical of many defense wastes, creates problems in all of the processes. The salt has a melting point of 307°C and, upon melting, forms a viscous, sticky mass that resists further decomposition. The addition of finely divided metallic iron to the wastes helps with this problem. PNL has also used powdered silica additives in the feed stream so that the crusts accumulating on the interior walls of the calciner are at least partially silicates, which are more readily knocked loose by vibrating hammers acting on the outside of the walls. Processess have also been developed, particularly in Europe, where much of the nitrate in the wastes is destroyed by pretreatment with formaldehyde or formic acid. Most calcines as originally produced will still contain traces to substantial amounts of undecomposed nitrate salts as well as small amounts of residual water. A brief "bake-out" of the product at 700–900°C is generally carried out if the material is to be stored for any length of time in order to avoid problems of radiolytic decomposition.

The calcines produced at ICCP by the Fluidized Bed Process are typically either high in aluminum or high in zirconium content. Table 5.15 presents data relevant to these two types. Table 5.16, assembled from several sources, gives some other properties as reported in the literature for the four major types of calcines. These properties are obviously highly dependent on the compositions of the original wastes, which can vary considerably. Most of the numbers in Table 5.16 are given as ranges.

5.2.4 Glasses

Phosphate glass is being studied at the Eurochemic Corporation at Mol for incorporation into metal matrices (Vitromet) as described above. The Russians apparently are also planning to use phosphate glass as the final disposal form for at least part of their wastes.[73] Research on this type of glass has been carried out in

TABLE 5.15 Fluidized Bed Calcines

	Aluminum Waste	Zirconium Waste
Mass mean particle diameter (mm)	0.5–0.7	0.3–0.7
Bulk density (g/cm^3)	1.1	1.7
Thermal conductivity (kcal/hr m °C)		
40°C	0.12	0.48
800°C	0.37	1.07
Chemical composition (wt%)		
Al as Al_2O_3	88.8	21.9
Zr as ZrO_2	—	21.3
Na as Na_2O	1.7	—
Ca as CaF_2	—	54.2
N as N_2O_5	4.0	1.9
Hg as HgO	2.9	—
H_2O	2.0	0.6
F.P. oxides	0.6	0.1

SOURCE: Lakey and Wheeler, Reference 112.

TABLE 5.16 Some Properties of Calcines

Property	Calcine Type			
	Pot	Fluidized Bed	Spray	Rotary Kiln
Bulk density (g/cm^3)	1.1–1.4	2.0–2.4	1.0–2.4	1.0–1.3
Particle density (g/cm^3)	—	—	3.8–5.1	—
Maximum operating temperature (°C)	420	500–600	700	500
Thermal conductivity (W/m-K)	0.35–1	0.2–0.3	0.2	0.2
Specific area (m^2/g)	0.1–5	0.1–5	10–20	0.1–5
Porosity (%)	40–85	45–80	30–75	70–80
Volume (m^3/MTU)	0.044–0.058	0.032–0.40	0.03–0.06	0.045
Compressibility (%)	—	—	31–40	—

SOURCES: References 67, 83, and 91.

this country in the past,[90] but there are no sizable investigative efforts being undertaken at present.

Phosphate glasses will incorporate about the same waste loadings as borosilicate glasses and are better at incorporating molybdates and sulfates.[79a] Phosphate glass as a disposal form has particular attraction for solidification of wastes high in aluminum and sodium content, utilizing the Na_2O-Al_2O_3-P_2O_5 system. For low-to-moderate melting temperatures, the optimum range of the Na to P ratio is 1.0:1.3. This ratio can be increased at higher temperatures, and at 1400–1500°C, glasses can be made with up to 40% Al_2O_3.[77b] The chief disadvantages of the phosphate glass process are the extreme corrosiveness of the molten mixture, necessitating a platinum melter, and the readiness with which the product vitrifies during storage, causing a thousandfold increase in leachability.[79a]

The wastes plus phosphoric acid or phosphorus pentoxide plus akalai metal oxides are evaporated to a thick, syrupy slurry. This material is fed to the melter where the remaining volatile components are volatilized and removed. Heating to 1000–1200°C then produces a molten glass that is drained into mild steel canisters for cooling and solidification.

Table 5.17 includes some data on the properties of phosphate glasses. In addition, Reference 83 gives the following numbers:

Specific volume	0.036–0.078 m^3/MTU
Specific area	0.005–0.05 m^2/kg
Coefficient of linear expansion	8–10 $\times 10^{-6}$/K
Heat capacity	1100–1200 J/kg-K

TABLE 5.17 Glasses and Glass Ceramics Composition and Properties of Phosphate and Borosilicate Glasses and Glass Ceramics

		Phosphate Glasses (wt%)	Borosilicate Glasses (wt%)	Glass Ceramic	
				Celsian-Types (wt%)	Fresnoite-Types (wt%)
P_2O_5		30–60	—	—	—
Fe_2O_3		5–40	—	—	—
SiO_2		0–20	27–50	28–38	22–28
Al_2O_3		5–41	0–1	10–18	0–2
B_2O_3		0–17	9–22	2–7	0–3
CaO		—	0–4	0–6	0–4
BaO		—	—	13–16	28–36
Na_2O		0–30	4–20	0–2	—
Li_2O		—	0–4	1–3	—
TiO_2		—	0–6	3–4	14–23
ZnO		0–5	0–22	3–5	0–6
PbO		0–30	—	0–3	—
Waste oxides		$\leqslant 40$	20–30	20	20
Melting temperature	K	1050–1400	1200–1500	1450 ± 50	1475 ± 25
100 poise temperature	K	1050–1400	1200–1500	1350–1450	1400–1500
Electrical conductivity $[\Omega\ cm]^{-1}$ at prepared temperature		0.5–1.0	0.5–1.0	0.040–0.09	—
Nucleation temperature	K	—	—	900 ± 25	975 ± 25
Nucleation time	h	—	—	3	3
Crystal temperature	K	—	—	1075–1125	1125
Crystal time	h	—	—	10–15	10–15
Heat conductivity W/m/K (500–900 K)		0.8–1.2	1.2–1.4	1.2–1.4	1.2–1.4
Transformation temperature	K	650–750	775–875	775–875	925–950

Dilatomic softening point	K	$\leqslant 800$	$\leqslant 800$	$\geqslant 1000$	$\geqslant 1000$
Coefficient of thermal expansion $\times 10^7$		—	80–120	80–100	80–100
Crystal phases		Fe-phosphate Na-Zr-Phosphate	Ru-Oxide, Pd, Ce-Oxide, spinel Zn-Silicate Sr-Molybdate Gd-Titanate Ca-RE-Silicate Gd-Ca-Phosphate and others	Cymrit or h-Celsian m-Celsian (Ba) $BaAl_2Si_2O_8$ RE-Titanate (RE, Ac, Sr) $RE_2Ti_2O_7$ Ba-Molybdate (Mo, Ba) $BaMoO_4$ Pollucite (Cs, Rb) $(Cs, Na)AlSi_2O_6$	Fresnoite (Ba, Sr) $Ba_2TiSi_2O_8$ Priderite (Ba) $K_2Fe_2Ti_6O_{16}$ RE-Titanate (RE, Ac, Sr) $RE_2Ti_2O_7$ Ba-Molybdate (Ba, Mo) $BaMoO_4$
Leach resistance:					
Hydrolytic class [German grain titration test]		Not applicable	1–5	1–3	1–2
Soxhlet test–3 days ($g/cm^2/d^1$)					
Powder		—	10^{-3}–10^{-6}	10^{-4}–10^{-5}	10^{-4}–10^{-5}
Block		—	10^{-2}–10^{-5}	10^{-3}–10^{-4}	—
Bead		1.5–10^{-5}	—	—	—
Impacted block (grain sizes $\leqslant 0.1$ cm in %)		—	$\geqslant 50$	<30	<30
Density (g/cm^3)		—	2.6–2.8	3.1	3.7
Density change (%) after $10^{18} \alpha/g$		—	$\leqslant \pm 1\%$	—	—
Stored energy (J/g) after $10^{18} \alpha/g$		—	80–400	—	—

SOURCE: Lutze et al., Reference 77a.

Worldwide, borosilicate glass and calcine have received more development effort than any other potential final disposal forms for HLLW. Since simple calcine has now been largely eliminated from consideration for repository disposal, borosilicate glass will probably be the first form utilized on a routine basis, at least for the lower-activity content defense wastes.

Figure 5.2 diagrams the major processes that have been developed here and abroad for making borosilicate waste glass. In the PNL In-Can-Melter technique, the spray calciner described in the previous section is joined directly to the disposal canister, which also acts as the glass-manufacturing vessel. Glass frit is added to the calcine as it drops into the container, which is surrounded by a zoned furnace. The calcine-frit mixture is heated to 1000–1100°C, and a molten glass is formed. As the canister fills, the heating band of the furnace is raised to match the reaction zone. The containers are generally filled to the 90% level, at which time the calcine-frit feed is moved to a second canister already in place. The filled "can" is removed and replaced by an empty one to be ready for the next switching operation. The process can be made essentially continuous by alternating filling stations.

A lesser amount of development has also been carried out at PNL aimed at using a fluidized bed calciner in conjunction with the in-can production technique.

Both the rotary kiln and superheated steam processes shown in Figure 5.2 require a separate melter. In the rotary kiln method, the melter is connected directly to the rotating barrel of the calciner and the melter, in turn, drains directly into the storage canister. The superheated steam approach follows the same pattern. The denitrated wastes are calcined at 600°C by being sprayed as a mixture with superheated steam into the reaction vessel. Glass frit is then mixed with the calcine, and the mixture is passed into the melter. The glass produced then drains into the storage container. The use of two filling positions again allows either process to be applied on a continuous basis.

The British Rising Glass process can be considered as an extension of the Pot Calcine procedure. The HLLW plus glass-forming additives are fed into the final storage container, which is surrounded by a zoned furnace. Heating begins at the lowest portion of the furnace, and three layers form—molten glass at the bottom, covered by calcine, and with a layer of boiling liquid on top. As the glass level rises, the higher heating elements of the furnace are activated. The procedure as shown in Figure 5.2 is "FINGAL." Newer developments include the use of an annular container, with an additional heater in the annulus, and utilization of internal fins in the container for heat removal. This is the "HARVEST" process. The Indians and the British plan to use the Rising Glass method for solidification of their wastes. The French depend on the Rotary Kiln in combination with a continuous melter, as shown in Figure 5.2, or in combination with a Continuous Ceramic Melter, described below. The Germans expect to employ the Figure 5.2 Superheated Steam approach, while the Japanese are still undecided.[73]

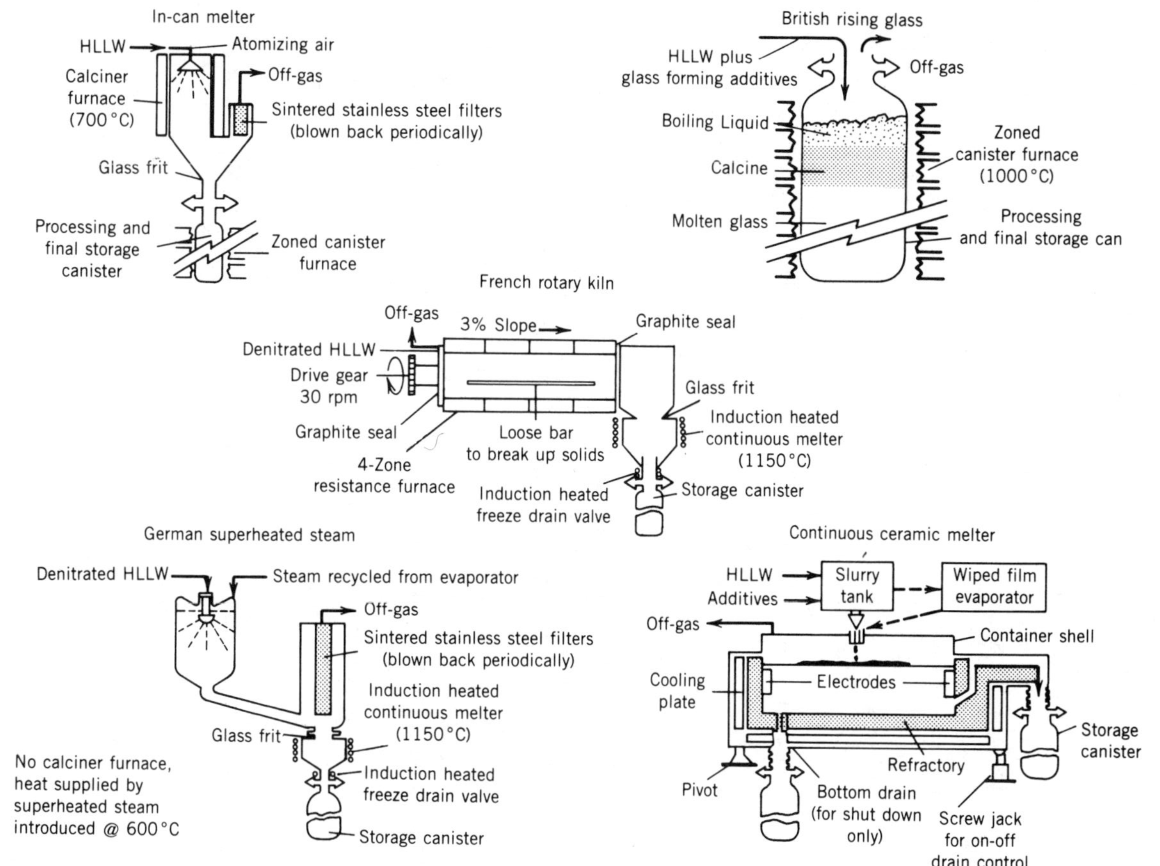

FIGURE 5.2. Current waste-glass production processes. (Reference 73.)

The last of the processes shown in Figure 5.2, the Continuous Ceramic Melter, is an adaptation of a successful commercial glassmaking technique. The reaction vessel is a large ceramic-lined tank. Heating to produce a glass melt is obtained by passing alternating current through the glass between electrodes placed at either end of the tank. The passage of the current forces mobile ions (primarily sodium) to migrate through the melt. Resistance to this migration causes the dissipation of energy known as joule-heating. Application of this approach to HLLW solidification has been used again at PNL, originally with calcine from the spray calciner (plus glass additives) as the feed. An alternative to the use of calcine is to eliminate most of the water from HLLW by means of a wiped-film evaporator and to use this material for admixture with the additives. The most recent development work has been aimed at introducing the liquid wastes plus additives directly into the melter without a predrying or preconcentration step. As in the Rising Glass process, three layers result in the tank with this approach—glass covered with calcine covered with boiling liquid. The ceramic melter has been successfully used with all three feed variations.

In this process, the storage canister is filled by tilting the entire melter tank to a 5° angle. The second container shown in Figure 5.2 is for receipt of any scrap glass left in the tank at times when the unit is to shut down.

Use of the Ceramic Melter allows higher temperatures to be used than in the In-Can-Melter process, which in turn permits a higher Si content and a better glass. While final decisions have not been made, it is highly probable that the Ceramic Melter will be chosen for solidifying both the West Valley and Savannah River HLLW.

One of the major attractive features of glass as a waste disposal form is its ability to contain practically any element in the Periodic Table over a range of concentrations. But there are exceptions: The halogens and mercury compounds are volatilized, and sulfate in more than 1% concentration causes the formation of a separate water-soluble phase. Sulfate can be volatilized as SO_2 by addition of a suitable reducing agent. Fluorine is also volatile, but can be retained in the melt to a large extent by the addition of calcium to form calcium fluoride. Ruthenium dioxide, if present, is not soluble and is encapsulated as particles.[73]

Glasses are made of network formers and network modifiers. Silicon, boron, aluminum, and phosphorus are network formers because they bond to oxygen with high bond strengths to form an extended network. Boron, for example, lowers the melting temperatures; aluminum and boron increase chemical stability and slow the rate of devitrification of glass; too much aluminum and boron make the viscosity of the melt too high.[73]

Other elements fit into interstices of the glass network and are called network modifiers. Each has a specific effect on the overall properties of the glass. The alkali-metal oxides lower the melting temperature of glass by decreasing the viscosity, but excessive amounts reduce chemical stability. Cesium is volatile at glass-forming temperature, but its volatility is effectively suppressed by oxides of boron, titanium, and molybdenum. Calcium and magnesium oxides tend to improve chemical durability and/or prevent crystallization. Generally, no more than two parts by weight of glassmaking additives are required to convert one part of radioactive waste, as oxide, to a satisfactory glass.[73]

Semiempirical studies[73, 112c, 114] have provided guidelines for formulation of acceptable glasses. The important atom ratios are

Designation	Atom Ratio
A	Si/Al B
B	Oxygen/Network formers
C	Network formers/Network modifiers
D	Network modifiers/Oxygen-2 (network formers)

When Ratio A is more than 1.5, leachability is low. The melting point of the glass increases as A increases. Glass formation is best and leachability low when Ratio B is between 2.2 and 2.4. Excess oxygen is needed to balance the electrical charge of the network modifiers. Density increases as A decreases and B increases. Ratio C should be about 2 for good glass formation. As C decreases, the glass network is disrupted and the tendency for glass formation decreases. Ratio D represents the ratio of network modifiers to excess oxygen not required for the network formers to be four-coordinate and should have a value as close as possible to 2.

A low-viscosity glass formation is preferred to minimize blending problems and component volatilities and resultant off-gas cleanup requirements. Commercial soda-lime glass (72 w/o SiO_2, 15 w/o Na_2O, 10 w/o CaO-MgO, 2 w/o Al_2O_3 1 w/o misc.) is melted at 1475–1500°C. Formulations that have been demonstrated for radioactive waste glasses melt at about 1000–1250°C.

Table 5.17 includes material relevant to the composition ranges and physical properties of borosilicate waste glass. Tables 5.18–5.21 extend this information, partially to demonstrate that glass formulations can be varied over a very considerable range and partially to present numerical data.

Table 5.18b gives compositions of glasses developed at PNL. The code used for the "waste type" line refers to the waste calcine, with PW indicating Purex Waste. The PW-7 series, for example, was developed in connection with the projected output of the Barnwell Reprocessing Plant where the calcine would be "dirty," being high in Na and phosphate from admixtured intermediate-level waste. The PW-4 series was for a "clean" waste, that is, small amounts of residual U and Pu plus only fission and corrosion products. The PW-8 series of formulations was generated in connection with the projected output of the West Valley Plant at the time when it was still being planned to modernize and reopen

TABLE 5.18 Waste in Borosilicate Glass

(a) Borosilicate Glass Characteristics

Composition	
SiO_2	25–40 wt%
B_2O_3	10–15 wt%
Alkali metal oxides	5–10 wt%
ZnO	0–20 wt%
Waste oxides	20–35 wt%
Typical volume	60–80 *l*/MTU
Density	3.0–3.6 g/cm^3
Thermal conductivity	0.9–1.3 w/m-°C
Leach rate	10^{-4}–10^{-7} g/cm^2-day
Processing temperature	1000–1400°C

(b) Typical Waste Glass Compositions (wt%) Weight Percent

Glass code	72-68	76-68	77-260	77-10
Waste type	PW-4b	PW-8a	PW-7c	PW-9
Frit code	73-1	76-101	77-269	77-268
SiO_2	27.3	40.0	36.0	38.0
B_2O_3	11.1	9.5	9.0	13.0
Na_2O	4.0	7.5	8.0	2.0
K_2O	4.0	—	2.0	4.0
ZnO	21.3	5.0	—	5.0
CaO	1.5	2.0	1.0	2.0
MgO	1.5	—	—	—
SrO	1.5	—	—	—
BaO	1.5	—	—	—
TiO_2	—	3.0	6.0	3.0
Al_2O_3	—	—	2.0	—
CuO	—	—	3.0	—
Waste	26.3	33.0	33.0	33.0

SOURCE: Report DOE/EIS-0046D, Reference 67.

TABLE 5.19 Additional Borosilicate Glass Formulations

Element as Oxide	PNL76-68	GLR	GLQ	GLP
Composition of Radwaste Glasses (wt%)				
Network formers				
SiO_2	39.80	40.00	38.10	60.60
B_2O_3	9.47	9.50	17.30	—
Al_2O_3	—	—	13.80	17.50
P_2O_5	0.48	0.50	0.06	0.30

TABLE 5.19 (Continued)

Element as Oxide	PNL76-68	GLR	GLQ	GLP
Composition of Radwaste Glasses (wt%)				
Alkali metals				
Na_2O	12.51	13.00	17.30	3.70
K_2O	—	0.07	—	2.10
Rb_2O	0.13	—	0.03	—
Cs_2O	1.03	0.46	0.25	—
Alkaline earth metals				
MgO	—	—	1.60	2.80
CaO	2.00	2.00	—	5.90
SrO	0.38	0.41	0.90	—
BaO	0.56	0.61	0.13	—
Filled d-shell ions				
ZnO	4.97	5.00	—	—
CdO	0.03	0.04	0.01	—
Ag_2O	0.03	0.03	0.01	—
Third-row transition metals				
TiO_2	2.97	3.00	—	0.80
Cr_2O_3	0.41	0.45	0.90	—
MnO_2	—	—	—	0.20
Fe_2O_3	9.77	11.30	5.20	6.10
CoO	—	0.12	—	—
NiO	0.20	0.60	0.60	—
Fourth- and fifth-row transition metals				
ZrO_2	1.77	1.90	0.42	—
MoO_3	2.28	2.50	0.54	—
RuO_2	1.07	—	0.26	—
Rh_2O_3	0.17	—	0.04	—
PdO	0.53	—	0.04	—
Lanthanides				
CeO_2	1.19	3.92	0.25	—
La_2O_3	0.53	2.00	0.28	—
Pr_6O_{11}	0.53	0.40	0.06	—
Nd_2O_3	1.67	1.40	0.20	—
Sm_2O_3	0.33	0.25	0.03	—
Eu_2O_3	0.07	0.07	0.01	—
Gd_2O_3	0.05	0.16	0.02	—
Y_2O_3	0.21	0.02	0.002	—
Other elements				
TeO_2	0.26	0.28	0.06	—
U_3O_8	4.58	—	—	—
Cm_2O_3	—	—	0.14	—

SOURCE: Houser et al., Reference 77c.

TABLE 5.20 **Proposed Savannah River Glass, (Composition of Typical SRP Borosilicate Glass)**[a, b]

Calcine	Composition (wt %)	Frit	Composition (wt %)
Fe_2O_3	42.0	SiO_2	52.5
Al_2O_3	8.5	B_2O_3	10.0
MnO_2	11.8	Na_2O	18.5
U_3O_8	3.9	Li_2O	4.0
NiO	5.2	CaO	4.0
SiO_2	3.8	TiO_2	10.0
Na_2O	4.7		
Zeolite	8.8		
$NaNO_3$	2.6		
$NaNO_2$	0.2		
$NaAlO_2$	0.2		
NaOH	3.9		
Na_2SO_4	1.3		

SOURCE: Report DOE/EIS-0023, Reference 82.
[a]Glass will contain 28 wt % calcine.
[b]Average density of glass will be 2.7 g/cm^3.

the facility. That waste would have been relatively clean, but unusually high in iron content.

Glasses PNL 72-68 and PNL 76-68 have received much study and have essentially achieved the status of standards for glasses produced from commercial wastes. Table 5.19 gives an unusually complete summary of the composition of PNL 76-68. The other glasses in the table were developed by Corning Research Laboratories, with GLR being a soda-zinc-borosilicate type, GLQ, a soda-alumina-borosilicate, and GLP, a synthetic obsidian.

Table 5.20 gives the composition of a glass for incorporation of Savannah River sludge, a material very high in iron, aluminum, manganese, and sodium. Table 5.21 shows compositions of glasses being developed by a European group, along with an unusually extensive listing of physical properties. These glasses are of higher than normal Si content for waste glasses.

The literature on borosilicate waste glass, and indeed, for most of the solidification forms, often presents conflicting numbers for specific properties, particularly for leachability. Much of this confusion has arisen because different groups use different testing methods. As an effort towards standardization, the U.S. Department of Energy has established a Materials Characterization Center (MCC) at PNL. The MCC has published two extensive reports, the first[81] being a survey of properties of solidified HLLW forms, and the second,[115] a start on a handbook of standardized testing methods.

Apparently, no direct measurements have been made of the gamma ray atten-

TABLE 5.21 Compositions and Properties of Selected Glass Products

Composition (wt %)	Glass Product GP 98/12	GP 98/18	GP 98/26
SiO_2	48.20	45.60	46.09
TiO_2	3.91	3.70	3.74
Al_2O_3	2.21	2.10	2.11
B_2O_3	10.54	10.00	10.08
MgO	1.80	1.70	1.71
CaO	3.48	3.30	3.33
Na_2O	14.88	14.10	14.23
Gd_2O_3	—	4.50	3.71
HAW Oxides[a]	15.00	15.00	15.00
Product Data			
Specific gravity (g/cm^3)	2.83	2.86	
Viscosity at 1150°C (Poise)	125.00	75.00	
Thermal expansion (100–400°C) ($K^{-1} \times 10^{-6}$)	9.5	9.7	
Electrical conductivity at 1150°C (Ω^{-1}/cm)	2.56	2.37	
Thermal conductivity at 1150°C (W/m/K)	1.10	1.36	
Evaporation velocity at 1150°C ($m/s \times 10^{-9}$)	2.60	3.30	
Impact resistance (cm^2/J)	6.6	9.33	
Soxhlet leach rate in 21st day at 70°C ($g/cm^2\ d \times 10^{-4}$)	0.87	0.46	
Crystallization after 360 hr at 800°C	None	Part	
Formation of yellow phase	None	None	
Specific heat ($J/kg/K \times 10^{-3}$)	1.13	1.30	

SOURCE: Guber et al., Reference 77d.

[a] 15 wt % of HAW oxides contain 0.91 wt % of phosphate.

uation coefficients (μ) of waste glasses. Data from the Schott catalogue[116] for hot-cell-shielding window glasses of comparable densities are

Radiation (MeV)	Glass Density (g/cm^3)	$\mu(cm^{-1})$
^{60}Co (1.25)	2.53	0.141
	3.53	0.181
^{137}Ce (0.67)	2.53	0.194
	3.53	0.270

High Silica Glass is the last product in Table 5.7 to be discussed. An intimate mixture of porous glass frit and waste calcine is sintered in the form of a cylinder at 1100°C, much lower than the melting temperature of 1800°C. This sintered core is then placed in an inactive glass tube, which is collapsed around the core by heating at 1300°C. It is estimated that the waste loading would be about 18 w/o on an equivalent oxide basis. The leachability of the core would be 5×10^{-10} g/cm^2-day, and with the outer skin in place, about 10^{-12} g/cm^2-day. Thermal conductivity is projected as being ca. 0.8 Btu/hr-ft-°F at 100°C and the final density to be 150 lb/ft^3.

5.3 LEACHABILITY

One of the chief concerns in disposing of solidifed waste in an underground repository is that future intrusion of water could gradually dissolve the solid, thus providing the most plausible path for reintroduction of radioactivity into the biosphere. The question of leachability of the various solid waste forms has accordingly received more attention than the determination of any other physical property. The mechanisms by which leaching occurs are still not completely understood, but are known to be affected by many variables: the compositions of the waste and the host matrix, temperature, the chemical nature of the element being measured to determine leachability, the chemical nature of the leaching fluid (distilled water, groundwater, brine), radiation effects, and so forth. Mendel and his associates[81] review this problem, and a number of groups worldwide are studying the phenomenon on an increasingly fundamental basis.

Values reported in the literature for leachability are often conflicting for similar solidified waste forms, making intercomparisons difficult unless the studies have been made by the same group of experimenters, using the same measuring techniques throughout. It is probably this last factor—the use of a large variety of measurement methods by different investigators—that accounts for many of the contradictory values in the literature. This situation is gradually being improved by the availability of standardized measurement techniques. The IAEA has recommended[118] one such method, and the MCC has published[115] several provisional procedures.

Just as there have been many experimental approaches for measuring leachability, a number of mathematical approaches have also been developed for calculating and expressing the results. These techniques are discussed in Reference 81 where the most commonly used expression is as follows:

$$(\mathrm{LR})_i = \frac{(A_i)}{(A_0)} \times \frac{(W_0)}{(SA \times t)} \tag{5.1}$$

where $(LR)_i$ = g/cm^2-day, normalized to component i
A_i = amount of i leached in time t
A_0 = original amount of i in the sample
W_0 = original weight of i in the sample (grams)
SA = surface area of the sample in cm^2
t = days

A_i and A_0, of course, must be expressed in the same units. The leaching rate is also sometimes seen as g/m^2-day, that is, a factor of 10^4 larger than the $(LR)_i$ value as derived above. Equation 5.1 applies to a static testing technique. If testing is with a flowing stream of leaching solution, the situation is more complicated. Barney[119] (as quoted by Jardine and Steindler) has an interesting approach. He argues that an acceptable degree of waste fixation can be said to exist if the level of a specific radionuclide in the leaching medium never exceeds the maximum permissible concentration (MPC) in drinking water for that nuclide as set by the NRC and other regulatory agencies. (Some such MPC values are given in the Health Physics chapter.) The maximum permissible leach rate (MPLR—expressed as g/m^2-sec) then becomes

$$\text{MPLR} = \frac{\rho}{r} \times \frac{(F)}{(A)} \tag{5.2}$$

where $r = (K_i)/(\text{MPC})_i$ (5.3)
ρ = density of the solid (kg/m^3)
K_i = concentration of i in the solid (Ci/m^3)
F = flow rate of the leachant (m^3/sec)
A = surface area of the sample (m^2)
$(\text{MPC})_i$ = maximum permissible concentration of i in drinking water (Ci/m^3)

One of the major problems in using this approach is in accurately determining the flow rate of the leaching medium. In Equations 5.1 and 5.2, it is implicitly assumed that the surface area of the solid remains relatively unchanged with time and that the component being measured is uniformly distributed in the solid and remains so with time.

Table 5.22 (from Reference 106) lists some leach rates that have been reported in the literature. (This table is also given in Reference 81. Either of these sources should be consulted for original citations.) Figures 5.3 and 5.4 present similar information, probably more realistically as ranges rather than as specific numbers. Figure 5.3 indicates that leachability of the rare earths and plutonium from all of the solidification forms to be appreciably less than that of the alkali and alkaline earth elements.

TABLE 5.22 Leach Rates for Various Waste Forms

Waste Form	Leach Rate (g/cm^2-day)	Test Conditions
Pot calcine	10^{-1}	25°C-H_2O
Fluid bed calcine	10^{-1}	25°C-H_2O
Pot calcine	10^{-2}	25°C-H_2O
Fluid bed calcine	10^{-2}	25°C-H_2O
Cement	10^{-2}-10^{-3}	25°C-H_2O
Al metal matrix-sintered	3×10^{-3}	25°C-H_2O
Concrete	10^{-3}	25°C-H_2O
Al metal matrix-cast	10^{-3}-10^{-4}	25°C-H_2O
Aqueous silicates (clay)	10^{-4}-10^{-5}	25°C-H_2O
Bottle glass[a]	5×10^{-5}	100°C-H_2O
Borosilicate glass	3×10^{-5}	100°C-H_2O
Lead matrix[a]	2×10^{-5}	100°C-H_2O
Vitromet (63 vol % phosphate glass/37 vol % lead)	2×10^{-5}	100°C-H_2O
Phosphate glass	1×10^{-5}	100°C-H_2O
Lead matrix[a]	7×10^{-6}	20°C-H_2O
Glass (devitrified)	5×10^{-6}	25°C-H_2O
Asphalt	4×10^{-6}	25°C-H_2O
Zn borosilicate glass (devitrified)	2×10^{-6}	25°C-H_2O
Borosilicate glass (in-can melted)	1.3×10^{-6}	25°C-H_2O
Borosilicate glass	10^{-5}-10^{-7}	25°C-H_2O
Phosphate glass	10^{-6}-10^{-7}	25°C-H_2O
Titanates	5×10^{-7}	25°C-H_2O
Alumina phosphate glass	5×10^{-7}	25°C-H_2O
Phosphate glass	5×10^{-7}	20°C-H_2O
Vitromet (63 vol % phosphate glass/37 vol % lead)	5×10^{-7}	20°C-H_2O
Zn borosilicate glass (as formed)	3×10^{-7}	25°C-H_2O
Glass (as formed)	2×10^{-7}	25°C-H_2O
Borosilicate glass	10^{-4}-10^{-7}	25°C-H_2O
Silicate glass (Canadian)	10^{-6}-10^{-7}	25°C-H_2O
Sintered glass ceramics	10^{-5}-10^{-8}	25°C-H_2O
Industrial glass[a]	10^{-6}-10^{-7}	25°C-H_2O
Industrial glass[a]	10^{-5}-10^{-8}	25°C-H_2O
Silicate melts (fired clay)	10^{-6}-10^{-8}	25°C-H_2O
Silicate glass (U.S.A.)	2×10^{-7}-6×10^{-8}	25°C-H_2O

SOURCE: Jardine and Steindler, Reference 106.

[a]Contain no simulated waste. For comparison only.

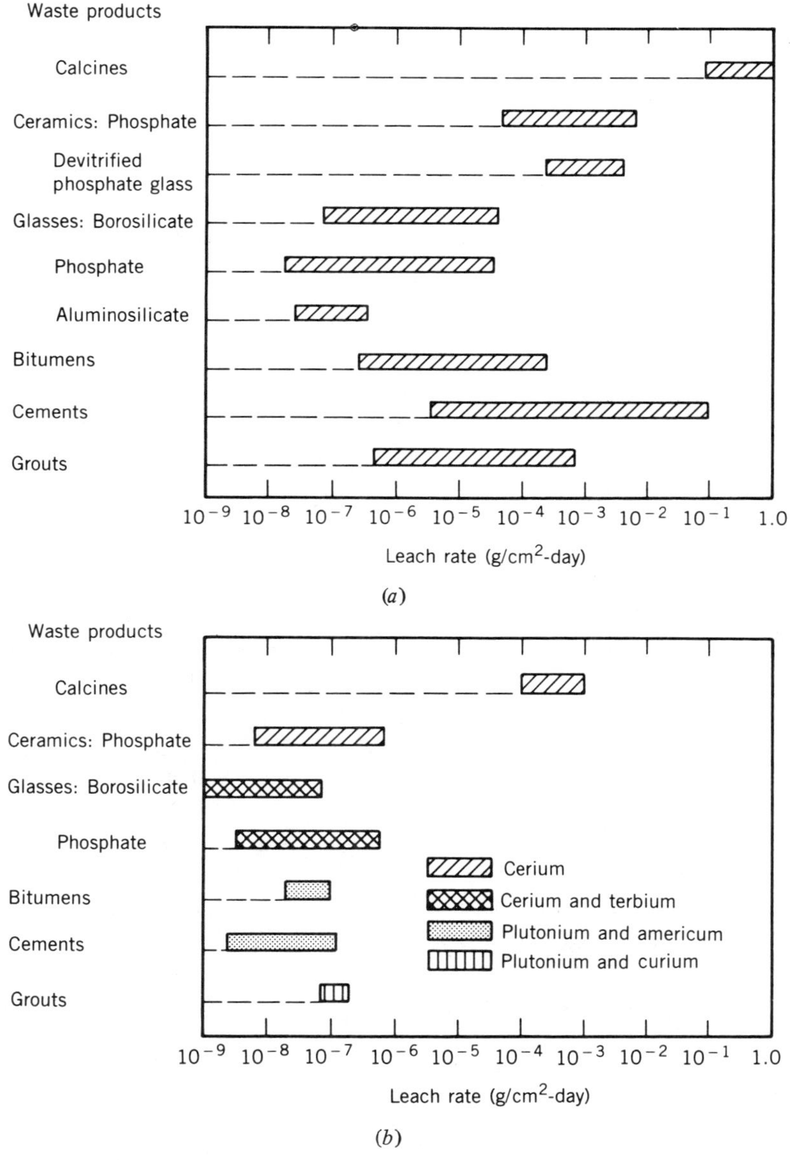

FIGURE 5.3. Comparison of leach rates for (*a*) alkali and alkaline-earth elements and (*b*) rare-earth and actinide elements from various waste products. (Reference 73.)

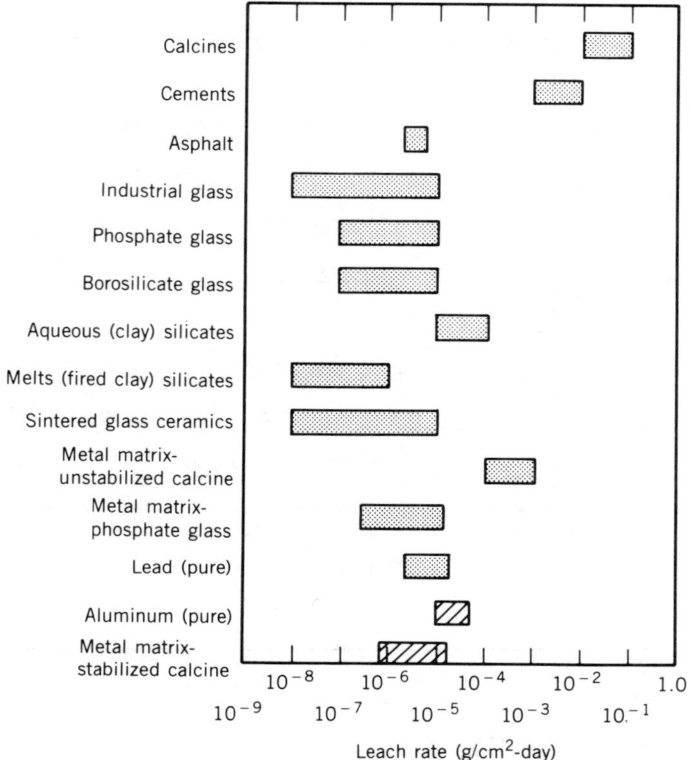

FIGURE 5.4. Water leaching of various waste forms. (Reference 106.)

Table 5.23 represents one of the few cases where leachability tests have been made by the same group of investigators on the same wastes (SRL sludges) incorporated into several different forms, using the same testing technique (MCC #1) throughout. The studies also considered the leachability of several different elements (Cs, U, Ce), several different leaching solutions were used and tests were made at two different temperatures. Enough test specimens were run for each solid-leachant combination to permit assignment of uncertainty ranges for each value given in Table 5.23. These limits are not reproduced here, but appear in the original paper. The variability is usually in the 5–25% range.

Figure 5.5 graphs data from the same study and shows the change in the leach rate of cesium in distilled water at 40°C and of uranium in the same leachant at 90°C as a function of time. Most of the curves are typical of many leaching experiments—the rate is higher in the beginning then levels off to a more constant value after the first few days.

TABLE 5.23 Leach Rates of SRL Sludges in Various Solid Forms

	Leach Rates (g/m^2-day)			
Solidification Form	DI Water (40°C)	DI Water (90°C)	SI Water (90°C)	Brine (90°C)
		Cesium		
B-Si glass	0.055	1.54	0.76	0.44
Hi-Si glass	0.004	0.03	0.12	0.04
Tailored ceramic	0.521	5.09	1.08	1.36
SYNROC	0.90	0.49	0.39	0.24
		Uranium		
B-Si glass	0.011	0.055	0.236	<0.019
Hi-Si glass	0.061	0.052	0.081	<0.020
Tailored ceramic	<0.002	0.005	0.014	<0.014
SYNROC	<0.002	<0.002	0.001	<0.014
		Cerium		
B-Si glass	<0.001	<0.004	0.003	<0.029
Hi-Si glass	0.027	0.014	0.015	<0.004
Tailored ceramic	<0.003	<0.005	<0.006	<0.011
SYNROC	0.001	<0.001	0.0009	<0.007

SOURCE: Stone, Reference 117.

Abbreviations: DI, distilled; SI, silicate; B-Si, borosilicate; Hi-Si, high-silica.

Wallace and his associates[120] examined the question of the time needed to destroy a solidified solid as a function of the leach rate. They assumed long cylinders of a solid having a density of 2 g/cm^3, and end effects were ignored. The problem then became the number of years needed to completely dissolve such cylinders of differing radii (r_0) as a function of leach rate. The results of their calculations are shown graphically in Figure 5.6, which is also given in Reference 106. It will be seen that even a cylinder having only a 1-cm radius would not be entirely destroyed after 100,000 years if the leach rate could be held to 10^{-7} g/cm^2-day or less. Leachabilities of that order have already been demonstrated for several of the proposed waste solidification forms.

5.4 ELEMENT VOLATILITY

Mercury and the halogens will volatilize and be transferred out through the off-gas in any waste-solidification process requiring higher temperatures. Other

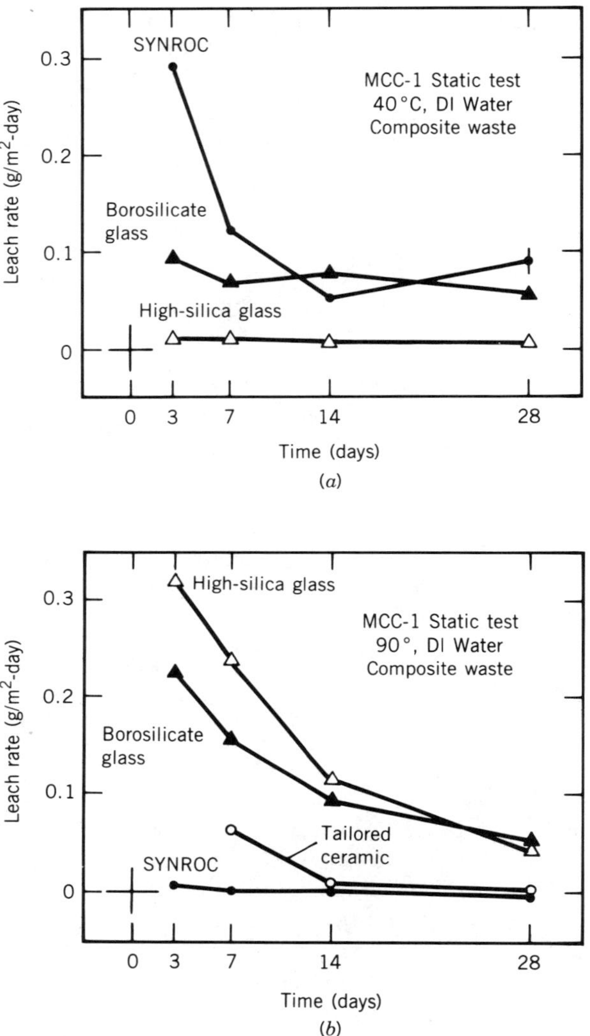

FIGURE 5.5. Leachability of solidified Savannah River sludge. (*a*) Cesium and (*b*) uranium leach rates. (Reference 117.)

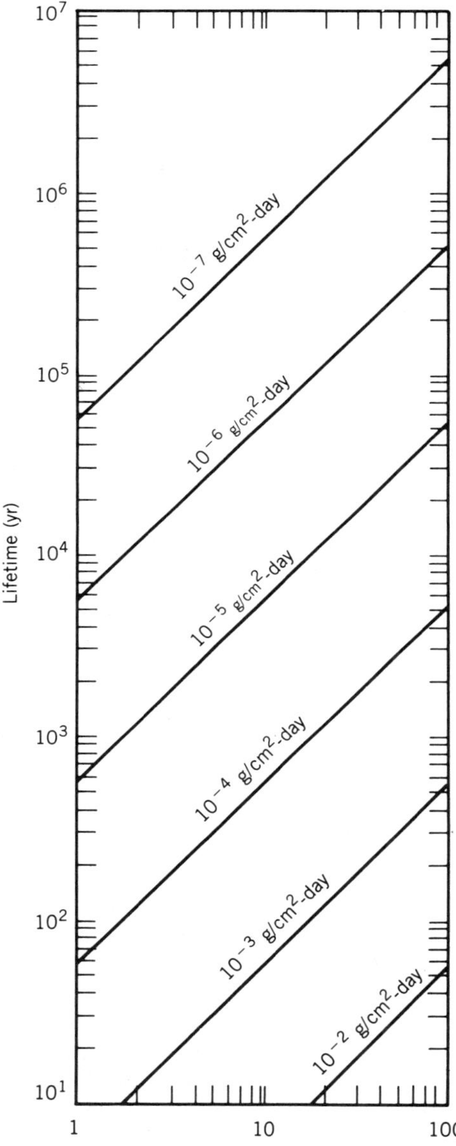

FIGURE 5.6. Effect of leach rate on cylinder lifetime. (References 120, 106.)

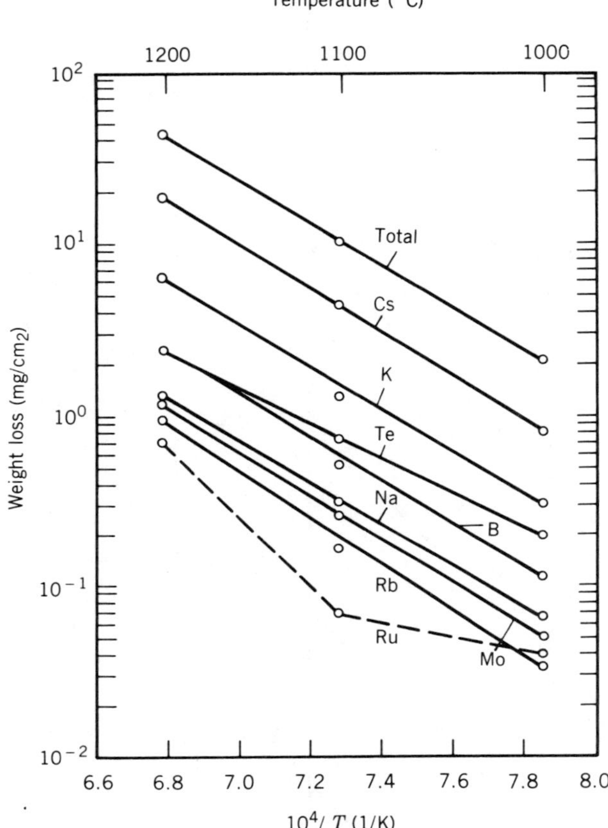

FIGURE 5.7. Weight loss from glass vs. temperature. (Reference 96.)

elements may also be lost to varying degrees as shown in Figure 5.7, which graphs weight losses from a simulated waste glass as a function of temperature. The two elements in the group that have received the most attention are ruthenium and cesium because of the ^{103}Ru, ^{106}Ru, and ^{137}Cs radioactivities. Inactive ruthenium is also produced in large quantities in fission and, if volatilized, may subsequently condense and generate plugging problems.

As much as 50% of the ruthenium in spent oxide fuel elements may remain in insoluble form in the reprocessing plant dissolver.[121] This refractory material is probably in the form of colloidal RuO_2. The fraction that does go into solution apparently does so in the form of nitrosyl complexes having the general formula $RuNO(NO_3)_x(OH)_{3-x}(H_2O)_2$. As the acidity increases, the proportion of the trinitrato also increases, as does ruthenium volatility as shown in Figure 5.8.

These nitrosyl complexes are thought to be converted in the high-temperature processes to higher ruthenium oxides such as RuO_4, generally assumed to be responsible for the ruthenium losses, although it has been demonstrated that RuO_2 will also volatilize at 1100°C and above.

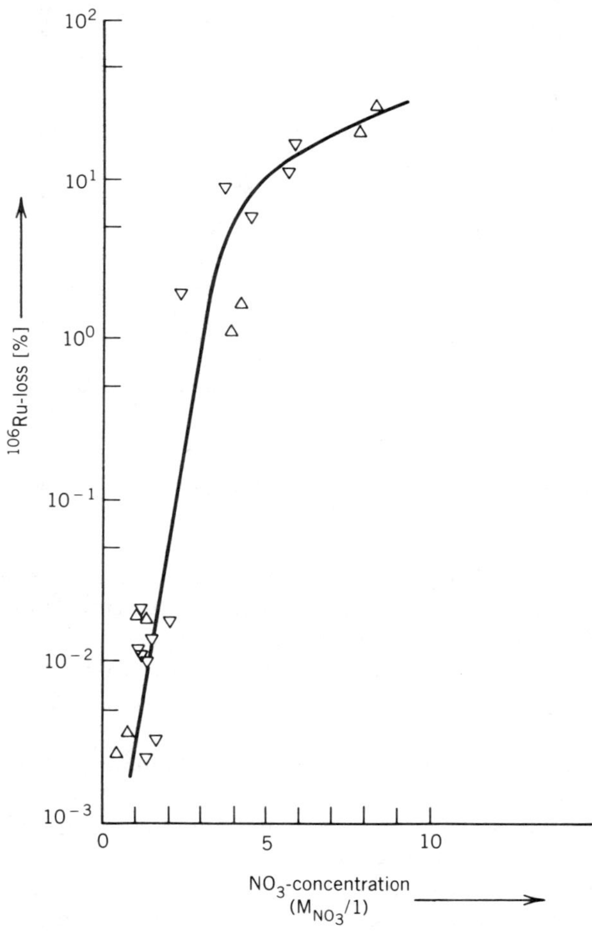

FIGURE 5.8. Loss of ^{106}Ru as a function of the NO_3^- concentration in the melt. (Reference 112b.)

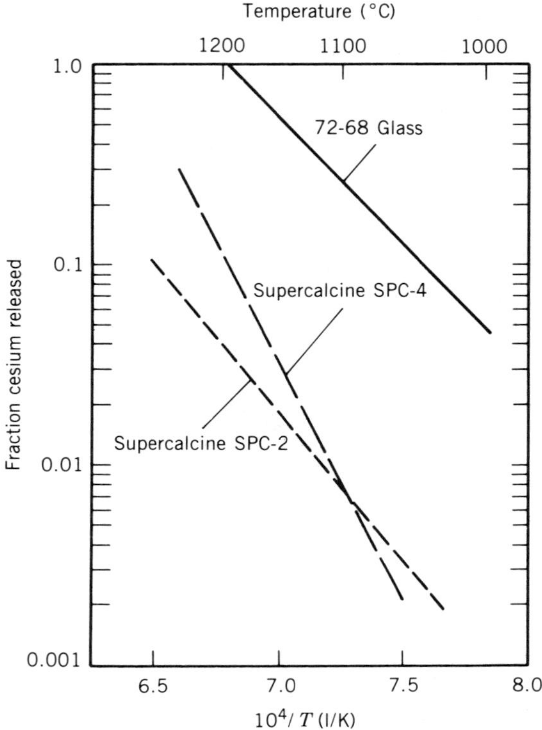

FIGURE 5.9. Cesium loss from glass and supercalcine. (Reference 107.)

Much of the above is reported[79c] in connection with the German waste-glass development program at Julich. The approach of that group to the ruthenium problem is to pretreat the HLLW with formaldehyde in order to destroy the nitrate. An alternative is to place special filters in the off-gas system. Silica gel, a mixture of Fe_2O_3 and silica,[122] and hydrous ZrO_2 at 100°C[123] have all been used.

Cesium losses can be appreciable from solid-waste forms held at high temperatures for protracted periods as shown in Figure 5.9. Cesium (and Ru) losses are very low or do not occur in the lower temperatures of the Idaho Fluidized Bed calcination process.[112a] Any cesium leaving the PNL spray calciner apparently passes through the sintered stainless steel filters of the unit as particulates that can be collected downstream and returned to the feed stream. Again, however, only minor amounts of Cs, Te, and Se were seen in the Waste Solidification

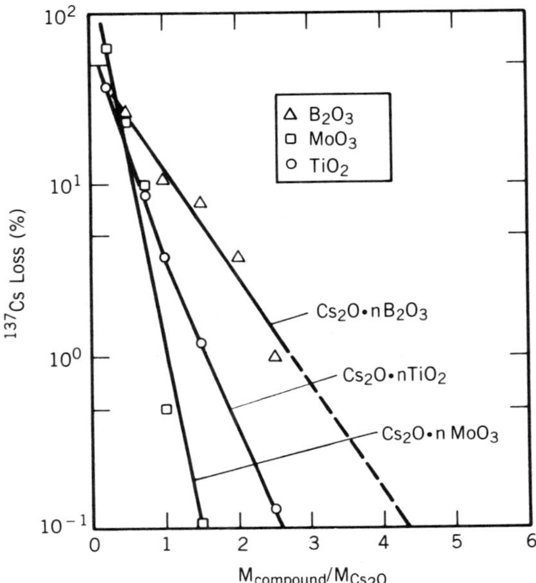

FIGURE 5.10. Suppression of ^{137}Cs loss. (Reference 112b.)

Engineering Prototypes (WESP) Program at Hanford prior to 1971, and essentially none in the later Waste Fixation Program (WFP).[91]

Not unexpectedly, the composition of the HLLW has a definite effect on cesium volatility. As shown in Figure 5.10, B_2O_3, MoO_3, and TiO_2 in the waste help to suppress cesium losses.

S I X

NON-HIGH-LEVEL WASTES

6.1 BACKGROUND AND LITERATURE

Whereas it has been possible to hold HLLW in underground tanks and spent fuel assemblies in storage pools until a repository becomes available, the types of wastes discussed in this section are produced routinely and in such bulk as to require reasonably immediate treatment, although it was decided in 1970 that TRU wastes (those contaminated with plutonium or other alpha emitters) should be held in temporary surface storage until a decision is made on means for their ultimate disposal. In general, however, the problem of handling increasing volumes of non-HLW has existed for the last 40 years. A considerable literature has developed, including various reviews[73,75,124–126] of disposal techniques. Since these surveys are reasonably accessible, the emphasis in this chapter will be primarily on definitions and regulatory aspects.

6.2 LOW-LEVEL WASTES (LLW)

A considerable increase in the quantity of contaminated waste naturally came about with the start-up of plutonium production plants for military purposes in 1943. Land was set aside at most of the major production and research sites for disposal of these wastes by shallow land burial. The techniques used have changed little with time. Solid wastes in drums, barrels, or boxes are placed in relatively shallow (30 feet being typical) trenches filled to about one meter from the top and backfilled with the excavated soil. Liquid wastes are converted to solid form and similarly treated.

(In the early days, some of the government facilities located near the coasts and with limited space for land burial disposed of solidified wastes by dumping them at sea. This, while still practiced abroad to a limited extent, was specifically banned in this country by Congress in 1972[127]).

Until the early 1960s, the AEC accepted the relatively small amount of LLW then being generated commercially at either Oak Ridge or at the Idaho reservation. As the quantity needing disposal grew, Congress responded in 1963 by establishing new rules allowing LLW disposal facilities to be privately operated, although still required to be on state or federally owned land. Six such sites were established: two in the East (New York, South Carolina), two in the Midwest (Illinois, Kentucky), and two in the West (Nevada, Washington). Figure 6.1 shows the locations of these commercial sites, open or closed. DOE operates burial grounds at 14 government-owned sites, five of which are considered major operations. These are also shown in the figure as well as the locations of government facilities generating different types of waste.

Of the commercial sites, the New York facility closed in 1975 due to leakage

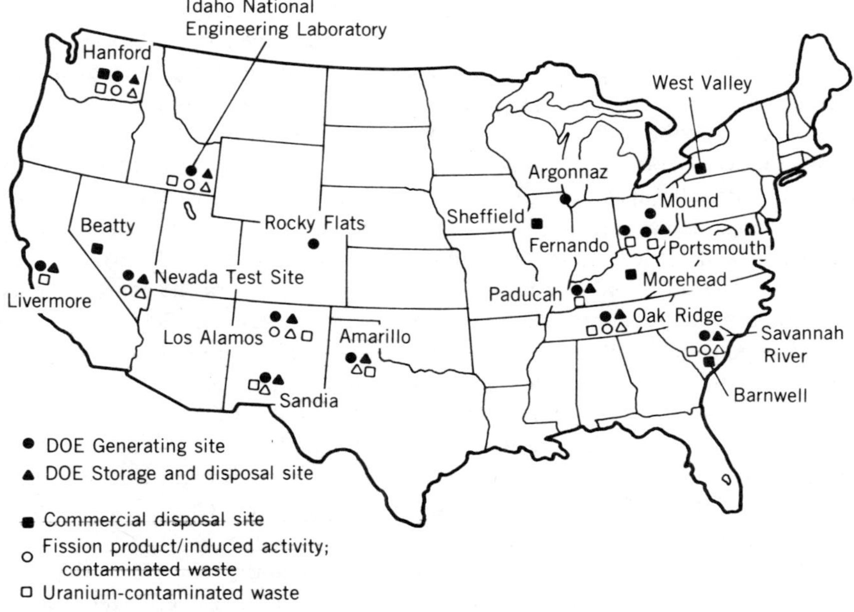

FIGURE 6.1. Low-level radioactive waste burial sites. (Reference 128.)

problems from the burial trenches. The Kentucky site was closed in 1977, and the site in Illinois in 1978 when it reached its licensed capacity limit. Barnwell in South Carolina now accepts only 49,500 m^3 of waste per year. The western sites continue to operate normally,[128] but are far from the major generators of LLW, most of which are in the Northeast.

Late in the 1970s, universities, hospitals, research laboratories, and industrial sources were producing about 28,000 m^3 of solid low-level waste annually; nuclear power plants an additional 64,000 m^3/yr. During the 1980s, the quantity generated by medical, academic, and industrial institutions is expected to increase to 56,000 m^3/yr, and the total to 295,000 m^3.[129]

The question of LLW disposal was obviously becoming a serious problem. Congress responded by passing the Low-Level Waste Policy Act of 1980, placing the responsibility on each state for disposal of wastes generated within its borders. A single site serving several states is acceptable if a compact is agreed to between the responsible authorities of the states involved. As of January 1, 1986, no state or compact-member state will be required to accept LLW from outside their own jurisdiction.[130]

The USNRC responded to this situation by new regulations governing the established disposal areas as well as the required new LLW areas. These rules appear in a new addition to the Code of Federal Regulations (10CFR61).[131] Most of this deals with the mechanics of the licensing process and sets forth monitoring and security rules, financial and maintenance responsibilities, and so on. Subsection 61.7 discusses the concepts used in preparing the regulations, and subsections 61.55–61.58, the classification and characteristics of the wastes. The brief discussion following is based on these subsections.

Wastes acceptable for disposal by near-surface burial are classified as types A, B, or C. Broadly speaking, Class A wastes contain only low levels of radioactivity and, in effect, are simply contaminated trash. Such materials are to be segregated and buried separately unless they are in as stable a form as that required for Classes B and C wastes, where either the form or the container must maintain gross physical properties and identity over a period of 300 years.

Institutional control of access to the site is required for up to 100 years. Class B wastes contain nuclides that will decay to innocuous levels within that control period and can be buried without provision of special protection against subsequent intrusion. Class C wastes contain longer-lived species that could still present a hazard after the 100-year period. Such materials require deeper burial or the installation of intrusion barriers, such as concrete covers, having an effective life of 500 years or more.

Classification of the wastes is thus based on the quantity of radioactivity present and the specific nuclides involved. Table 6.1 considers the longer-lived activities. If the concentration does not exceed 0.1 times the value given in the table, the waste is classified as Class A. If the amount is between 0.1 and 1 of

TABLE 6.1 Longer-Lived Activities in LLW

Radionuclide	Concentration (Ci/m^3)
C-14	8
C-14 in activated metal	80
Ni-59 in activated metal	220
Nb-94 in activated metal	0.2
Tc-99	3
I-129	0.08
Alpha-emitting transuranic nuclides with half-life greater than 5 years	100[a]
Pu-241	3,500[a]
Cm-242	20,000[a]

SOURCE: 10CFR61.55, Reference 131.
[a]Units are nanocuries per gram.

the value, the waste is C; and if the concentration is over the table value, the waste is not generally acceptable for near-surface disposal without specific authorization by the NRC.

Similarly, Table 6.2 considers shorter-lived activities. If the concentration does not exceed the value in Column 1, it is a Class A waste; if the concentration is between the numbers in Column 1 and Column 2, Class B; between Column 2 and Column 3, Class C; and more than Column 3, not acceptable for near-surface burial.

TABLE 6.2 Shorter-Lived Activities in LLW

	Concentration (Ci/m^3)		
Radionuclide	Column 1	Column 2	Column 3
Total of all nuclides with less than 5-year half-life	700	a	a
H-3	40	a	a
Co-60	700	a	a
Ni-63	3.5	70	700
Ni-63 in activated metal	35	700	7000
Sr-90	0.04	150	7000
Cs-137	1	44	4600

SOURCE: 10CFR61.55, Reference 131.
[a]There are no limits established for these radionuclides in Class B or C wastes. Practical considerations such as the effects of external radiation and internal-heat generation on transportation, handling, and disposal limit the concentrations for these wastes.

The balance of subsection 61.55 is mostly devoted to methods (not given here) of classifying wastes containing a mixture of radioactivities. It is also stated that nuclide concentrations may be determined by indirect methods, such as the use of scaling factors relating the inferred concentration of one isotope to another actually measured. The concentration may be averaged over the volume of the waste, or weight of the waste in the cases where the limits are expressed as nanograms per gram in Table 6.1.

Subsection 61.56 deals with the characteristics required of wastes acceptable for disposal. These characteristics (slightly abbreviated) are:

1. Wastes must not be packaged in cardboard or fiberboard boxes.
2. Liquids must be solidified or packaged in sufficient absorbent material to absorb twice the volume of the liquid.
3. Solid-waste-containing liquid shall contain as little freestanding and noncorrosive liquid as is reasonably achievable, but in no case over 1% of the total volume of the waste if it is packaged in a container designed for stability, or over 0.5% otherwise.
4. The wastes must not be capable of detonation, explosive decomposition, or explosive reaction with water.
5. The wastes must not contain sources of, or be capable of generating, toxic fumes, although gases containing not over 100 Ci of activity and packaged in containers at a pressure that does not exceed 1.5 atmospheres at 20°C can be accepted.
6. Wastes must not be pyrophoric or contain pyrophoric materials.
7. Wastes containing hazardous biological, pathogenic, or infectious material must be treated to reduce the potential hazard to the maximum extent practicable.

Classes B and C wastes must be structurally stable, that is, able to maintain physical dimensions and form under disposal conditions such as the weight of overburden and compacting equipment, the presence of moisture, and microbiological activity, and able to withstand internal factors, such as radiation effects and chemical changes. The stability can be provided by the waste form itself, by processing to a stable form, or by placing the waste in a suitably designed container.

If the wastes are beyond the Class C limits, the NRC can specifically authorize their disposal mode. A proposal must be submitted to the Commission, and if it is decided that the characteristics of the wastes, the adequacy of the proposed disposal site, the planned techniques of disposal, and future surveillance programs are all within the safety performance objectives of 10CFR61, the NRC can give the necessary authorization on a special-case basis.

If the wastes contain only radionuclides not listed in either Tables 6.1 or 6.2, they are given a Class A designation (10CFR61.55). An Environmental Impact Statement has been prepared[132] in support of the 10CFR61 regulations.

6.2.1 10CFR20 Regulations

Part 20 of 10 CFR is titled "Standards for Protection Against Radiation." Subsections 20.303-20.306 are concerned with the disposal of radioactive waste, although 20.304, which formerly allowed a licensee to bury minor amounts of activity on his own premises, has now been withdrawn, the implication being that all buriable wastes are to go to authorized burial grounds in accordance with 10CFR61 except for trivial amounts of C-14 and tritium (see below).

Appendix B of 10CFR20 lists a large number of radionuclides and establishes the maximum concentration of each (MPCs) that cannot be exceeded in releases to the atmosphere or in water supplies. A person or an institution holding a license for handling radioactivity can dispose of liquids or readily dispersible or soluble solids into a sanitary system if the quantity per day, when diluted by the average amount of sewage generated per day, does not result in a concentration exceeding the Appendix B MPC for the isotope established for liquids released to unrestricted areas. Part 20 also includes another list of nuclides with limits expressed in terms of microcuries. This list applies primarily to the labeling of shipping containers and work areas, but, as an alternative to the MPC approach, also to the disposal of liquids to sanitary systems. Ten times the Appendix C quantity can be so released at a time, but the concentration, when averaged over a month's time, cannot exceed the MPC. Up to 5 curies of tritium and 1 curie of C-14 per year can go into the sewage system, but the total of all other radionuclides cannot exceed one curie per year. Further exceptions are given for tritium and C-14—0.05 microcuries or less of each isotope per gram of the medium used for scintillation counting, or the same quantity per gram of animal tissue, averaged over the weight of the entire animal, can be disposed of by any convenient means without regard to its radioactivity.

A selected number of MPC limits are given in the Health Physics chapter. Table 6.3 gives the Appendix C microcuries limits for all the isotopes listed.

Subsection 10CFR20.305 states that the licensee cannot dispose of active materials by incineration, the only exception being for tritium and C-14 below the 0.05 microcurie limit as described above. Incineration of any other active materials requires specific authorization by the NRC.

6.3 TRU WASTES

TRU wastes were originally defined as those contaminated with U-233 or transuranic nuclides. The AEC proposed changes[133] to 10CFR20 in 1974 to require segregation and retrievable storage of such materials if their average radioactivity

TABLE 6.3 Data from 10CFR20, Appendix C

Isotopes	Microcuries
Am-241, Pu-239, Ra-226, U-233, U-234, U-235, any other alpha emitter or unknown mixture of alpha emitters	0.01
I-129, Po-210, Sr-90, any unlisted beta emitter or unknown mixture of beta emitters	0.1
Bi-210, Ce-144, Cs-134, Co-60, Eu-152 (13y), Eu-154, I-125, I-126, I-131, I-133, Ru-106, Ag-110m, Sr-89	1
Sb-124, Sb-125, As-74, As-76, Ba-131, Ba-133, Ba-140, Br-82, Cd-109, Cd-115m, Ca-45, Cs-135, Cs-136, Cs-137, Cl-36, Cl-38, Co-58m, Co-58, Dy-165, Eu-155, Gd-153, Ga-72, Hf-181, In-114m, In-115, I-132, I-134, I-135, Ir-192, Fe-55, Fe-59, Kr-87, La-140, Mn-52, Mn-54, Hg-203, Ni-63, Nb-93m, Nb-95, Nb-97, Os-185, P-32, K-42, Pm-147, Pm-149, Rb-86, Rb-87, Ru-103, Rh-105, Sm-151, Sc-46, Sc-48, Se-75, Ag-105, Na-24, Sr-85, Sr-91, Sr-92, Ta-182, Tc-96, Tc-99, Te-125m, Te-127m, Te-129m, Te-131m, Te-132, Tb-160, Tl-204, Tm-170, Tm-171, Sn-113, Sn-125, W-181, W-185, V-48, Y-90, Y-91, Zn-65, Zr-93, Zr-95, Zr-97	10
Sb-122, As-73, As-77, Cd-115, C-14, Ce-141, Ce-143, Cs-134m, Cu-64, Dy-166, Er-169, Er-171, Eu-152 (9.2h), Gd-159, Ge-71, Au-198, Au-199, Ho-166, In-113m, In-115m, Ir-194, Fe-55, Kr-85, Lu-177, Hg-197m, Hg-197, Mo-99, Nd-147, Nd-149, Ni-59, Ni-65, Os-191m, Os-191, Os-193, Pd-103, Pd-109, Pt-191, Pt-193m, Pt-193, Pt-197m, Pt-197, Pr-142, Pr-143, Re-186, Re-188, Rh-103m, Rh-105, Sm-153, Sc-47, Si-31, Ag-111, S-35, Tc-97m, Tc-97, Tc-99m, Te-127, Te-129, Tl-200, Tl-201, Tl-202, Th (natl.), W-187, U (natl.), Xe-133, Xe-135, Yb-175, Y-92, Y-93, Zn-69m	100
Cs-131, Cr-51, F-18, H-3, Xe-131m, Zn-69	1000

SOURCE: U.S. Code Federal Regulations, Title 10, Part 20, Appendix C.

level was 10 nCi/g or higher. These modifications were never incorporated into 10CFR20, but AEC itself (and successors ERDA and DOE and their contractors) accepted these rules. The resulting accumulation of alpha wastes is now stored at the Idaho site awaiting disposal in the Waste Isolation Pilot Plant (WIPP) now being constructed in southern New Mexico.[134]

The definition of TRU wastes was effectively changed with the adoption of 10CFR61, and they are now designated as alpha wastes, since they could include isotopes such as Ra-226, Ac-227, and Pa-231. As shown in Table 6.1, wastes containing less than 10 nCi/g of alpha emitters having half-lives of greater than 5 years are defined as Class A, and those to 100 nCi/g as Class C. Commercially generated wastes in this range (if the Class C type were placed in proper form) presumably could undergo near-surface burial in privately operated facilities. Wastes contaminated with alphas at any but low levels are, however, predominantly generated in connection with government research and military programs. These materials would go into WIPP.

Table 6.1 shows values for setting the upper limit of Class C wastes at 3500 nCi/g for beta-emitting Pu-241 and 20,000 nCi/g for short-lived Cm-242. The concern in these two cases is not so much with the parent activities as with keeping the concentrations of the respective daughters (Am-241 and Pu-238) within the 100 nCi/g limit at all times.

6.4 INTERMEDIATE-LEVEL WASTES

The dividing lines between low- and intermediate-level wastes at one end of the scale and high- and intermediate-level at the other are fuzzy, as shown by a survey[135] conducted by IAEA wherein various nuclear installations around the world were asked for definitions of the different waste categories. For medium-level liquid wastes, some of the responses were

Installation	μCi/ml
Poland	10^{-4} to 10^{-2}
Sweden	10^{-3} to 10^{2}
United States	3×10^{-3} to 3×10^{3}
United Kingdom	10^{-2} to 10^{3}
U.S.S.R	10^{-1} to 10^{4}

Oak Ridge[87] uses a definition of 10^4 to 10^6 times the MPC values to classify wastes as being intermediate-level. The bulk of liquids generated at this level of radioactivity come from the nuclear fuel cycle as indicated in Table 5.1. ORNL has estimated[48] that 200 gallons of liquid ILW is produced for each MTU of fuel processed. On that basis, it was projected that the annual generation rate by year 2000 would be 3.2×10^6 gallons, with a total accumulation to that date of 4.9×10^7 gallons. These figures will, of course, only be valid if fuel reprocessing is once again undertaken in this country. The Oak Ridge shale-fracture technique for disposing of such liquids is noted in Section 5.2.2.

Class B and Class C commercial wastes as defined in 10CFR61 (Tables 6.1 and 6.2) can be considered as intermediate-level solid wastes towards the bottom end of the scale. At the other end, as shown in Table 5.1, are the cladding hulls and assembly hardware left behind when irradiated fuel is put into solution as the first step in reprocessing. Many of these materials are too radioactive for acceptance at commercial burial sites. Since the only accumulation of hulls to date from commercial fuel reprocessing in this country was generated at West Valley, the AEC-NRC took care of the problem by licensing a separate burial ground on the site, a solution that presumably could be used in connection with other privately operated fuel reprocessing plants if these are ever authorized. Intermediate-level military wastes at Hanford, Idaho and Savannah River can be buried on their respective sites without licensing by the NRC.

If commercial reprocessing is again allowed, it seems probable that the more radioactive intermediate-level liquids will be mixed with the high-level materials and solidified for repository disposal. Very bulky, highly radioactive, obsolete equipment and structures from reactors and accelerators may present a problem. Fortunately, the nuclides present in such materials generally have relatively short half-lives, which simplifies matters to some degree.

6.5 IMMOBILIZATION OF NON-HLW

During the period when all six of the U.S. privately operated radioactive-waste burial sites were functioning, the Atomic Industrial Forum reviewed and published[136] the detailed rules developed by the operators at each location. Most of the facilities would not accept liquids absorbed on vermiculite, or solid wastes bonded with bitumen. All liquids had to be solidified, although several sites carried out the solidification themselves for a fee. Bottled gases were accepted at several sites, not at others. Charges to the customer were predominantly based on the radioactivity level of the package and its size, with surcharges for odd shapes. The fact that each operator could establish his own rules, over and above those of the NRC, should be kept in mind.

The following brief discussion of immobilization techniques for non-HLW is based primarily on Reference 73.

> Absorbents have been widely used in the nuclear industry to immobilize liquids for transportation and burial. They are favored for oils, solvents and other organic and inorganic liquids that are difficult to solidify by other means. The practice is limited to radioactive liquids that qualify as "low specific activity material" (49CFR 173.309) unless the disposable container is overpacked with a DOT-acceptable container.

The above CFR reference is to the Department of Transportation regulations for shipment of hazardous materials. These rules are highly involved, appearing primarily in 49CFR, Parts 170–189. (DOT has prepared a booklet[137] that summarizes the rules. The regulations are based on a classification system developed by IAEA[138]). The DOT definition of low specific activity material is also somewhat complicated, but basically includes uranium and thorium ores, unirradiated uranium and thorium, tritium in aqueous solutions containing less than 5 mCi/ml, materials containing alpha emitters at less than 100 nCi/g or beta emitters where the concentration limit depends on the specific isotope, and contaminated equipment where the activity level on the surface is less than 1 μCi/m^2.

Absorbents are usually granular or powdered materials and include vermiculite, various clays, diatomite (natural calcium silicate), calcium sulfate desiccants, silica gel, portland cement, and plaster of paris. The amount of liquid waste that can be absorbed is typically in the range of $\frac{1}{3}$ to $\frac{1}{2}$ of the final volume, giving a product in the 0.4 to 1.0 g/cm^3 density range. While the liquids are generally satisfactorily immobilized, the radioactivity is easily displaced if the product comes in contact with water.

Waste-management methods for immobilizing radioactive liquids at Hanford and Savannah River have included solidification in the form of salt cakes. In essence, water is evaporated from the liquids *in situ* in the storage tanks to the point where the contained salts crystallize to form a solid monolith, with the radioactive species largely immobilized in the entrained water. This technique has limited application in nongovernment facilities, although proprietary processes have been developed wherein a molten neutral salt is used to incorporate wastes. The mixture formed is transferred into containers where it hardens upon cooling. Dispersible solids such as ashes, zeolites, and cladding hulls, can thus be stabilized.

Urea-formaldehyde and portland cement are the two immobilization techniques that have been most widely used in this country for immobilization of radioactive wastes above the low specific activity level, whereas bitumen is widely used in Europe as a binding agent to accomplish the same purpose. Urea-formaldehyde immobilization procedures have been largely developed as proprietary processes, using commercially available U-F adhesives and grouts. Urea and formaldehyde react together under alkaline or neutral conditions to form monomethylol and dimethylol urea. Addition of an acid catalyst causes polymerization to produce a gel. Excess water is removed by vacuum distillation, and further water loss occurs on exposure to air. Curing to a reasonably hard, freestanding solid continues over a period of several hours, giving a monolithic product in the 1.0–1.3 g/cm^3 density range. The product has fair mechanical strength and moderate leach resistance, but may become friable if completely dehydrated.

Cement solidification systems are widely used throughout the world to solidify or encapsulate virtually all types on non-HWL generated in fuel cycle operations, and several firms supply total radwaste solidification systems based on cement technology. Table 6.4 shows some typical radwaste formulations.

TABLE 6.4 Typical Cement–Radwaste Formulations

Waste Type	Weight Ratio, Waste Constituent/Cement: Dry Waste Salt	Total Water
Solutions and Slurries		
25 wt% Na_2SO_4	0.12	0.37
70 wt% $NaNO_3$ slurry	0.52	0.22
30 wt% $NaNO_3$ solution	0.15	0.35
Neutralized HNO_3-$Al(NO_3)_3$	0.37	0.37
20–25% Water treatment sludge[a]	0.09–0.17	0.27–0.51
30–40% Evaporator sludge	0.28–0.35	0.42–0.53
Concentrated BWR reactor waste[b]	0.17	0.51
400 g/l evaporator concentrate	0.33[c]	0.70

Waste Type	Weight Ratio, Waste Constituent/Cement: Dry Solid Basis		Wet Solid Basis		Wt% Water in Wet Sludge
	Solid	Water	Solid	Water	
Solids and Wet Sludges					
Acid digestion process residue	0.12	0.37	–	–	–
Al_2O_3-ZrO_2 calcine	2.4	0.25	–	–	–
Fly ash	0.33	0.30	–	–	–
Fe/Al hydroxide sludge	3.1	2.0	4.9	0.22	58
Diatomaceous earth	0.42	1.1	1.3	0.25	67
Linde AW-500 zeolite	1.21	1.0	2.0	0.23	39
Amberlite 200 cation resin	0.83	1.1	1.7	0.23	~50
BWR bead resins, recommended[d]	0.25	0.56	0.51	0.31	50
BWR bead resins, range[d]	0.25–1.0	0.45–1.3	–	–	50
BWR filter sludge, recommended[e]	0.26	1.19	0.51	0.94	50
BWR filter sludge, range[e]	0.04–0.28	0.48–1.3	–	–	50

SOURCE: Report ERDA-76-43, Vol. 2, Reference 73.

[a]Miscellaneous flocculating and scavenging agents.

[b]Assumed to be ~25% Na_2SO_4.

[c]Plus 0.13 kg of vermiculite/kg cement.

[d]Preneutralized with NaOH to prevent swelling and crumbling.

[e]Cellulose fiber filter aid, powdered resins, and so on.

Hydraulic cements are usually used, which harden under water and consist primarily of calcium silicates and aluminates. The most common of these products are the portland cements that have a density of about 3.1 g/cm^3, the tapped density of the bulk powder being about 1.5. Cement blocks prepared from waste concentrates and sludges will generally have less strength than conventional concrete, dependent on the salt content of the wastes. Maximum waste contents of the solidified products are typically 75 w/o for solids, 33 w/o for aqueous solutions and slurries. Solidified product densities range from about 1.5-2.0 g/cm^3. Limited amounts of organic liquids can be incorporated by using dispersing agents, as can small amounts of solvents absorbed on vermiculites. Since the chemistry of the setting of concrete is still incompletely understood, but known to be affected by a number of variables, cold laboratory tests should be carried out in advance for any new waste in order to determine the optimum formulation.

As has been previously indicated, waste immobilization with bitumen (asphalt) is widely used abroad,[139] but not in this country, presumably because the end product is combustible, particularly in the presence of nitrates and nitrites. Bitumenization processes operate in the 150-230°C range, which is sufficient to drive off most of the water in the wastes, although the procedure generates a low-activity condensate requiring further treatment. The solids content of a wide variety of wastes has been immobilized with bitumen, including neutralized evaporator concentrates and sludges, ion-exchange minerals and resins, incinerator ashes, sand, and vermiculite. Up to 25 w/o tributyl phosphate has been incorporated into a marginally acceptable product by the addition of appropriate fillers. In general, bitumen products with satisfactorily high mechanical strength may contain 40-60 w/o solids and have densities in the 1.0-1.3 g/cm^3 range.

Reference 73 lists other immobilization techniques for non-HLW where investigations have been made, but where further development is needed. These technologies include alumino-silicate mineral absorption (clay fixation), glass, metal matrix, pelletization, polyethylene, and water-extensible polymers. The first three of these techniques have been discussed in Chapter 5 in connection with high-level wastes.

Several of the processes described in Chapter 5 for solidification of HLW involve pelletization or compaction steps, but none of these have received full-scale, hot-cell demonstration. Sandia Laboratory has conducted smaller-scale studies on vacuum hot-pressing of inorganic hydrous-oxide ion-exchange minerals, the results of which indicate that the technology is probably suitable for the routine handling of a variety of wastes. The technique is attractive because calcining, pressing, and sintering are all accomplished in one step, but equipment and procedures for remote operation in high-activity backgrounds require much more development. Cold-pressing methods are widely used in industry, but development for waste-management applications (outside of trash compaction) has been limited. More sophisticated applications are believed to be well within the limits of current technology.

Batch-type laboratory experiments have demonstrated that both organic and inorganic waste materials can be successfully incorporated into polyethylene. The equipment and procedures used are similar to or identical with those used for bitumenization. The water and other volatiles of liquid waste can be driven off by addition at an appropriate rate to molten polyethylene to build up a product containing as much as 40 w/o waste solids. This mixture is then drained into containers at 180°C, where it hardens. Ion-exchange resins have also been successfully incorporated, as have up to 40 w/o of organic liquids.[140]

Readily emulsifiable water-extensible unsaturated polyester resins are available commercially which, when catalyzed, harden to a plasterlike substance that can hold up to 65% water and still remain a solid. Laboratory studies have shown that various non-HLW liquids from the LWR fuel cycle can be solidified, but more development is needed.

SEVEN

PACKAGED RADIOACTIVE WASTES

Three important items of information are needed before filled canisters of solidified waste are stored or shipped: the maximum internal temperature at the vertical centerline, the temperature at the surface halfway down the cylinder, and the exposure rate in R/h at the same point. Techniques that can be used for hand calculation of these data at generally acceptable levels of precision are given below.

7.1 CENTERLINE AND SURFACE TEMPERATURE

Cylindrical containers are assumed, having fill height-to-diameter ratios such that heat losses through the ends can be ignored. Only steady-state temperatures are considered, and variations in thermal conductivity with temperature are not considered. No temperature gradient in the canister wall nor thermal resistance between the solidified waste and the container wall are assumed.

The temperature difference between the surface (T_s) and center (T_c) of the cylinder may then be calculated by[141]

$$T_c - T_s = \frac{qa^2}{4k} \tag{7.1}$$

where q = volumetric heat-generation rate (Btu-hr/ft^3)
a = cylinder radius (ft)
k = thermal conductivity (Btu-hr/ft-°F)

If it is assumed that the waste cylinders are to be cooled by natural convection of air, the temperature difference between the cylinder surface and ambient air (T_a) may be calculated by

$$T_s - T_a = \frac{Q}{h} \tag{7.2}$$

where Q = surface heat flux (Btu-hr/ft^2)
h = film coefficient (Btu-hr/ft^2-°F)

The film coefficient is itself a function of the temperature difference between the surface and the air and is given by

$$h = 0.29 \left(\frac{T_s - T_a}{L}\right)^{1/4} \tag{7.3}$$

where L is the fill height of the waste in the cylinder in feet.

These last two equations contain two unknowns and, therefore, must be solved by trial-and-error procedures to find values for the film coefficient and temperature difference that will satisfy both expressions. The determination of $(T_s - T_a)$ sets the surface temperature (T_s). Addition of the $(T_c - T_s)$ differential establishes the centerline temperature, or

$$T_c = T_a + (T_s - T_a) + (T_c - T_s) \tag{7.4}$$

EXAMPLE: An aged waste is encapsulated in glass and placed in a 2′ × 10′ canister to the 90%-fill level and then stored in air at an ambient temperature of 80°F. The total heat generation of the waste is 332 Btu-hr.

The volumetric heat generation rate q and the surface heat flux Q are calculated, respectively, by dividing the total heat output by the waste volume and by the external surface area of a cylinder 9′ in height and 2′ in diameter. These numbers come out to be 11.7 Btu-hr/ft^3 and 5.9 Btu-hr/ft^2. A thermal conductivity value for glass of 0.6 Btu-hr/ft-°F is assumed. Then

$$T_s - T_c = \frac{(11.7)(1)^2}{(4)(0.6)} = 4.9°\text{F}$$

Equations 7.2 and 7.3 then are

$$T_s - T_a = \frac{5.9}{h}$$

and

$$h = 0.29\left(\frac{T_s - T_a}{9}\right)^{1/4}$$

By trial and error, it is found that a value of 0.34 Btu-hr/ft^2-°F satisfies both equations and produces a $T_s - T_a$ difference of 17.3°F. T_s is thus 80 + 17.3 or 97.3°F and T_c is 97.3 + 4.9 or 102.2°F.

The discussion above is based on the use of English units. Some convenient conversions to metric units are

Multiply	By	To Obtain
Btu-hr	0.293	W
Btu-hr/ft^2	3.15	W/m^2
Btu-hr/ft^3	10.4	W/m^3
Btu-hr/ft-°F	0.578	W/m-°C
Ft	0.305	m
Ft^2	0.0929	m^2
Ft^3	0.0283	m^3
$\frac{5}{9}$(°F-32)		°C

7.2 CANISTER MIDPOINT EXPOSURE RATES

A number of literature sources[15,21,141,142] consider the shielding situation shown in Figure 7.1, wherein exposure measurements are made at point P for a uniformly gamma-radiating cylinder with a slab shield interposed between the cylinder surface and the monitoring instruments. The situation is treated as if all of the radioactivity is concentrated in a hypothetical vertical line source that will be displaced from the cylinder axis towards the observer due to the effects of distance and self-attenuation in the cylinder contents. The problem considered here, that is, no interposing shield, and with the measuring point P directly on the cylinder surface at the midline of the waste volume, can be treated as a special case of the Figure 7.1 geometry.

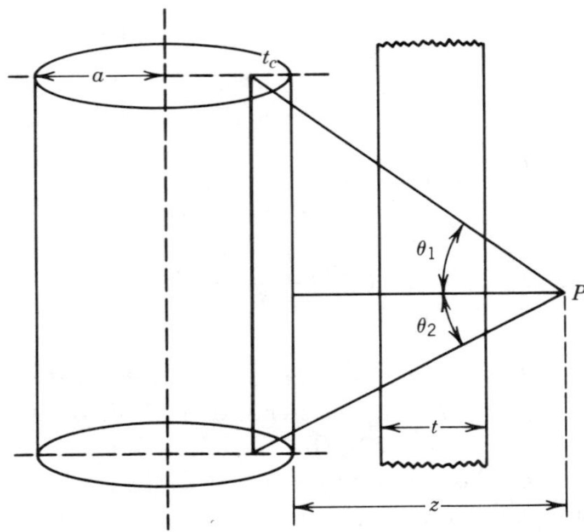

FIGURE 7.1. Cylinder shielding configuration. (Reference 15.)

(The procedure as given below was developed by Drs. L. E. Trevorrow and S. Vogler of Argonne National Laboratory, based on the more general approach of Blizard.[142] In this special case analysis, the thickness and composition of the canister wall were ignored.)

Blizard's notation is used in the following discussion:

a = radius of the cylinder

t_c = horizontal distance in from the cylinder surface of the hypothetical line source

z = horizontal distance between the cylinder surface and the observation point, P, (equal to zero in the special case)

t = thickness of the slab shield (equal to zero in the special case)

θ_1 and θ_2 = angles between the top and bottom of the cylinder contents and the observation point (these angles are the same in the special case)

μ_c = absorption coefficient of the pertinent gamma ray in the cylinder contents

μ = absorption coefficient of the ray in the slab shield (zero in the special case)

p = volumetric source strength: MeV/cm^3-sec

ϕ = energy flux at P: MeV/sec-cm^2

ϕ_{eq} = factor for converting ϕ to mR/hr

The basic applicable equation as given by Blizard is then

$$\phi = \frac{pa^2}{4(z + t_c)} [F(\theta_2, \mu t + \mu t_c) + F(\theta_1, \mu t + \mu t_c)] \tag{7.5}$$

While the steps to be taken are given below in cookbook fashion, the location of the hypothetical line source (the value of t_c) is first determined, the integrated flux at P from all points along the source is evaluated (the function portions of the equation), the energy flux at P is then calculated, and this, in turn, is converted to exposure units.

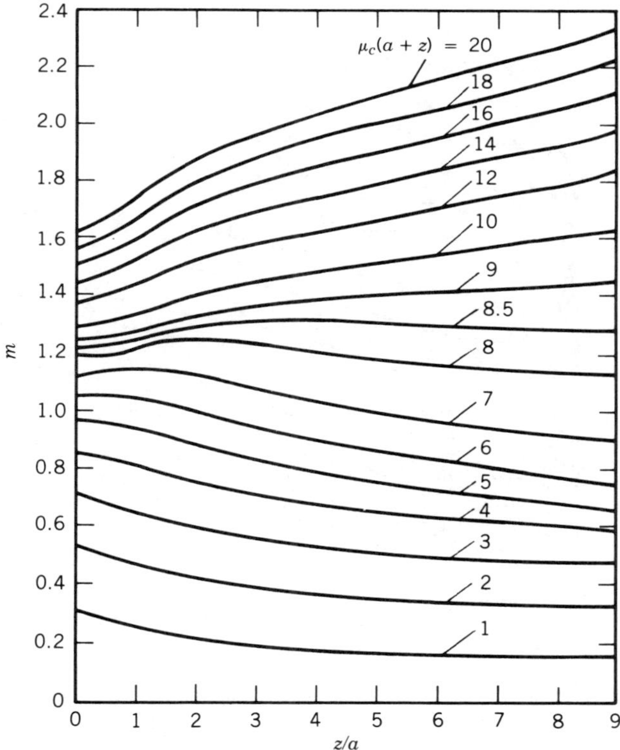

FIGURE 7.2. The self-absorption parameter, m. These curves apply to cylinders when z/a is less than 10. (Reprinted by permission of the publisher from *Nuclear Engineering Handbook*, H. E. Etherington, ed., Section 7.3. Copyright 1958 by McGraw-Hill Book Company. Reference 142.)

1. Derive the value of m from Figure 7.2. Since z/a is zero in the special case, this will be at the ordinate on the appropriate curve determined by $\mu_c a$.
2. From Figure 7.3, obtain the value of $\mu_c t_c/m$, remembering that in the special case, both z/a and μt are equal to zero.
3. Using the known value of μ_c and the derived value of m, calculate the distance t_c, thus locating the position of the hypothetical line source.
4. From the height of the waste in the cylinder and the derived value of t_c,

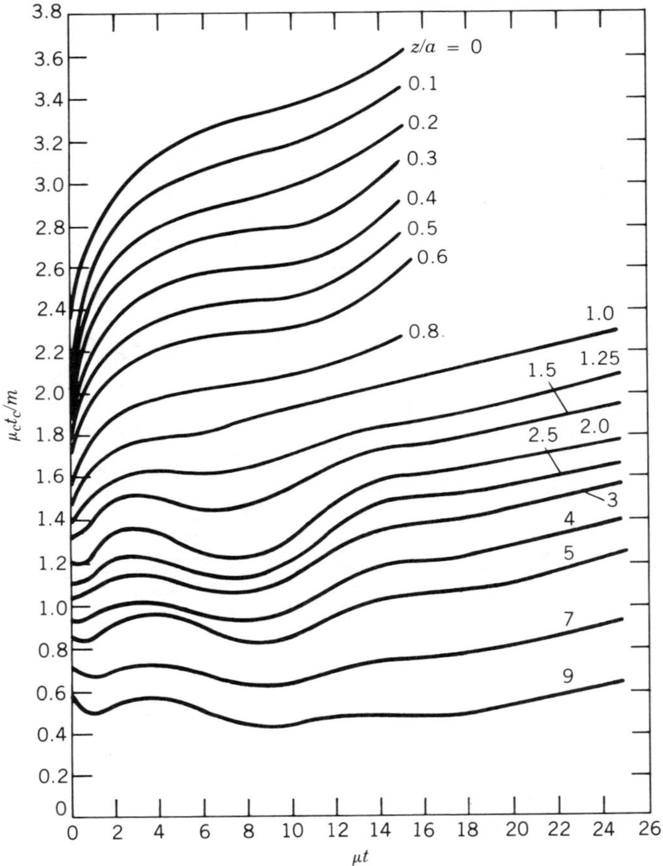

FIGURE 7.3. Self-absorption exponent divided by self-absorption parameter, $\mu_c t_c/m$, as a function of μt for $z/a < 10$. (Reprinted by permission of the publisher from *Nuclear Engineering Handbook*, H. E. Etherington, ed., Section 7.3. Copyright 1958 by McGraw-Hill Book Company. Reference 142.)

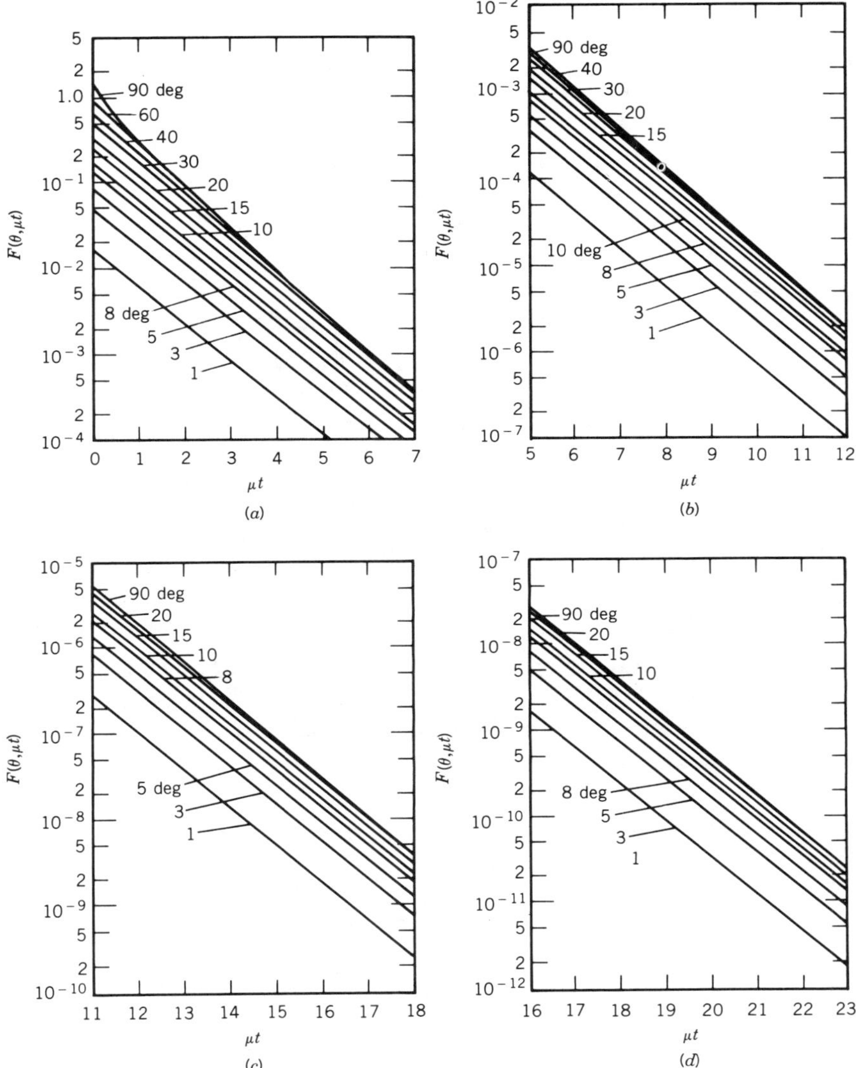

FIGURE 7.4. The $F(\theta, \mu t)$ function, (*a*) μt = 0–7; (*b*) μt = 5–12; (*c*) μt = 11–18; (*d*) μt = 16–23. (Reprinted by permission of the publisher from *Nuclear Engineering Handbook*, H. E. Etherington, ed., Section 7.3. Copyright 1958 by McGraw-Hill Book Company. Reference 142.)

calculate the tangent of θ_1 (the same as for θ_2), and the values of the angles.

5. Assuming that the curves of Figure 7.4 can be used on the basis of the abscissa values being for $\mu t + \mu_c t_c$ and the ordinate for $F(\theta, \mu t + \mu_c t_c)$, determine this latter quantity.
6. Calculate the volumetric source strength p:

$$\frac{\text{Total Ci}}{\text{Volume}} \times 3.7 \times 10^{10} \text{ dis/s} \times \text{MeV/dis}$$

7. Calculate the MeV flux ϕ at point P from Equation 7.5.
8. Determine the flux equivalent to 1 mR/hr (ϕ_{eq}), from the MeV/cm^2-sec conversion curve of Figure 7.5.
9. Calculate the exposure rate in mR/hr from the ratio of ϕ to ϕ_{eq}.

EXAMPLE: A 2′(OD) × 10′ canister is filled to the 90% level with an aged waste containing a total of 7.2×10^4 Ci of 0.66 MeV gamma from ^{137m}Ba, the predominant remaining activity and the only one that need be considered. The known parameters are then: a, 30.5 cm; height of waste, 274 cm; volume of waste, 8.0×10^5 cm^3; μ_c, taken as 0.2 cm^{-1}; total Ci, 7.2×10^4; and effective gamma energy, 0.66 MeV.

Steps 1-3 above place the position of the hypothetical line source 12.3 cm in from the cylinder surface. This allows calculation of a value of 84° 52′ for each of the angles θ_1 and θ_2. Figure 7.4 then gives a value of 6×10^{-2} for each of the $F(\theta, \mu t + \mu_c t_c)$ terms.

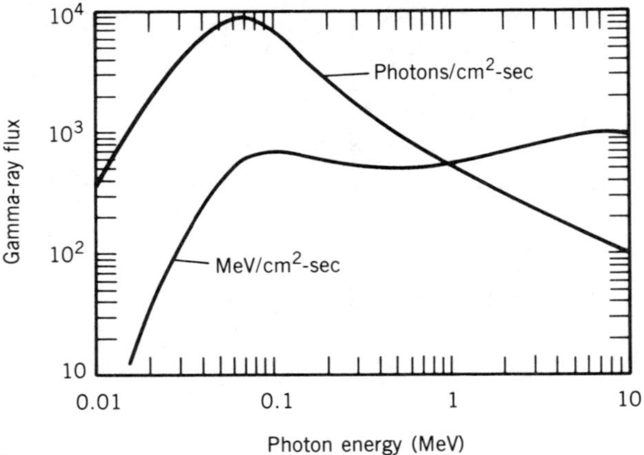

FIGURE 7.5. Flux equivalent of 1 mR/h. (Reference 15.)

TABLE 7.1 Data Relating to Possible Waste Containers[a]

English				Metric			
ID X Height (ft)	Volume (ft^3)	Capacity (gal)	Surface (ft^2)	ID X Height (m)	Volume (m^3)	Capacity (l)	Surface (m^2)
1 X 10	7.1	53	28	0.31 X 3.1	0.20	200	2.6
2 X 10	28	212	57	0.61 X 3.1	0.80	800	5.3
3 X 10	64	480	85	0.92 X 3.1	1.8	1800	7.9
4 X 10	113	845	113	1.22 X 3.1	3.2	3200	10.5
2 X 12	34	250	68	0.61 X 3.7	0.96	960	6.3
4 X 12	136	1015	136	1.22 X 3.7	3.8	3800	12.6
2 X 15	42	320	85	0.61 X 4.6	1.2	1200	7.9
3 X 15	95	710	127	0.92 X 4.6	2.7	2700	12
4 X 15	170	1270	170	1.22 X 4.6	4.8	4800	16
Steel drums							
30 gal	3.6	27	6.2	See text	0.10	100	0.58
55 gal	6.6	50	7.6	See text	0.19	190	0.71
83 gal	10	75	–	See text	0.28	280	–

SOURCE: Calculated figures.

[a]Nominal sizes are given in Column 1. All other numbers are for 90% fill.

TABLE 7.2 Corrosion Rates (mm/yr) at 250°C for Various Candidate Overpack Alloys in Simulated Isolation Environments (Based on Weight Loss After 14–50 Days)

Alloy	Deoxygenated Seawater + Sediments	Oxygenated Seawater	Deoxygenated Brine	Oxygenated Brine	Remarks
1018	0.33	11.3	1.72	6.9	
90-10 CuNi	0.065	0.7	0.137	0.44	
Cu	0.047	4.6	0.065	0.2	
Pb	0.27	1.0	0.41	1.2	
304L	0.009[a]	–	0.018[a]	–	Susceptible to SCC
20Cb3	0.006	–	0.007	0.1[b]	Questionable SCC resistance
Ebrite 26-1	0.005[a]	–	0.016[a]	0.24	Susceptible to H_2 embrittlement
Monel 400	0.11	–	0.029	–	
Hastelloy C-276	0.002	0.2[c]	0.007	0.06[c]	
Ticode 12	0.002	–	0.006	0.004	

SOURCE: Reference 77e

[a]Minor superficial pitting.

[b]Due to crevice corrosion.

[c]Due to pitting and crevice corrosion; deepest pit 1 mm, average 0.25 mm in 14 days.

The volumetric source strength (Step 6) is 2.2×10^9 MeV/cm^3-sec. Solution of Equation 7.5 gives a value of 4.99×10^9 MeV/cm^2-sec for the flux at point *P*. This is converted to an exposure estimate of 9.9×10^6 mR/hr through use of Figure 7.5.

The above example is for aged wastes where ^{137m}Ba radiation predominates. The situation could be quite different for solidified fresh waste, where a number of gammas from shorter-lived nuclides would have to be considered. In hand calculation, a separate evaluation would be made for each activity, and the aggregate exposure levels summed. This type of problem is, of course, susceptible to computer treatment. The SDC code used by Arnold[144] was established to handle up to 12 different gamma energies. This computer code, or its descendants, is available from the Oak Ridge Radiation Shielding Center.[44]

Rules regarding measurements to be made on waste canisters being sent to federal repositories have not as yet been set, so the requirement of an exposure evaluation at the cylinder surface may instead be for a midline estimate at one foot or some other distance from the package. The treatment given above is still applicable, but z would no longer be zero. Another modification might be to consider the canister wall as being the slab shield shown in Figure 7.1.

7.3 WASTE CONTAINER SIZES

Different authors make different assumptions as to the size of the canisters for packaging solidified HLW, with a $2' \times 10'$ cylinder being perhaps the most popular choice for handling aged defense wastes. Table 7.1 provides data relating to possible canister sizes. Commercially available steel drums are also included in the table.

7.4 CORROSION OF CANISTER MATERIALS

Burial of canisters of solidified HLW in bedded salt beds or in subseabed sediments are possibilities for disposal. Both of these are highly corrosive environments, and an overpack would probably be required, that is, the canister would be placed in a secondary container of resistant alloy before interment. Table 7.2 shows the results of corrosion tests made at Sandia Laboratory[77e] on candidate alloys. (The term "SCC" was not defined in the original paper.)

EIGHT

REPOSITORY DATA

8.1 BACKGROUND

Numerous suggestions have been made[73] of techniques for final disposal of high-level radioactive wastes. The passage by Congress of the Nuclear Policy Act of 1982[145] has now determined that the first direction taken will be by construction of a deep-underground repository. The act called for a suitable site to be selected by March 31, 1987, and the facility to be operational by 1998. The responsibility for locating the site was given to the Department of Energy, which announced at the end of 1983 that they are likely to fall three years behind the indicated schedule. Potential sites are being considered in six states. DOE hopes to narrow the choice to three by January 1985, then after an exploratory drilling and testing program at each location that would last until December 1989, to make a final decision in early 1990.[146] The facility itself would require licensing by the Nuclear Regulatory Commission at various stages of construction, operation, and eventual shutdown and closure.

In line with this responsibility, the NRC proposed technical criteria for the repository in the *Federal Register* in 1981.[147] After assessing the comments received to the proposal, the Commission issued, as a final rule, the addition to the federal regulations of 10CFR60, "Disposal of High-Level Radioactive Wastes in Geologic Repositories, Technical Criteria."[148]

The Nuclear Policy Act also affected the Environmental Protection Agency, which has the responsibility for establishing radiation-release limits to the environment for all parts of the nuclear fuel cycle. The EPA has accordingly submitted[149] the text of a proposed new regulation for comment—40CFR191, "Environmental Protection Standards for Management and Disposal of Spent Nuclear Fuel, High-Level and Transuranic Radioactive Wastes." One of the standards would be that the facility design would guarantee that the annual dosage by direct radiation from the operation would be limited for any member of the public to 25 mrem, whole body, 75 mrem to the thyroid, and 25 mrem to any other organ.

A second standard that would be established in 40CFR191 deals with allowable radiation releases from the repository to the external environment. Two levels would be considered, "reasonably foreseeable releases" and "very unlikely releases." The former would include cases where it could be estimated that a release had more than 1 chance in 100 of occurring in 10,000 years; the latter, where the estimated chance of a release would be between 1 in 100 and 1 in 10,000 over the same time period.

Table 8.1 lists these release limits in terms of curies per: (a) HLW generated from 1000 metric tons of irradiated heavy metal (MTHM), or (b) to an amount of transuranic waste containing one million curies of alpha-emitting transuranics. The figures as given would apply to "reasonably foreseeable releases." Ten times the quantities shown would be permitted for "very unlikely releases."

TABLE 8.1 EPA Proposed Repository Release Limits

Nuclide	Release Limit (Ci per 1000 MTHM)
Am-241	10
Am-243	4
C-14	200
Cs-135	2,000
Cs-137	500
Np-237	20
Pu-238	400
Pu-239	100
Pu-240	100
Pu-242	100
Ra-226	3
Sr-90	80
Tc-99	10,000
Sn-126	80
Any other alpha emitter	10
Any other non-alpha emitter	500

SOURCE: *Federal Register*, 47, 58204-58206 (Dec. 29, 1982).[149]

8.2 HOST ROCK PROPERTIES

Bedded rock salt as a suitable location for a deep-underground HLW repository has been under investigation since the 1960s.[73] This type of geologic structure is now being used for the WIPP facility,[150] and is a viable candidate for disposal of commercial wastes. In recent years, there has also been a steadily increasing interest in other possibilities for repository siting, such as large formations of basalt, granite, tuff, or shale. A very considerable literature has accordingly accumulated. Isherwood of Lawrence Livermore Laboratory has made a two-volume survey[151] of rock properties. Among the many other sources of information are articles,[152,153] reports,[154,155] textbooks,[156,157] symposia proceedings,[77,79,150] etc. The political aspects of site selection, of which there are many, have been reviewed.[158]

The tables and figures in the following pages are only representative of the large mass of data available, although an effort was made to present items of general, rather than specific interest. Ranges of values are again most meaningful, since generic terms such as basalt or granite include rocks of widely varying composition and corresponding differences in physical and chemical properties.

8.2.1 Permeability and Porosity

Essentially, the only credible mechanism for radioactivity to escape from a properly designed waste repository placed deep in the earth is through intrusion of water. Conceivably, such an intrusion could result in destruction of waste canisters and attack on the solidified HLW itself, with subsequent movement of dissolved radioactive species into channels leading to the outside environment. The permeability of the host rock, which is inversely related to its porosity (percent of void space), is thus a property of major importance. Tables 8.2 and 8.3 present data pertinent to permeability (expressed as water conductivity in cm/s) and porosity.

The conductivity ranges for soils given in Table 8.2 are important to the choice of backfill material in multiple-barrier repository installations.

8.2.2 Thermal Properties

A major concern in repository design is the effect of the heat from the encapsulated waste on the physical and chemical properties of the surrounding rock. Thermal properties are, therefore, of major interest. Figures 8.1 and 8.2 show thermal conductivity values and ranges for various rock types. Figure 8.3 and Table 8.4 give specific heat data, and, while repository temperatures are unlikely to reach into the rock-melting range, fusion temperatures are given in Table 8.5.

TABLE 8.2 Water Conductivity of Rocks and Soils at 20°C

Conductivity, cm/s	10^2	10	1	10^{-1}	10^{-2}	10^{-3}	10^{-4}	10^{-5}	10^{-6}	10^{-7}	10^{-8}	10^{-9}
Degree of conductivity	Very high			High		Moderate			Low		Very low	
Soil type	Clean gravel			Clean sands; clean sand and gravel mixtures				Very fine sands, silts, mixtures of sand, silt and clay; glacial till; stratified clay deposits; etc.			Homogeneous clays below zone of weathering	
Rock type	←—Shale——											
	←—(Fractured)——Sandstone——→											
	——(Solution cavities)——Limestone and dolomite——(Unfractured)——→											
	←—(Fractured or weathered)——Volcanic rocks excluding basalts——											
	——(Cavernous and fractured)——Basalt——(Dense)——→											
	←—(Weathered)——Metamorphic rocks——											
	←—Bedded salt——											
	←—(Weathered)——Granitic-type rocks——											

SOURCE: Reference 159 as quoted in Reference 151.

TABLE 8.3 Conductivity and Porosity Ranges

Rock Type	Total Porosity (%)	Interstitial Conductivity (cm/sec)
Volcanic	<1	10
Granite	0.05–3	5×10^{-11} to 2×10^{-10}
Metamorphic	0.02–2.4	5×10^{-11} to 2×10^{-6}
Basalt	0.9	2×10^{-9} to 5×10^{-6}
Sandstone	0–51	10^{-7} to 10^{-2}
Shale	0.7–45	4×10^{-11} to 2×10^{4}
Salt	<1	7×10^{-9} to 4×10^{-6}

SOURCE: Reference 160 as quoted in Reference 151.

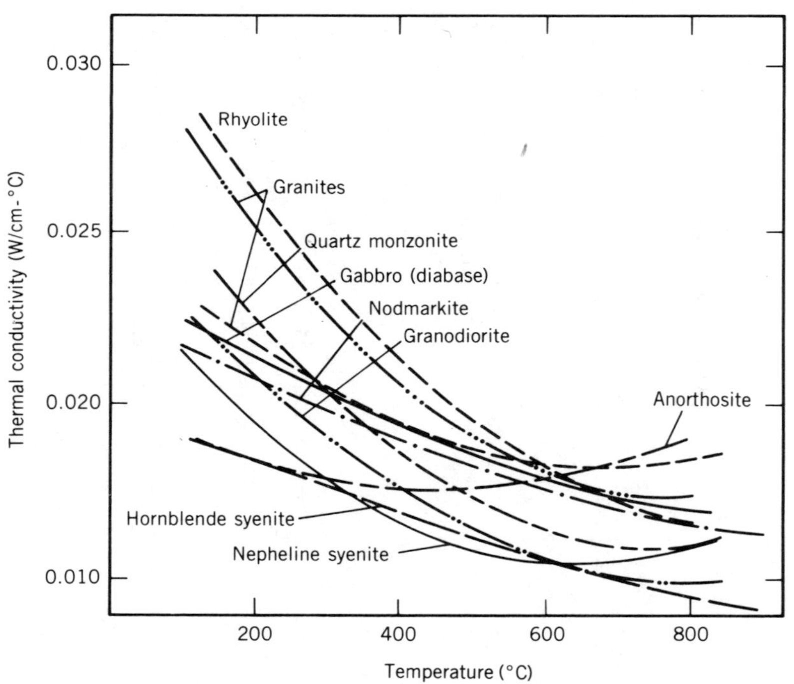

FIGURE 8.1. Thermal conductivity of igneous rocks. (Reference 161, quoted in Reference 151.)

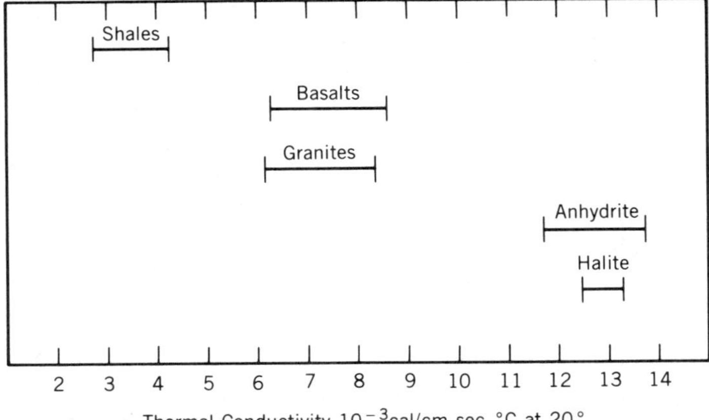

FIGURE 8.2. Thermal conductivities. (Reference 77f.)

8.2.3 Shear Strengths

Shear-strength data for various rocks and minerals are given in Figure 8.4

8.2.4 Specific Rock Types

The geological formations receiving the most attention as possible waste repositories are those composed of bedded salt, basalt, granite or volcanic tuff, and there has been some investigation of shales. Reference 151 contains extensive information regarding basalts, granites, and shales. Reference 153 presents an excellent summary of the status of current investigations of possible repository sites.

8.2.4.1 Basalts

Tables 8.6 and 8.7 summarize the thermal and physical properties of representative basalts. The "intact value" heading of Table 8.6 refers to a solid, unfractured sample, whereas the "rock-mass value" is for the material as it exists in the field. The data of Table 8.7 are for intact samples.

8.2.4.2 Granites

The Office of Nuclear Waste Isolation of the Department of Energy is located at Battelle-Columbus Laboratories in Ohio. The Office has prepared a Granitic Briefing Book[165] that identifies eight major formations of granitic rock throughout the United States that are possibilities for a repository site.

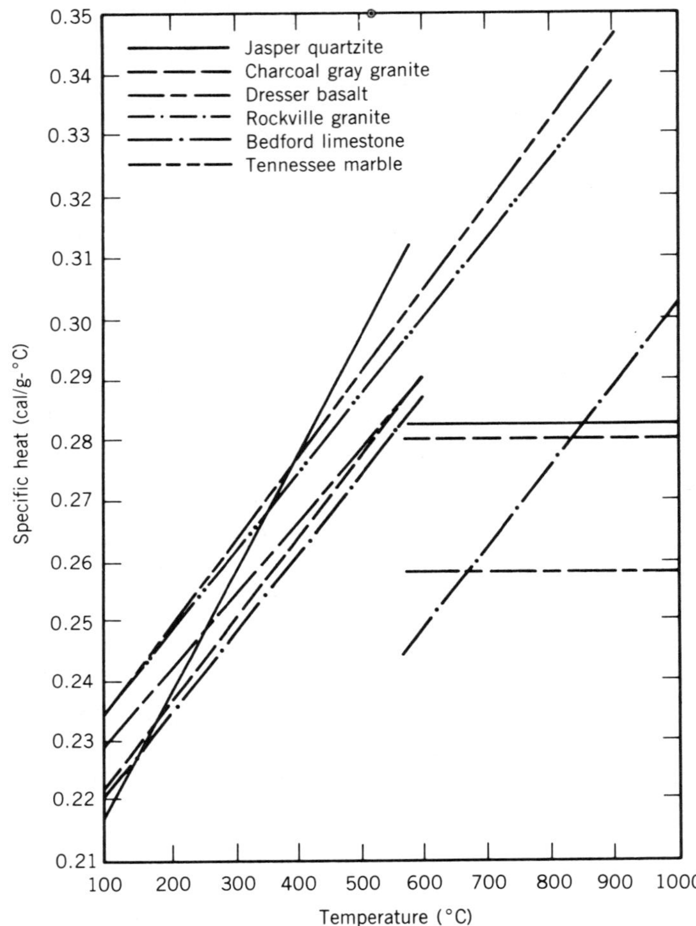

FIGURE 8.3. Specific heats for six rock types. (Reference 162, quoted in Reference 151.)

In an earlier report,[155] the office assumed various properties for a generic granite for modeling purposes. Their assumptions are presented in Table 8.8. A similar summary, including ranges, is given in Table 8.9.

8.2.4.3 Volcanic Tuffs

These are pyroclastic flow deposits, that is, solidified lava. Two varieties of tuff, welded and zeolitic, are of interest in connection with repository siting.[153] Welded tuffs are volcanic ash flows that were solidified at the time of formation, and are dense, low in porosity, and able to accept high thermal loads. Zeolitic

TABLE 8.4 Specific Heats of Rock Materials

Common Designation	Specific Heat (cal/g-°C)
Granite	0.241
Anorthosite (Peribonca black)	0.234
"Nepean sandstone"	0.246
"Diabase"	0.226
Nepheline syenite	0.237
"Basalt"	0.236
"Rhyolite"	0.237
Syenite (Sienna red)	0.229
Nordmarkite (Scots green)	0.231

SOURCE: Reference 163 as quoted in Reference 151.

Note: Values are means of two runs between 25 and 625°C.

TABLE 8.5 Fusion Temperatures for Various Rocks

Rock Type and Origin	Fusion Temperature (°C)	Pyrometric Cone Equivalent (P.C.E.) No.
Crystalline Igneous Rocks		
Basalt (traprock), N.J.	1152–1168	C-3
Diabase traprock ("black granite"), French Creek Granite Co.	1168–1186	C-4
Medium red granite, Leeds County, Ontario	1168–1186	C-4
White granite (high mica content), Chelmsford, Mass.	1168–1186	C-4
Opalescent granite, Cold Springs, Minn.	1201–1222	C-6
Grey granite, Norway	1215–1240	C-7
Grey granite, Salisbury, N.C.	1236–1263	C-8
Collins pink granite, Salisbury, N.C.	1236–1263	C-8
"Syenite," Lake Asbestos, Quebec	1236–1263	C-8
Red granite, Wausau, Wisc.	1236–1263	C-8
Melrose pink granite, Guenette, Quebec	1236–1263	C-8
Granite gneiss, Mount Wright, Quebec	1236–1263	C-8
Grey granite, Chelmsford, Mass.	1285–1305	C-10
Quartz-Rich Crystalline Igneous Rocks (More Than 10% Quartz)		
Granite, Stanstead, Quebec	1198–1221	C-5 1/3
"Granite" (Stanstead grey), Stanstead, Quebec	1201–1222	C-6
Granite (Saguenay red), Lake St. John, Quebec	1177–1196	C-5
Granite (Vermilion pink), Vermilion Bay, Ontario	1207–1232	C-6 1/2
"Rhyolite," Havelock, Ontario	1215–1240	C-7
Quartz-Poor Crystalline Igneous Rocks (Less Than 10% Quartz)		
"Basalt," Havelock, Ontario	1149–1165	C-2 2/3
"Diabase," Bell's Corners, Ontario	1164–1179	C-3 2/3
Anorthosite (Peribonca black), Peribonca River, Quebec	1297–1319	C-11 1/3
Syenite (Sienna red), Rawcliffe, Quebec	1168–1186	C-4
Nordmarkite (Scots green), Mount Megantic, Quebec	1167–1184	C-3 4/5
Nepheline syenite, Methen Township, Ontario	1168–1186	C-4

SOURCE: Reference 163 as quoted in Reference 151.

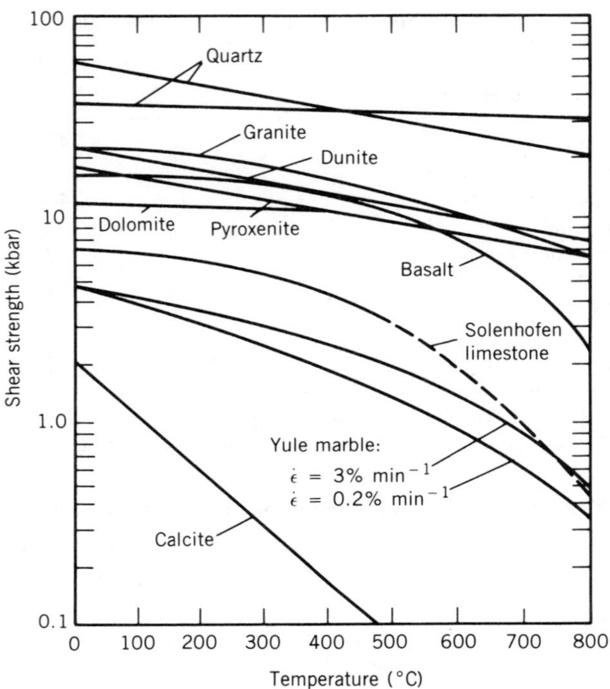

FIGURE 8.4. Shear strengths of rocks and minerals. (Reference 64, quoted in Reference 151.)

TABLE 8.6 Properties of a Typical Dense Basalt

	Intact Value	Rock-Mass Value
Index properties		
Density, g/cm^3	3.01	2.88
Porosity, %	2.0	0.6
Stress-strain properties		
Young's modulus	69	12.4
Poisson's ratio	0.26	0.26
Strength properties		
Uniaxial compressive strength, MPa	276	124
Tensile strength, MPa	16	0
Thermal properties		
Coefficient of linear thermal expansion, 10^{-6} °C^{-1}	1.7	1.7
Heat capacity, J/g-°C, at		
0°C	0.71	0.71
100°C	0.80	0.80

TABLE 8.6 (Continued)

	Intact Value	Rock-Mass Value
Heat capacity, J/g-°C, at		
200°C	0.92	0.92
300°C	0.96	0.96
Thermal conductivity, W/M-°C, at		
0°C	1.12	1.12
50°C	1.19	1.19
100°C	1.26	1.26
150°C	1.32	1.32
200°C	1.38	1.38
300°C	1.47	1.47
400°C	1.56	1.56

SOURCE: Reference 155 as quoted in Reference 151.

TABLE 8.7 Intact Properties for Four Basalts

	Origin of Sample			
	Dresser Basalt	Amchitka Island	Nevada Test Site	Columbia River Basalt
Index properties				
Unit weight; kg/m^3	3000	2700	2700	2403–3085
Natural moisture content, %	–	–	–	–
Porosity (rock-mass), %	0.19	2.8	4.6	–
Stress-strain properties:				
Young's modulus, GPa	93	61.1	34.9	55.2–110
Poisson's ratio	0.264	0.19	0.32	0.22–0.30
Bulk modulus, GPa	–	32.8	–	–
Shear modulus, GPa	4.15	25.7	–	–
Strength properties				
Uniaxial compressive strength, MPa	440.7	250	148	193.1–400.0
Tensile strength, MPa	22.1	15.5	13.1	12.41–24.14
Thermal properties				
Coefficient of linear thermal expansion, 10^{-7} °C^{-1}	8.9	8.90	–	16.7
Heat capacity, J/g-°C	0.96	0.96	–	0.72–0.96
Thermal conductivity, W/m-°C	1.30	1.30	–	1.00–1.35

SOURCE: Reference 155 as quoted in Reference 151.

TABLE 8.8 Properties of a Generic Granite

	Intact Rock	Rock Mass
Index properties		
Unit weight, kg/m^3	2643.3	2643.3
Natural moisture content, %	–	–
Porosity, %	0.4	0.4
Stress-strain properties		
Young's modulus, GPa	50.3	17.2
Poisson's ratio	0.18	0.18
Strength properties		
Strength parameters: A	4.5	4.5
k	0.75	0.75
Uniaxial compressive strength, MPa	182.8	131
Tensile strength	6.9	0
Thermal properties		
Coefficient of linear thermal expansion, $10^{-6}\ °C^{-1}$	31.0	31.0
Heat capacity, J/g-°C, at		
0°C	0.88	0.88
100°C	0.92	0.92
200°C	0.96	0.96
300°C	1.05	1.05
400°C	1.09	1.09
Thermal conductivity, W/m-°C, at		
0°C	2.86	2.86
50°C	2.70	2.70
100°C	2.56	2.56
150°C	2.44	2.44
200°C	2.34	2.34
300°C	2.15	2.15
400°C	1.99	1.99
Hydrologic properties		
Horizontal permeability, 10^{-8} m/min	3.05	3.05
Vertical permeability, 10^{-8} m/min	3.05	3.05

SOURCE: Reference 155 as quoted in Reference 151.

TABLE 8.9 Properties of Intact Granite

	Mean Value	Range
Index properties		
Unit weight, kg/m^3	2646.5	2306.9–3043.8
Natural moisture content (intact), %	–	0–0.32
Porosity (rock-mass), %	1.6	0.05–11.2
Stress-strain properties		
Young's modulus, GPa	50.3	15.9–83.4
Poisson's ratio	0.18	0.045–0.39
Strength properties		
Uniaxial compressive strength, MPa	175.1	35.2–353.1
Tensile strength, MPa	6.3	3.4–55.9
Thermal properties		
Coefficient of linear thermal expansion, $°C^{-1}$	2.5×10^{-6}	1.67×10^{-6} to 3.34×10^{-6}
Heat capacity, J/g-°C	0.92	0.67–1.38
Thermal conductivity, W/m-°C, at		
0°C	2.86	
50°C	2.70	
100°C	2.56	
150°C	2.44	
200°C	2.34	
300°C	2.15	
400°C	1.99	

SOURCE: Reference 155 as quoted in Reference 151.

tuffs have a high content of hydrous silicate minerals (zeolites) and a low density and high permeability and porosity. They are, however, very effective in retarding water-borne movement of radioactivity leaked from a repository. The hope is to find a thick deposit of welded tuff as the repository site, located between layers of the zeolitic type to act as a buffer against radioactivity movement.

8.3 NUCLIDE MIGRATION THROUGH ROCK

While ion filtration and species solubility will act to some extent as geochemical barriers to the migration of radionuclides from a repository, the most important factor is probably ion exchange, that is, the degree of sorption on the accessible rock surface. The effectiveness of sorption for a particular ion is generally expressed as the ion-exchange distribution coefficient K_d (sometimes seen as R_d), defined as

$$K_d = \frac{(\text{g of sorbed nuclide})/(\text{g of solid})}{(\text{g nuclide remaining in solution})/(\text{ml solution})}$$

The higher the K_d, of course, the more of the radionuclide is retained on the rock.

Ion-exchange K_d's are usually determined by the static method—a known amount of an element dissolved in a known amount of solution is equilibrated with a known amount of crushed rock. After equilibrium has been reached, analysis is made for the amount of element remaining in solution and the amount on the rock obtained by difference. The exchange process is, however, affected by many variables, and values reported in the literature for supposedly similar systems can vary considerably.

The obvious importance of being able to estimate the rate at which a radionuclide might leak from a repository has in recent years engendered a tremendous amount of research on the migration problem. The first volume of the survey by Isherwood[151] includes a thorough review of reported results through 1979.

The most critical parameter in determining the migration rate (save for the chemistry of the ion itself) is possibly the composition of the groundwater or brine associated with the repository rock or salt. Table 8.10 shows the composition of some brines, and Table 8.11 of some groundwaters. The brine data illus-

TABLE 8.10 Chemical Analyses of Bittern Brines—Concentrations in mg/L

	Michigan	Mississippi	New Mexico	Utah
Na	28,000	79,000	44,400	18,800
K	9,000	7,080	30,250	5,990
Ca	80,000	34,000	500	52,700
Mg	16,000	3,920	62,600	39,200
Sr	2,000	1,520	–	2,000
Cl	250,000	198,700	251,500	241,000
Br	–	2,040	530	3,080
SO_4	–	176	3,300	4
TDS	385,000	326,714[a]	395,090[b]	366,608[c]

SOURCE: Reference 77g.

[a]Includes: 17 SiO_2, 79 Fe, 47 Mn, 6 Zn, 25 Ba, 49 Li, 55 NH_4. Density 1.220.

[b]Includes: 1170 HCO_3 and 840 BO_3.

[c]Includes: 66 Al, 750 Fe, 260 Mn, 6 Pb, 60 Zn, 8 Ba, 6 Cu, 0.4 Ag, 66 Li, 20 Rb, 849 NH_4, 1,010 HCO_3, 25 F, 42 I, 660 B. Density 1.331.

TABLE 8.11 Groundwater Compositions (mg/L)

Solute	Basalt	Granite	Shale	Tuff
Alkalinity (as $CaCO_3$)	146	140	530	98
Calcium	<0.1	300	100	10
Magnesium	<1	3	50	3
Sodium	300	300	700	50
Silica	100	10	10	70
Chloride	140	73	61	7
Fluoride	52	0.8	0.1	2.3
Sulfate	75	980	2000	19

SOURCE: With permission of the American Association for the Advancement of Science. Copyright 1983 by AAAS. Reference 166.

trate a feature applicable to all types of waters—there are wide differences in composition, depending on the source of the samples.

In any but trace amounts, competition exists between ions for the same exchange sites on the rock, affecting the K_d's. (Trace levels are hard to define—Figure 8.5 shows the effect of concentration down to 10^{-10} M.) In monovalent/monovalent ion competition, the K_d of the minor ion is roughly inversely proportional to the concentration of the major univalent competing ion.[168] Simi-

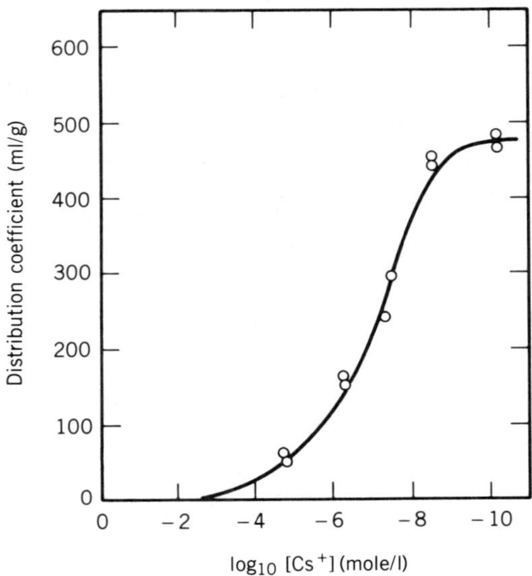

FIGURE 8.5. Cs absorption on basalt. (Reference 167, quoted in Reference 151.)

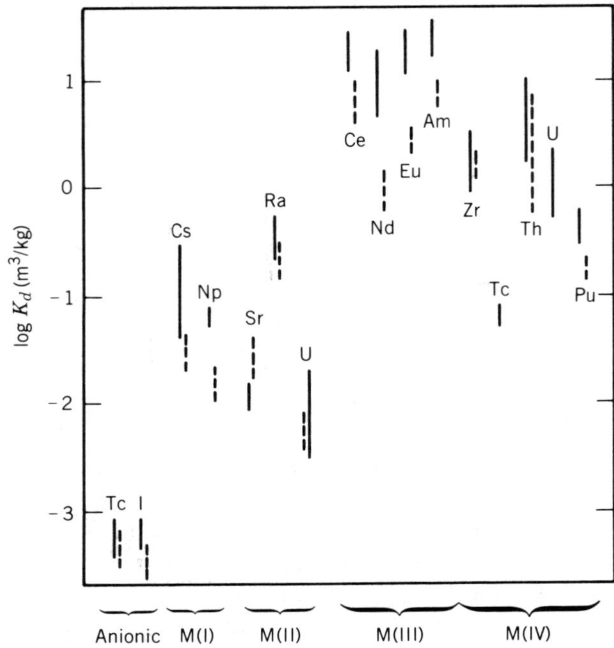

FIGURE 8.6. Mass related distribution coefficients (log K_d) for granite (solid lines) and bentonite quartz (dashed lines). (Reference 77h.)

TABLE 8.12 Average Sorption Ratios, R_d (ml/g)

Material	Element	22°C	70°C
Quartz monzonite	Sr	20	40
	Tc(VII)	<80	<16
	Cs	440	1,440
	Ba	160	730
	Ce	740	470
	Eu	960	540
	U(VI)	9	
	Pu	1,300	3,600
	Am	2,600	6,600
Argillite	Sr	130	290
	Tc(VII)	<40	<3
	Cs	2,500	1,900
	Ba	4,200	18,000
	Ce	>40,000	13,000
	Eu	>50,000	22,000

TABLE 8.12 (Continued)

Material	Element	22°C	70°C
Alluvium	Sr	200	
	Cs	7,000	
	Ba	5,000	
	Ce	>20,000	
	Eu	>20,000	
	U(VI)	10	
	Pu	>1,000	
Vitric tuff	Sr	13,000	14,000
	Cs	15,000	18,000
	Ba	5,000	50,000
	Ce	40	40
	Eu	30	80
	Pu	170	
	Am	170	220
Devitrified tuff	Sr	60	110
	Cs	120	100
	Ba	400	1,000
	Ce	80	80
	Eu	90	200
	Pu	110	
	Am	110	100
Zeolitized tuff	Sr	240	1,000
	Cs	600	1,400
	Ba	750	3,700
	Eu	6,000	4,000
	Pu	280	
	Am	590	700

SOURCE: Reference 77i.

larly, in monovalent/divalent ion competition, such as between Na^+ and Sr^{2+}, the divalent ion K_d is inversely proportional to the square of the monovalent ion concentration.[151]

Figure 8.6 and Table 8.12 present some K_d values for different radionuclides with several rock types. Figure 8.7 gives only Pu data, and is of interest because the results are from a cooperative program wherein nine different laboratories determined K_d's of cesium, strontium, and plutonium using common rock samples, standard solutions, and prescribed experimental methods. Consistent results were obtained for both Cs and Sr in brine for both limestone and basalt, and for basalt in basalt groundwater. There were wide variations for those two

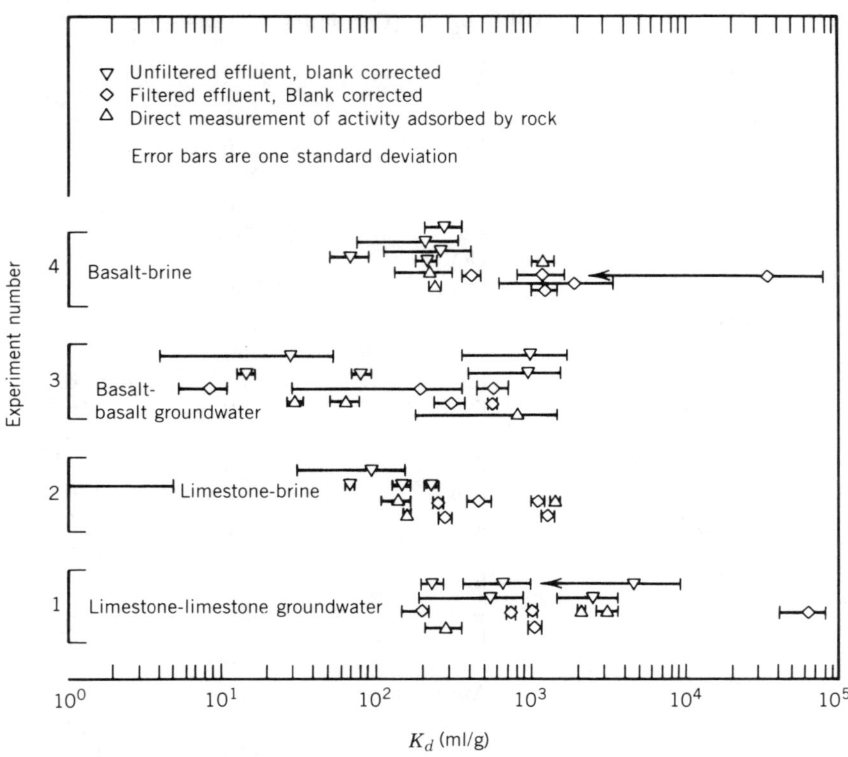

FIGURE 8.7. Pu absorption on basalt and limestone. (Reference 169, quoted in Reference 151.)

TABLE 8.13 Representative Batch Distribution Ratios for Backfill Candidates

	Distribution Ratio (R_d)							
Material	^{85}Sr	^{137}Cs	^{241}Am	^{237}Np	^{125}I	^{233}U	^{99}Tc	Final pH
Coconut charcoal	979	8	1.2E5	583	8.3E3	290	1.5E4	8.9
Iron (powder)	320	9	1.2E5	710	2.4	9.2E3	2.2E4	8.7
13-X, powder (Faujacite)	6.8E4	1.8E3	35	18	1.2	194	0.8	9.4
4A, powder (type A)	2.0E4	1.0E3	–	11	0.4	–	0.8	10.1

TABLE 8.13 (Continued)

Material	Distribution Ratio (R_d)							Final
	^{85}Sr	^{137}Cs	^{241}Am	^{237}Np	^{125}I	^{233}U	^{99}Tc	pH
Calcium Bentonite	865	1.1E5	9.2E3	78	1.5E-3	76	0.9	8.2
Sodium Bentonite	6.8E3	1.0E3	1.4E3	29	1.4E-3	8	–	8.6
AW-400 (erionite)	2.3E4	6.7E4	92	13	0.8	7	1.2	8.5
AW-500 (chabazite)	2.3E3	7.8E4	1.3E3	18	1.5	17	0.9	8.4
Zeolon-900 (mordenite)	1.5E4	8.3E4	24	12	1.8	35	1.6	8.6

SOURCE: Reprinted by permission of the publisher from *Scientific Basis for Nuclear Waste Management*, S. V. Topp, ed., pp. 329–336. Copyright 1982 by Elsevier Publishing Co. Inc. Reference 170.

elements in limestone water. In the plutonium case, the values obtained by the various investigators varied by two to three orders of magnitude for all of the systems studied.

Table 8.13 shows distribution data for some of the most important waste radionuclides when equilibrated with various materials under consideration as backfill in a multibarrier isolation scheme. The equilibration liquid used in this study was an artificial basalt groundwater having an initial pH of 9.85.[170]

PART FOUR

DATA FOR OPERATIONS

A number of topics having to do with day-by-day handling of radioactivity are touched upon in this part. The intent is to furnish quick reference information with a minimum of explanatory text. The areas are broad and generally complex, so citations are given to major literature sources in the event that a special situation requires more detail.

NINE

SHIELDING

9.1 THE SHIELDING LITERATURE

Shielding against radiation has been a major investigative effort over the last 40-plus years, and the available literature is extensive. The most comprehensive single source to date is probably the three volumes comprising the *Compendium on Radiation Shielding*.[21] Manuals have also been prepared by Rockwell,[171] Blizard,[142] Steigelmann,[172] Arnold,[173] Schaeffer,[174] Courtney,[175] and others. Fundamental and applied aspects of shielding are considered in textbooks such as Glasstone and Sesonski,[15] and in specialized multisubject and multivolume surveys such as the *Reactor Handbook*.[143] The Oak Ridge Radiation Shielding Information Center[44] is available for consultation for complicated situations.

9.2 SHIELDING TERMINOLOGY

Table 9.1 summarizes the relationships between the most frequently encountered shielding terms. These all derive from the basic exponential equation for radiation absorption:

$$I = I_0 \, e^{-\mu x} \tag{9.1}$$

where I_0 is the initial intensity of the radiation and I the intensity at depth x into the shield. The Greek μ is the absorption coefficient for the radiation in

TABLE 9.1 Shielding Terms

Quantity	Symbol	Unit	Relationships
Cross section	σ	cm^2	
Linear-absorption coefficient	μ	cm^{-1}	$\mu = n\sigma$
Density	ρ	g/cm^3	
Mass-absorption coefficient	μ_m	cm^2/g	$\mu_m = \mu/\rho$
Mean free path	mfp	cm	mfp = $1/\mu$ $I/I_0 = 0.367$
Relaxation length	λ	cm	$\lambda = 1/\mu$ $I/I_0 = 0.367$
Half-value layer	HVL	cm	HVL = $0.693/\mu$
Tenth-value layer	TVL	cm	TVL = $2.303/\mu$ TVL = 3.323 × HVL

SOURCE: Reference 176.

the material making up the shield. Since μ is usually expressed in terms of inverse centimeters, x must also be in centimeters.

This basic equation can be used with reasonable accuracy for calculating beta shielding and for gamma shielding if the shielding thickness is small, and particularly if the radiation beam is highly collimated. In most situations with gammas, however, the beam is not collimated, and there are numerous scattering episodes in the shield that result in an increased dose delivery outside the shield. This situation is covered by introduction of the buildup factor B:

$$I = BI_0 \, e^{-\mu x} \tag{9.2}$$

Most of the terms in Table 9.1 are defined there. The half-value layer HVL is the thickness of shield needed to cut the intensity in half, and the tenth-value layer TVL is the thickness needed to reduce the radiation by a factor of 10.

9.3 SHIELDING AGAINST ALPHAS AND BETAS

Both alphas and betas are charged particles and are stopped very rapidly by passage through matter. The ranges of isotopically generated alphas are less than the thickness of the human skin, so shielding is not a problem. The hazard with alphas develops if the generating nuclides are taken into the body, since most of

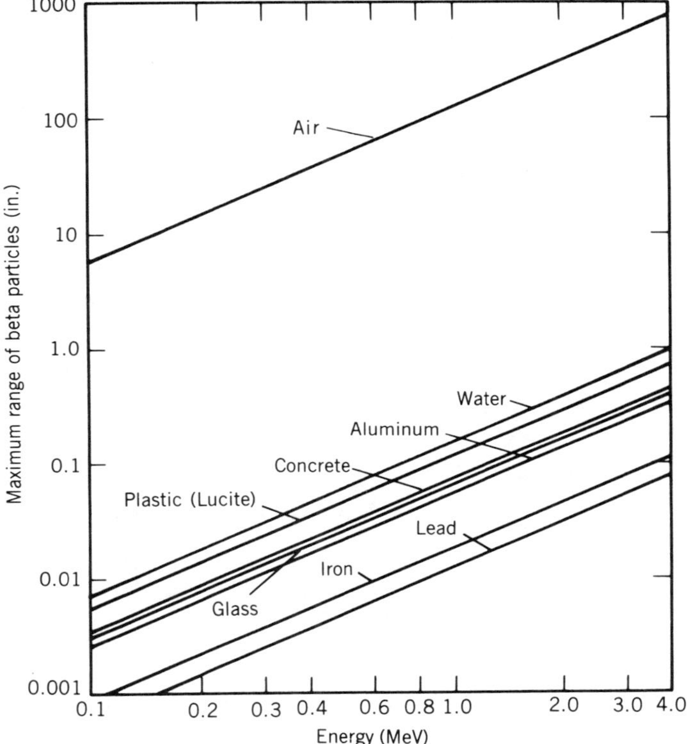

FIGURE 9.1. Beta radiation ranges. (Reference 177.)

these are very slowly eliminated, and continue to produce severe local damage over a lifetime in the bone or tissue in which they are deposited.

The maximum range of a beta particle in matter is roughly inversely proportional to the density of the absorbing material through which it passes. Figure 9.1 shows the range of betas with energies up to 4 MeV in a number of common materials. Very few isotopically produced betas have energies in excess of 3 MeV. The figures shows that 0.1 inch or less of almost any material provides adequate protection in these lower-energy ranges.

A complication in the shielding of betas is the production of *bremsstrahlung*, or x-rays created as the electron is slowed by its passage through matter. The radiation hazard from the ^{90}Sr–^{90}Y pair and from ^{147}Pm is essentially due to this braking radiation. Table 9.2 shows that normally 1% or less of the beta's energy is converted to bremsstrahlung, although this is partially dependent on the composition of the shield.

TABLE 9.2 Bremsstrahlung Production

Nuclide	Beta-Particle-Energy Maximum (MeV)	Beta-Particle-Energy Average (MeV)	Bremsstrahlung (MeV/beta)
^{106}Rh	3.54	1.515	1.29×10^{-1}
^{90}Y	2.27	0.944	2.81×10^{-2}
^{90}Sr	0.545	0.201	1.41×10^{-3}
^{147}Pm	0.23	0.067	2.02×10^{-4}
^{171}Tm	0.097	0.029	3.13×10^{-5}

SOURCE: Appendix 11 of Reference 34.

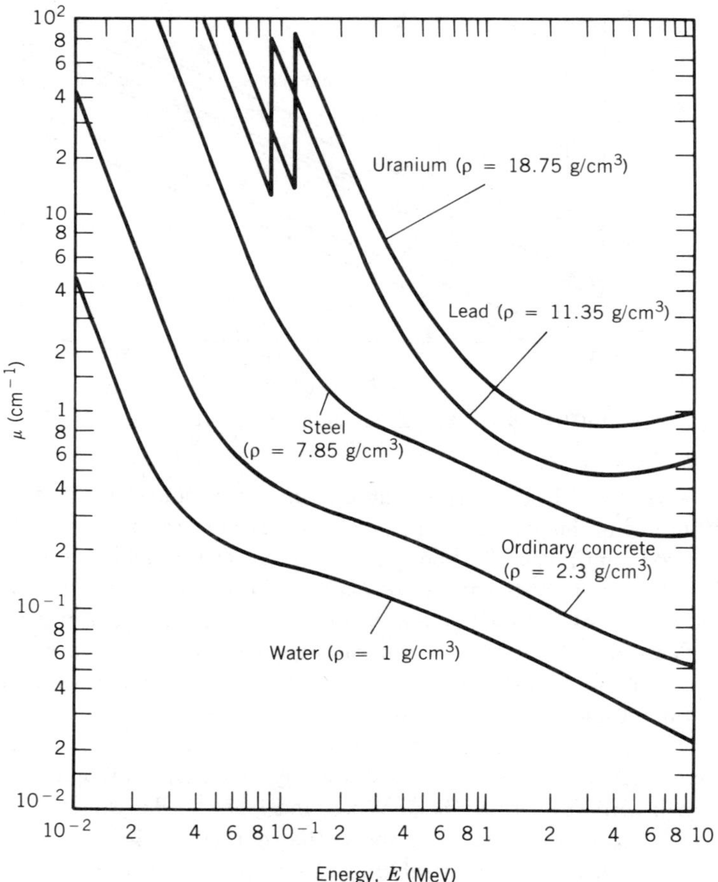

FIGURE 9.2. Gamma linear absorption coefficients. (Reference 172.)

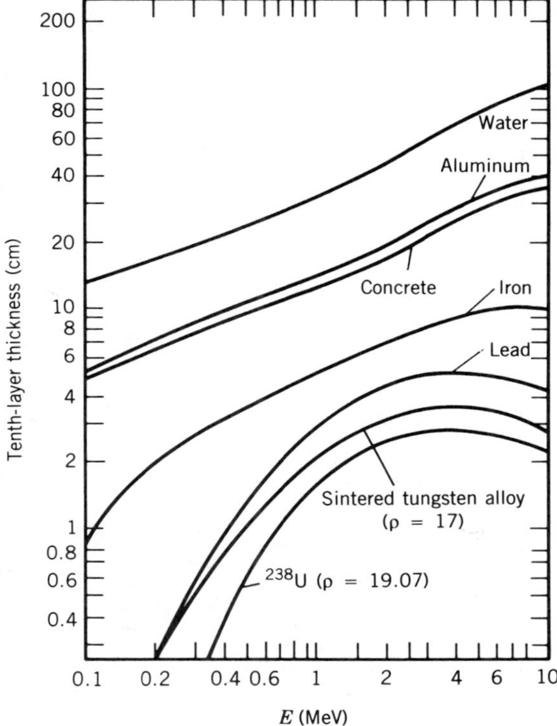

FIGURE 9.3. Tenth value layers for gammas. (Reference 174.)

9.4 SHIELDING AGAINST GAMMA RADIATION

Gammas, x-rays, and neutrons carry no charge, in contrast to alphas and betas. They are, therefore, highly penetrating, requiring a direct collision with an atom in the shield, rather than being slowed by interaction with the electrical fields of the atoms in the material being traversed. In the case of gammas, as stated above, the presence of scattering reactions also makes it necessary to introduce buildup factors into the basic absorption equations in most situations.

Figure 9.2 gives linear-absorption coefficients, and Figure 9.3 presents tenth-value-layer data for the most commonly used shielding materials. Both sets of curves are for the narrow-beam situation.

Buildup data are given in the "Radiological Health Handbook"[20] in tabular form and as graphs by Steigelmann.[172] The latter author's curves for water, ordinary concrete, iron, and lead are reproduced here as Figures 9.4–9.7. (Steigel-

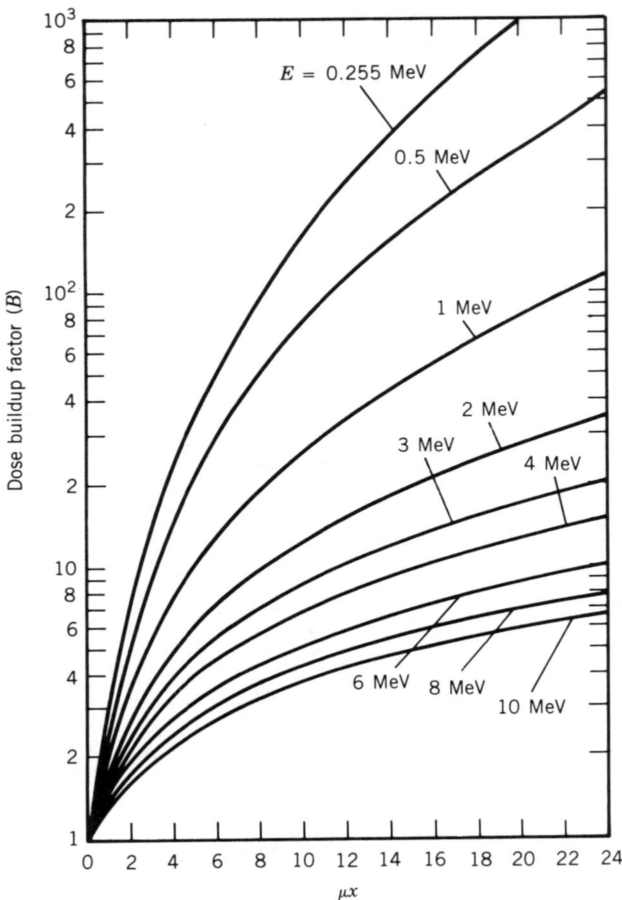

FIGURE 9.4. Buildup factors in water. (Reference 172.)

mann also gives curves for uranium metal and for lead for gammas having energies over 3 MeV. The *Radiological Health Handbook* has additional data for aluminum, tin, tungsten, and uranium.)

Buildup factors vary with both the composition of and the penetration depth into the shield. The abscissa values in the graphs are in terms of μx, in order to cover these variables, where μ is the absorption coefficient and x is the penetration depth in centimeters. The data are for shielding of a point isotropic source.

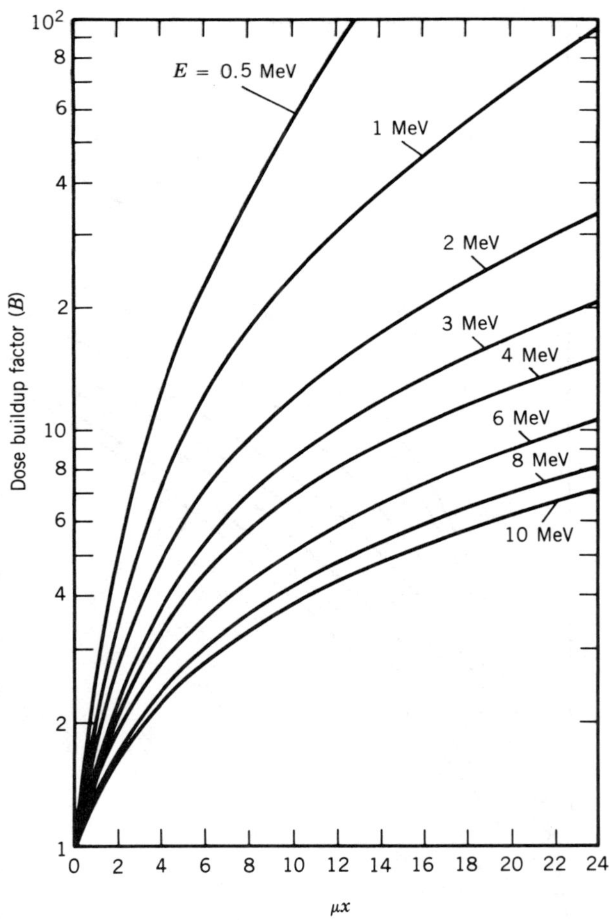

FIGURE 9.5. Buildup factors in ordinary concrete. (Reference 172.)

9.5 SHIELDING AGAINST NEUTRONS

Isotopically produced neutrons are usually generated with energies in the 1–10 MeV range. These fast neutrons have very small capture cross sections (Table 1.10), and must be slowed to the thermal range (0.025 eV at 22°C) where cross sections increase dramatically. Slowing of neutrons above the 0.1–0.5 MeV range is best accomplished by inelastic scattering processes in heavy metal shields. Below that threshold, energy is best reduced by elastic scattering, with materials

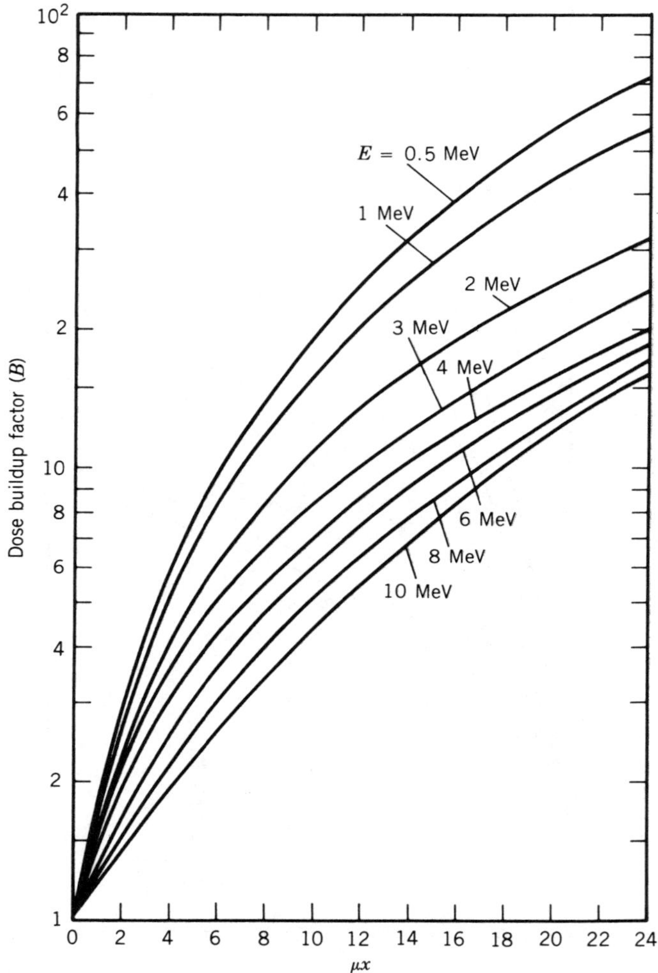

FIGURE 9.6. Buildup factors in iron. (Reference 172.)

with high hydrogen content being the most effective. Composite shields are often used because of these conflicting requirements.

Water contains 11.1% H by weight, roughly 6.7×10^{22} atoms/cm^3. Table 9.3 shows comparable data for other high-hydrogen-content materials.

Calculations for neutron shielding are very complex, and there is no relatively simple equation, graph, or data table available for application to all situations.

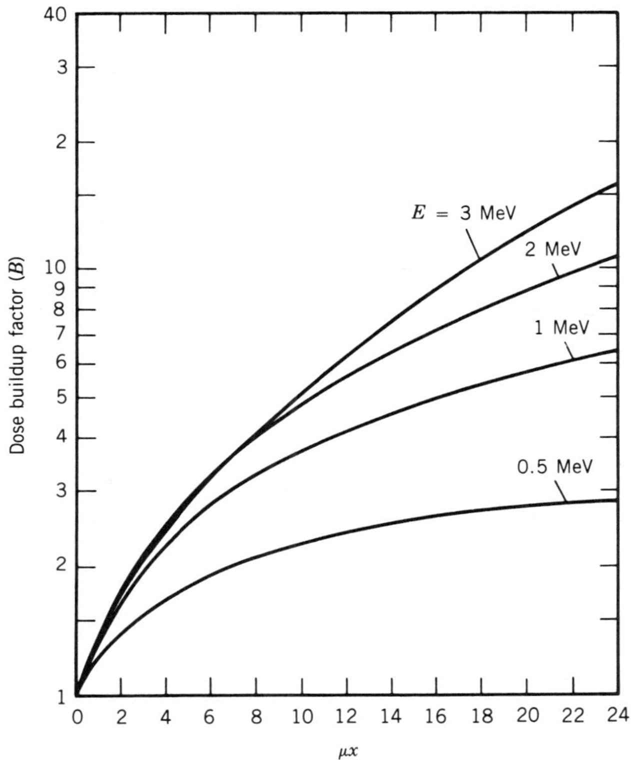

FIGURE 9.7. Buildup factors in lead. (Reference 172.)

TABLE 9.3 High-Hydrogen-Content Materials

Material	Density	Formula	Hydrogen Density (10^{22} atoms/cm^3)
Butyl rubber	0.96	$(C_4H_8)n$	7.90
Lucite	1.19	$(C_5H_8O_2)_n$	5.73
Nylon	1.11	$(C_{12}H_{22}O_2N_2)_n$	6.56
Polystyrene	1.05	$(C_8H_8)_n$	4.86
Polyethylene	0.93	$(CH_2)_n$	8.02
Natural rubber	0.93	$(C_5H_8)_n$	6.54

SOURCE: Reference 21.

TABLE 9.4 Shielding Materials—Density Data

Material	g/cm^3	lb/ft^3	Material	g/cm^3	lb/ft^3
Water	1.0	62.43	Concretes		
Metals, alloys			Barytes	3.5	219
Aluminum	2.70	169	Ferrophosphorus	4.8	300
Copper	8.96	556	Hematite	4.0	250
Iron	7.86	491	Ilmenite	3.8	237
Bagged shot	2.2	137	Iron shot	6.0	375
Cast iron	7.25	452	Magnetite	3.7	231
Pig, gray	6.7–7.6	417–453	Ordinary	2.3–2.5	144–156
Pig, white	7.0–7.8	436–486	Serpentine	2.1	130
Steel, ingots	7.87	491	Portland cement	0.82–1.95	51–121
Lead	11.4	718	Rocks, Minerals		
Magnesium	1.74	109	Basalt	2.7–3.2	169–200
Nickel	8.9	556	Dolomite	2.84	177
Uranium	19.1	1190	Granite	2.64–2.76	165–172
Zinc	7.14	446	Limestone	2.46–2.84	154–177
Zirconium	6.49	405	Marble	2.60–2.84	160–177
Brass	8.4	524	Quartz	2.65	165
Hevimet	17.1	1068	Sandstone	2.2–2.5	137–156
Mallory 1000	17.0	1061	Shale	1.54–2.81	96–175
Sintered W alloy	17.0	1061	Slate	2.6–3.3	162–205

Barite	4.3–4.6	268–287	Masonry brick, fresh	1.57–1.63	98–102
Borax	1.7	106	Ceramic tile	1.92	120
Galena	7.3–7.6	460–470	Cinder block	1.49	93
Hematite	4.9–5.3	306–330	Gypsum plaster	1.63	102
Halite	2.1–2.6	131–162	Linoleum	1.18	74
Ilmenite	4.5–5.1	281–318	Wood	0.31–1.33	19–83
Limonite	3.6–4.0	225–250	Fir	0.53	33
Magnetite	5.16–5.18	322–323	Oak	0.6–0.9	37–56
Serpentine	2.5–2.65	156–165	Pine	0.35–0.6	22–37
			Viewing Materials		
Soil Materials			Glass		
Clay	1.8–2.6	112–162	Flint	3.15–3.90	197–243
Earth, loamy, dry	1.6–1.9	100–116	Shielding	2.53–6.20	158–387
Gravel	1.8–2.0	112–125	Window	2.4–2.6	150–162
Loam, dry	1.5–1.6	92–100	$ZnBr_2$ solution, 78.4 w/o	2.54	158
Loam, fresh	1.7–1.85	104–115			
Sand, fine, dry	1.40–1.65	87–103	Mineral oil	0.86	54
Sand, fine, moist	1.90–2.05	118–128	Miscellaneous		
Building Materials			Asphalt	1.1–1.5	69–94
Fire brick	1.85–2.2	115–137	Slag	2.0–3.9	125–240
Masonry brick, dry	1.42–1.46	89–91	Blast furnace	2.5–3.0	156–187

SOURCES: Varied, please see text.

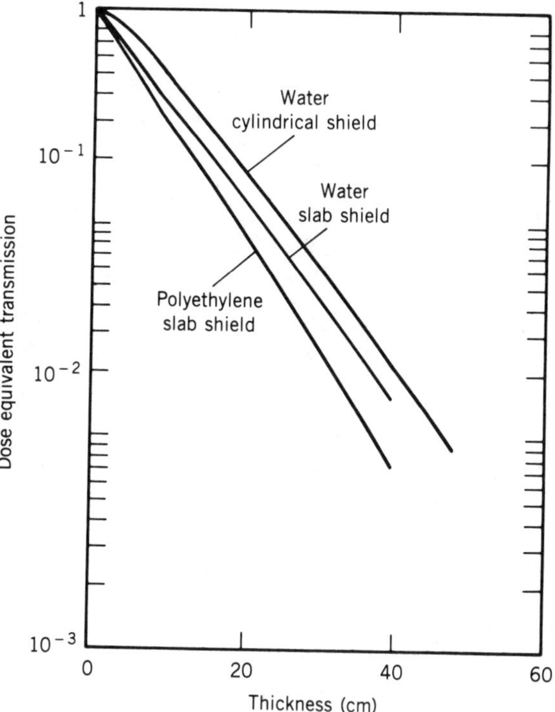

FIGURE 9.8. Shielding ^{241}Am-Be alpha-n neutrons. (Reference 34.)

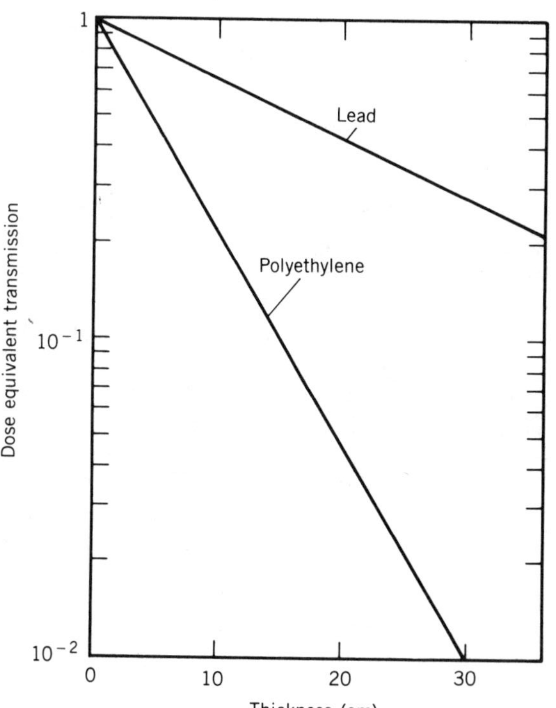

FIGURE 9.9. Shielding ^{252}Cf neutrons. (Reference 34.)

Figure 9.8 shows the transmission factors for ^{241}Am-Be alpha-*n* neutrons through water and polyethylene shields, and Figure 9.9 gives similar information for neutrons generated by the spontaneous fission of ^{252}Cf. Both figures are for broad-beam dose equivalent transmission.

9.6 SHIELDING MATERIALS

While the statement is an oversimplification, the effectiveness of a substance as a gamma shield is roughly proportional to its density. Table 9.4 gives density data for a number of materials. Ranges are given in many cases, and undoubtedly exist for others (the heavy concretes, for example).

Sources used include References 9, 10, 21, 116, 151, and 176.

TEN

HEALTH PHYSICS

10.1 BACKGROUND AND LITERATURE

The development of formal radiation protection standards has been an evolutionary process that is still continuing. The most influential of the agencies bringing about this development have been the International Commission on Radiation Protection (ICRP), the International Commission on Radiological Units and Measurements (ICRU), and the U.S. National Council on Radiation Protection (NCRP). The recommendations made by the specialist members of these commissions eventually become codified as legal standards, which, in the case of the United States, appear as Part 20 of Title 10 of the Code of Federal Regulations (10CFR20).

The reports prepared by the NCRP and ICRU are currently published from their shared headquarters in Washington, DC[179] whereas the ICRP's recommendations generally appear in hardback from Pergamon Press. To a considerable degree, the publications of these groups are the core of the radiological health physics literature. A few of the numerous available textbooks are cited,[180, 181, 182, 183] and a multipage summary listing of health physics sources has been published.[184] The "Radiological Health Handbook"[20] provides much useful information, as do many of the volumes in the Safety and Technical Series of books prepared by the International Atomic Energy Agency (IAEA).[185]

10.2 HEALTH PHYSICS TERMS

The ICRU in its Report 19[186] lists, defines, and explains the units and terms used in Health Physics application and research. Discussion of a few of those most commonly seen is given here.

Exposure Unit. This is the roentgen, defined as "the exposure of x- or gamma-radiation such that the associated corpuscular emission per kilogram of air produces, in air, ions carrying 2.58×10^{-4} coulombs of energy of either sign." The term applies only to x- or gamma-radiation and says nothing concerning the amount of energy absorbed by the target or of subsequent biological effects. The ICRU exposure unit is termed "X."

Absorption Unit. This is the rad (radiation-absorbed-dose) and is defined as the amount of energy imparted to matter by ionizing radiation per unit mass of irradiated material at the point of interest. One rad equals the absorption of 100 ergs of energy per gram. It is a broader term than the roentgen since it applies to any type of radiation and any type of material. The ICRU symbol is "D."

Figure 10.1 shows the dose rate produced by a one-curie point source of beta radiation, and Figure 10.2, the exposure rate from a similar one-curie source of gammas.

Dose Equivalent. While the absorbed dose quantifies the amount of energy imparted to a target, it still says nothing concerning biological effects. The unit in this case is the rem (roentgen-equivalent-man). The ICRU definition follows: "The dose equivalent, H, is the product of D, Q and N at the point of interest in tissue where D is the absorbed dose, Q is the quality factor and N is the product of other modifying factors:

$$H = DQN." \tag{10.1}$$

D is expressed in rads. N is a catchall factor that allows for unusual situations such as an organ receiving an uneven dose of radiation. Q, the quality factor, adjusts for the type of radiation and its energy and is determined by the rate of linear-energy transfer (LET) in the irradiated tissue. In cases where available information is incomplete, a good approximation for Q, is 1 for x-rays, gammas, and betas; 3 for thermal neutrons; 10 for fast neutrons, protons and isotope-produced alphas; and 20 for heavy recoil nuclei.[188] A more exact conversion of gamma and neutron fluxes to mrem/hr is shown in Figure 10.3.

Activity. The most widely used unit for expressing activity is the curie, defined as 3.7×10^{10} disintegrations per second.

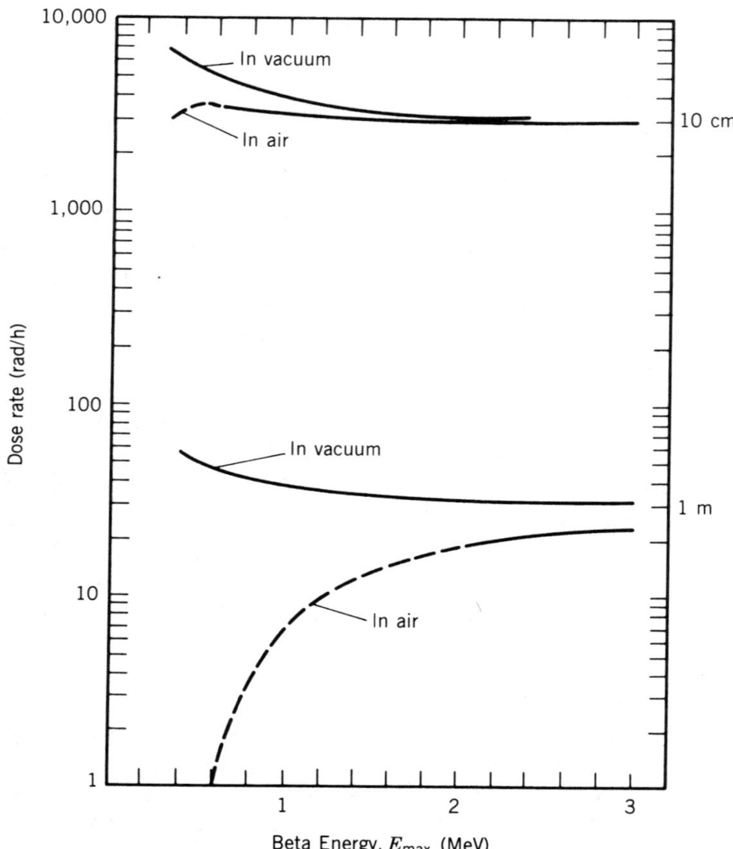

FIGURE 10.1. Dose rate from a 1-curie beta source. (Reference 187.)

10.2.1 SI Units

The above definitions developed over the last 50–60 years, and are still those in most general use. There is, however, considerable pressure for adoption of the International System of Units (abbreviated SI) generated by the Conférence Générale des Poids et Mesures as a coherent international set of definitions to cover all science and engineering, based on seven base units (meter, kilogram, second, ampere, kelvin, mole, and candela) plus two supplementary units, (radian and steradian). All other units are to be derived from these bases by simple multiplication, division, or the use of exponents. Thus, the unit of force becomes the newton (N) or kg · m/sec^2.

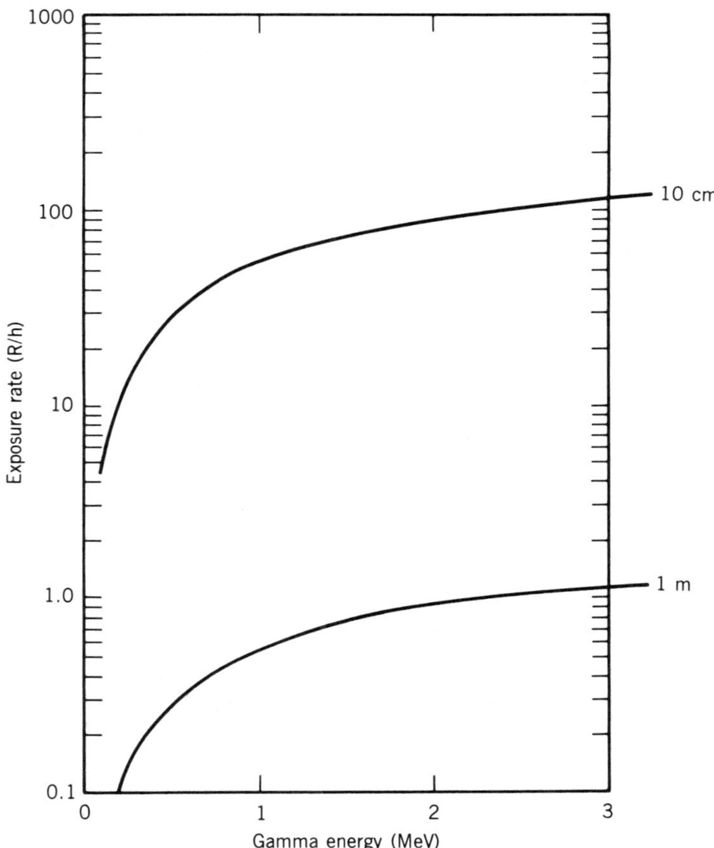

FIGURE 10.2. Exposure rate from a 1-curie gamma source. (Reference 187.)

Adoption of the system would affect the health physics units discussed in the previous section, the most startling change being in the unit of activity, which, rather than the curie, would become the becquerel (Bq), defined as 1 dis/sec. The absorbed dose would be expressed as joules/kg, termed the gray (Gy) and equivalent to 100 rads. The rem would change equally in magnitude and be termed the Sievert (Sv).

The ICRU accepted these recommendations in 1974, but asked for a 10-year transition period during which the old terms could still be used. That period of grace has now expired, but the older terms still seem to predominate in the literature.

Table 10.1 gives the relationships between present and SI units.

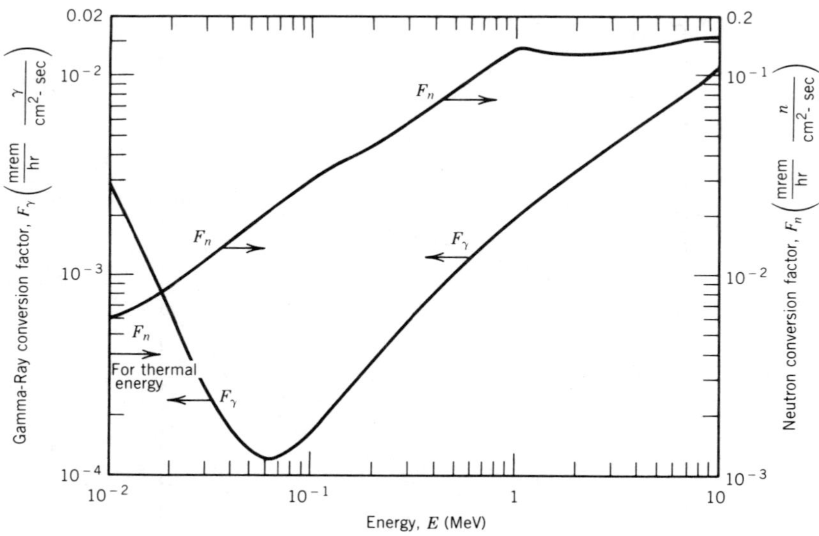

FIGURE 10.3. Gamma and neutron dose conversion factors. (Reference 172.)

TABLE 10.1 SI Units and Their Present Equivalents

Quantity	Derived SI Unit	SI Name and Symbol	Present Equivalent
Activity	Per second (sec^{-1})	Becquerel (Bq)	2.703×10^{-11} Ci
Absorbed dose	Joules/kilogram ($J \cdot kg^{-1}$)	Gray (Gy)	100 rad
Absorbed dose rate	Watts/kilogram ($W \cdot kg^{-1} = J \cdot kg^{-1}\ sec^{-1}$)	Gy/sec	100 rad/sec
Exposure	Coulombs/kilogram ($C \cdot kg^{-1}$)	—	3.876×10^{3} R
Exposure rate	Amperes/kilogram ($A \cdot kg^{-1} = C \cdot kg^{-1} \cdot sec^{-1}$)	—	3.876×10^{3} R/sec

SOURCE: Modified from Reference 189.

10.3 PROTECTION STANDARDS

Standards for protection against radiation are of two types, limits on exposure of the external parts of the body, and limits designed to prevent dangerous accumulation of radioactivity within the body.

Table 10.2 shows the recommendations developed by the NCRP for protection against external radiation. Most of these standards appear in 10CFR20, but not in such concise form. Differentiation between individuals exposed to radia-

TABLE 10.2 NCRP Exposure Recommendations

Occupational exposure limits	
Whole body (prospective)	5 rems in any one year
Whole body (retrospective)	10–15 rems in any one year
Whole body (accumulation to age N yr)	$(N - 18) \times 5$ rems
Skin	15 rems in any one year
Hands	75 rems in any one year (25/qtr)
Forearms	30 rems in any one year (10/qtr)
Other organs, tissues, and organ systems	15 rems in any one year (5/qtr)
Fertile women (with respect to fetus)	0.5 rem in gestation period
Dose limits for the public, or occasionally exposed individuals	
Individual or occasional	0.5 rem in any one year
Students	0.1 rem in any one year
Population dose limits	
Genetic	0.17 rem average per year
Somatic	0.17 rem average per year
Emergency dose limits—lifesaving	
Individual (older than 45 yr if possible)	100 rems
Hands and forearms	200 rems, additional (300 rems total)
Emergency dose limits—less urgent	
Individual	25 rems
Hands and forearms	100 rems, total
Family of radioactive patients	
Individual (under 45 yr)	0.5 rem in any one year
Individual (over 45 yr)	5 rems in any one year

SOURCE: NCRP Report 39, Reference 190.

tion as part of their jobs and the general public is handled in 10CFR20 by establishing rules for "restricted" (nuclear facilities) and "unrestricted" areas.

Table 10.3 deals with the hazard of introducing radioactive species into the body. The table was prepared from 10CFR20 as being the legal standards set by the NRC. While some of the numbers vary slightly (on the conservative side), they are essentially the same as those initially proposed by the ICRP[191] and the NCRP.[192]

The listings given by the original sources are much more elaborate than Table 10.3, which considers only the soluble forms of each isotope and lists only about one-quarter of the nuclides included in 10CFR20. The more official sources also give MPC values for insoluble forms, and ICRP and NCRP generally specify several limits for each isotope, depending on the organs (bone, GI tract, whole body, etc.) considered as being particularly susceptible to exposure to that particular nuclide.

The Maximum Permissible Concentration values in Table 10.3 are established

TABLE 10.3 Selected MPC Values (Above Natural Background)

Nuclide	Half-Life (Days)	Radiation Workers Air (μCi/ml)	Radiation Workers Water (μCi/ml)	General Public Air (μCi/ml)	General Public Water (μCi/ml)
Ac-227	8.00E 03	2E-12	6E-05	8E-14	2E-06
Am-241	1.70E 05	6E-12	1E-04	2E-13	4E-06
Am-242m	5.60E 04	6E-12	1E-04	2E-13	4E-06
Am-243	2.90E 06	6E-12	1E-04	2E-13	4E-06
Sb-124	6.00E 01	2E-07	7E-04	5E-09	2E-05
Sb-125	8.77E 02	5E-07	3E-03	2E-08	1E-04
A-41	7.63E-02	2E-06	—	4E-08	—
As-74	1.75E 01	3E-07	2E-03	1E-08	5E-05
As-76	1.10E 00	1E-07	6E-04	4E-09	2E-05
Ba-131	1.16E 01	1E-06	5E-03	4E-08	2E-04
Ba-140	1.28E 01	1E-07	8E-04	4E-09	3E-05
Bk-249	2.90E 02	9E-10	2E-02	3E-11	6E-04
Be-7	5.36E 01	6E-06	5E-02	2E-07	2E-03
Bi-207	2.90E 03	2E-07	2E-03	6E-09	6E-05
Bi-210	5.00E 00	6E-09	1E-03	2E-10	4E-05
Br-82	1.50E 00	1E-06	8E-03	4E-08	3E-04
Cd-109	4.75E 02	5E-08	5E-03	2E-09	2E-04
Cd-115m	4.30E 01	4E-08	7E-04	1E-09	3E-05
Ca-45	1.64E 02	3E-08	3E-04	1E-09	9E-06
Ca-47	4.90E 00	2E-07	1E-03	6E-09	5E-05
Cf-249	1.70E 05	2E-12	1E-04	5E-14	4E-06
Cf-252	8.03E 02	6E-12	2E-04	2E-13	7E-06
C-14	2.00E 06	4E-06	2E-02	1E-07	8E-04
Ce-141	3.20E 01	4E-07	3E-03	2E-08	9E-05
Ce-144	2.90 02	1E-08	3E-04	3E-10	1E-05
Cs-134	8.40E 02	4E-08	3E-04	1E-09	9E-06
Cs-137	1.10E 04	6E-08	4E-04	2E-09	2E-05
Cl-36	1.20E 08	4E-07	2E-03	1E-08	8E-05
Cr-51	2.78E 01	1E-05	5E-02	4E-07	2E-03
Co-60	1.90E 03	3E-07	1E-03	1E-08	5E-05
Cu-64	5.30E-01	2E-06	1E-02	7E-08	3E-04
Cm-242	1.62E 02	1E-10	7E-04	4E-12	2E-05
Cm-244	6.70E 03	9E-12	2E-04	3E-13	7E-06
Dy-166	3.40E 00	2E-07	1E-03	8E-09	4E-05
Es-253	2.00E 01	8E-10	7E-04	3E-11	2E-05
Es-254	4.80E 02	2E-11	4E-04	6E-13	1E-05
Er-169	9.40E 00	6E-07	3E-03	2E-08	9E-05
Eu-152	4.70E 03	2E-08	2E-03	6E-10	8E-05
Eu-154	5.80E 03	4E-09	6E-04	1E-10	2E-05
Fm-255	8.96E-01	2E-08	1E-03	6E-10	3E-05

TABLE 10.3 (Continued)

Nuclide	Half-Life (Days)	Radiation Workers Air (μCi/ml)	Radiation Workers Water (μCi/ml)	General Public Air (μCi/ml)	General Public Water (μCi/ml)
F-18	7.80E -02	5E -06	2E -02	2E -07	8E -04
Gd-153	2.62E 02	2E -07	6E -03	8E -09	2E -04
Ga-72	5.90E -01	2E -07	1E -03	8E -09	4E -05
Ge-71	1.20E 01	1E -05	5E -02	4E -07	2E -03
Au-198	2.70E 00	3E -07	2E -03	1E -08	5E -05
Hf-181	4.60E 01	4E -08	2E -03	1E -09	7E -05
Ho-166	1.10E 00	2E -07	9E -04	7E -09	3E -05
H-3	4.50E 03	5E -06	1E -01	2E -07	3E -03
In-114m	4.90E 01	1E -07	5E -04	4E -09	2E -05
I-126	1.33E 01	8E -09	5E -05	9E -11	3E -07
I-129	6.30E 09	2E -09	1E -05	2E -11	6E -08
I-131	8.00E 00	9E -09	6E -05	1E -10	3E -07
Ir-192	7.45E 01	1E -07	1E -03	4E -09	4E -05
Fe-55	1.10E 03	9E -07	2E -02	3E -08	8E -04
Fe-59	4.51E 01	1E -07	2E -03	5E -09	6E -05
Kr-85	3.93E 03	1E -05	—	3E -07	—
La-140	1.68E 00	2E -07	7E -04	5E -09	2E -05
Pb-203	2.17E 00	3E -06	1E -02	9E -08	4E -04
Pb-210	7.10E 03	1E -10	4E -06	4E -12	1E -07
Lu-177	6.80E 00	6E -07	3E -03	2E -08	1E -04
Mn-52	5.55E 00	2E -07	1E -03	7E -09	3E -05
Mn-54	3.00E 02	4E -07	4E -03	1E -08	1E -04
Hg-203	4.58E 01	7E -08	5E -04	2E -09	2E -05
Mo-99	2.79E 00	7E -07	5E -03	3E -08	2E -04
Nd-147	1.13E 01	4E -07	2E -03	1E -08	6E -05
Np-237	8.00E 08	4E -12	9E -05	1E -13	3E -06
Np-239	2.33E 00	8E -07	4E -03	3E -08	1E -04
Ni-59	2.90E 07	5E -07	6E -03	2E -08	2E -04
Ni-63	2.90E 04	6E -08	8E -04	2E -09	3E -05
Nb-93m	3.70E 03	1E -07	1E -02	4E -09	4E -04
Nb-95	3.50E 01	5E -07	3E -03	2E -08	1E -04
Os-191	1.60E 01	1E -06	5E -03	4E -08	2E -04
Os-193	1.30E 00	4E -07	2E -03	1E -08	6E -05
Pd-103	1.70E 01	1E -06	1E -02	5E -08	3E -04
P-32	1.43E 01	7E -08	5E -04	2E -09	2E -05
Pt-197	7.50E -01	8E -07	4E -03	3E -08	1E -04
Pu-238	3.30E 04	2E -12	1E -04	7E -14	5E -06
Pu-239	8.90E 06	2E -12	1E -04	6E -14	5E -06
Pu-241	4.80E 03	9E -11	7E -03	3E -12	2E -04
Pu-242	1.40E 08	2E -12	1E -04	6E -14	5E -06

TABLE 10.3 (Continued)

Nuclide	Half-Life (Days)	Radiation Workers Air (μCi/ml)	Radiation Workers Water (μCi/ml)	General Public Air (μCi/ml)	General Public Water (μCi/ml)
Po-210	1.38E 02	5E-10	2E-05	2E-11	7E-07
Pr-143	1.37E 01	3E-07	1E-03	1E-08	5E-05
Pm-147	9.20E 02	6E-08	6E-03	2E-09	2E-04
Pa-231	1.30E 07	1E-12	3E-05	4E-14	3E-07
Pa-233	2.74E 01	6E-07	4E-03	2E-08	1E-04
Ra-226	5.90E 05	3E-11	4E-07	3E-12	3E-08
Rn-222	3.82E 00	3E-08	—	3E-09	—
Re-186	3.79E 00	6E-07	3E-03	2E-08	9E-05
Rb-86	1.86E 01	3E-07	2E-03	1E-08	7E-05
Ru-103	4.10E 01	5E-07	2E-03	2E-08	8E-05
Ru-106	3.65E 02	8E-08	4E-04	3E-09	1E-05
Sm-151	3.70E 04	6E-08	1E-02	2E-09	4E-04
Sc-46	8.50E 01	2E-07	1E-03	8E-09	4E-05
Se-75	1.27E 02	1E-06	9E-03	4E-08	3E-04
Ag-110m	2.70E 02	2E-07	9E-04	7E-09	3E-05
Na-22	9.50E 02	2E-07	1E-03	6E-09	4E-05
Na-24	6.30E-01	1E-06	6E-03	4E-08	2E-04
Sr-89	5.05E 01	3E-08	3E-04	3E-10	3E-06
Sr-90	1.00E 04	1E-09	1E-05	3E-11	3E-07
S-35	8.71E 01	3E-07	2E-03	9E-09	6E-05
Ta-182	1.12E 02	4E-08	1E-03	1E-09	4E-05
Tc-99	7.70E 07	2E-06	1E-02	7E-08	3E-04
Te-127m	1.05E 02	1E-07	2E-03	5E-09	6E-05
Te-132	3.20E 00	2E-07	9E-04	7E-09	3E-05
Tb-160	7.30E 01	1E-07	1E-03	3E-09	4E-05
Tl-204	1.10E 03	6E-07	3E-03	2E-08	1E-04
Th-228	7.00E 02	9E-12	2E-04	3E-13	7E-06
Th-230	2.90E 07	2E-12	5E-05	8E-14	2E-06
Th-232	5.10E 12	3E-11	5E-05	1E-12	6E-06
Tm-170	1.27E 02	4E-08	1E-03	1E-09	5E-05
Sn-113	1.12E 02	4E-07	2E-03	1E-08	9E-05
Sn-125	9.50E 00	1E-07	5E-04	4E-09	2E-05
W-185	7.40E 01	8E-07	4E-03	3E-08	1E-04
U-232	2.70E 04	1E-10	8E-04	3E-12	3E-05
U-233	5.90E 07	5E-10	9E-04	2E-11	3E-05
U-235	2.60E 11	5E-10	8E-04	2E-11	3E-05
U-238	1.60E 12	7E-11	1E-03	3E-12	4E-05
V-48	1.61E 01	2E-07	9E-04	6E-09	3E-05
Xe-133	5.27E 00	1E-05	—	3E-07	—
Yb-175	4.10E 00	7E-07	3E-03	2E-08	1E-04

TABLE 10.3 (Continued)

		Radiation Workers		General Public	
Nuclide	Half-Life (Days)	Air (μCi/ml)	Water (μCi/ml)	Air (μCi/ml)	Water (μCi/ml)
Y-90	2.68E 00	1E -07	6E -04	4E -09	2E -05
Y-91	5.80E 01	4E -08	8E -04	1E -09	3E -05
Zn-65	2.45E 02	1E -07	3E -03	4E -09	1E -04
Zr-93	4.00E 08	1E -07	2E -02	4E -09	8E -04
Zr-95	6.33E 01	1E -07	2E -03	4E -09	6E -05

SOURCE: 10CFR20, Appendix B.

on calculations aimed at keeping radioactivity levels in breathing air or possible drinking water below the point where normal intakes could lead to harmful accumulation of activity anywhere within the body. The decay half-life must be taken into account, as well as the "biological half-life"—the time needed for half of the ingested material to be eliminated by normal metabolic processes. In the case of radiation workers, it is assumed that the individual is subject to exposure on the job during a 40-hour workweek over a period of 50 years. In the case of the general public, the figures are established on the basis of a 168-hour week over a lifetime. The fact that there are fetuses, small children, and other particularly susceptible individuals in the general population is also taken into account.

The MPC values in Table 10.3 are given in computer printout form. Considering the first entry, Ac-227, the notation 2E -12 indicates that the 2 preceding the E is to be multiplied by 10 taken to the -12 power. Therefore, the MPC for Ac-227 in a radioactivity-handling facility is 2×10^{-12} μCi/ml of air.

The values given for the rare gases (Xe, Kr, Ar, Rn) consider that the individual is submerged in an infinite half-hemisphere of air, so any exposure is to the whole body. Special rules, however, apply to radon since it generates radioactive daughters. A number of other special procedures are also given in 10CFR20 to handle mixtures of activities, nuclides not listed in their table, and so on. In some cases, such as with the uranium isotopes, the potential hazard may actually be greater from heavy metal poisoning than from radioactivity.

ELEVEN

RADIATION DAMAGE

11.1 BACKGROUND AND LITERATURE

Following World War II, the development of the embryonic nuclear power industry required numerous empirical studies of radiation damage to all sorts of materials, a program that tapered off to a certain extent by the early 1960s. The interest at present is more on radiation effects in space, in "hardening" of electronics against the electromagnetic surge accompanying a nuclear blast[193] and in selection of materials for construction of new accelerators.[194] More basic studies of radiation-effect mechanisms is, of course, a very active area of continuing research.

Citation of a single comprehensive and fairly recent literature source is difficult. Several textbooks can be listed[195–197] and various reports.[198,199] A journal on radiation effects chiefly concerned with basic studies is available.[200] Much of the earlier applied information is reviewed in the *Reactor Handbook*.[201]

Figure 11.1 shows the approximate levels of radiation used for medical and other applications, or where changes or damage are seen in various materials.

The brief presentation in Chapters 11 and 12 reproduces several tables and graphs prepared by the author for an earlier book.[176] The citations given here are to the original sources.

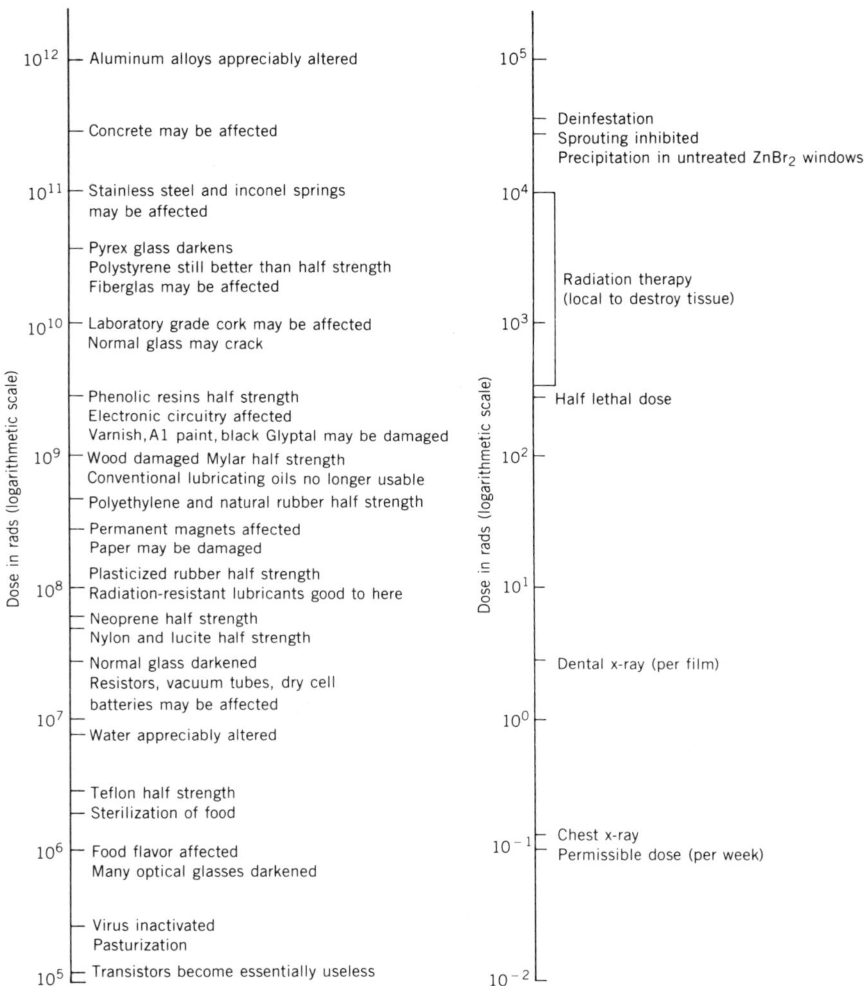

FIGURE 11.1. Approximate levels for radiation damage. (Reference 176.)

11.2 OILS AND LUBRICANTS

Oils generally show a marked increase in viscosity upon irradiation, while greases are softened. Therefore, viscosity is the chief criterion used in judging damage to such materials. The threshold values shown in Table 11.1 are at the point where viscosity changes are first observed and the limiting values at the point

TABLE 11.1 Radiation Effects on Oils and Lubricants

	Dosage (per cm^2 × 10^{-18})			
	Threshold		Limiting	
Compound Class	Neutrons	Photons	Neutrons	Photons
Polynuclear aromatics	1.8	12	15	90
Short-chain alkyl aromatics	1.5	2.5	7	30
Long-chain alkyl aromatics	0.5	2.0	2.5	10
Aliphatic polyethers	0.3	1.5	1.0	7.5
Aliphatic hydrocarbons	0.2	1.3	0.8	5
Aliphatic diesters	0.2	0.5	0.8	5
Aliphatic silicones	<0.1	<0.3	<0.5	<1

SOURCE: Reference 201.

where the sample solidifies. The neutron flux used was at thermal energies, and the gammas had an average energy of about 0.75 MeV.

The table was taken from Calkins and Shall,[201] who, in turn, quoted it from a classified document (Durand and Faris, NAA-SR-1304). Elsewhere, Calkins and Shall give a tentative rating of lubricating materials in order of increasing radiation stability: silicones, esters, mineral oils, ethers, and alkyl aromatics.

11.3 ELASTOMERS AND PLASTICS

A number of studies of radiation effects on elastomers and plastics have been undertaken in the past. Perhaps the most comprehensive review[202] of this early work was made by the Radiation Effects Information Center at Battelle Institute in Ohio as part of a series of similar reports dealing with a wide variety of materials. (The Institute is no longer functioning, but presumably these REIC reports could be obtained through the National Technical Information Service.)

Figure 11.2 was prepared from REIC Report 21.[202] The open portion of a particular bar is the dosage range over which damage is incipient to mild. The lighter crosshatched interval indicates mild-to-moderate damage, with the material still possibly useful in some applications. The darker end portion of the bar

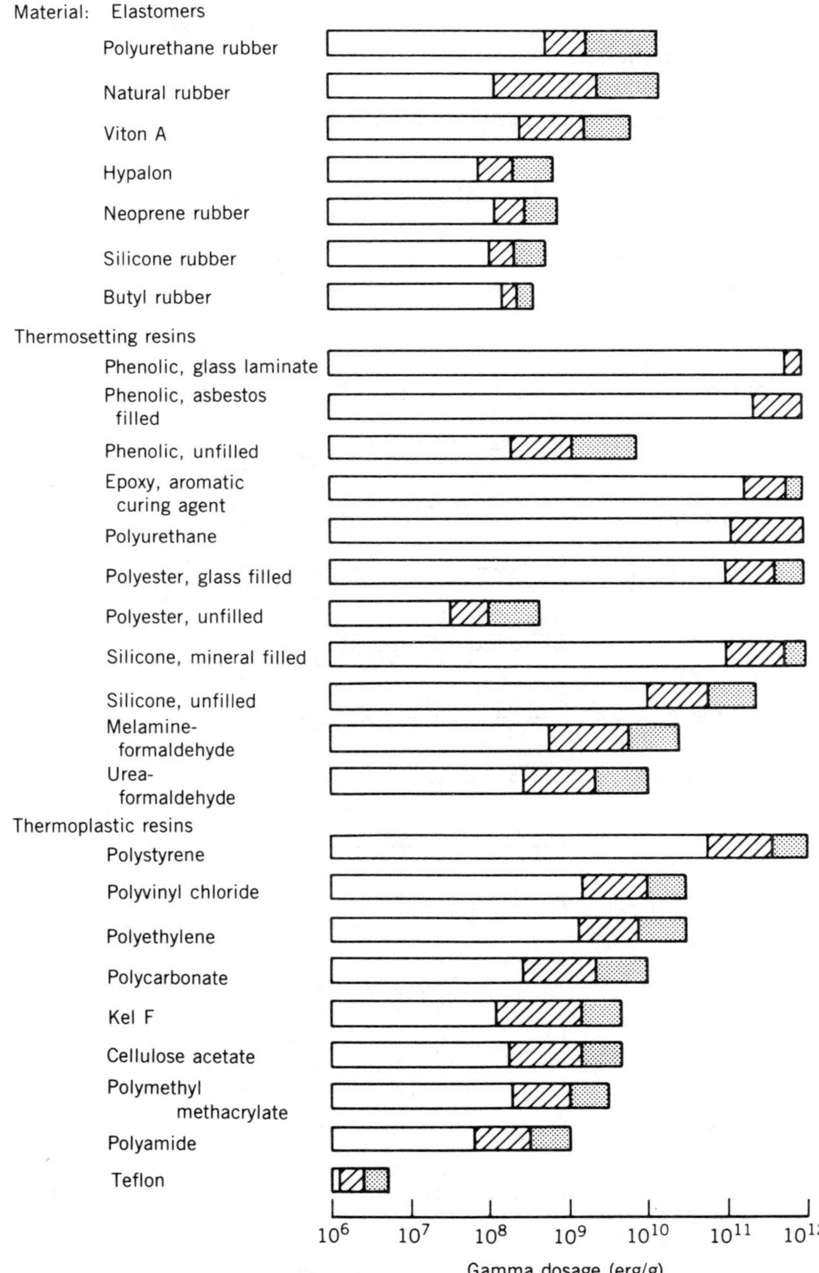

FIGURE 11.2. Radiation damage to elastomers and plastics. (Reference 202.)

gives the range where damage is severe. The intervals shown are approximations since the circumstances of the irradiation (temperature, cover-gas, dose rate, etc.) sometimes produce marked variations in observed effects.

The data in Figure 11.2 are for gamma radiation.

11.4 ELECTRONIC COMPONENTS

The designers and builders of large particle accelerators have a particular interest in the behavior of electronic components and circuitry in radiation fields. Figure 11.3 presents the results of tests carried out by CERN at its accelerator

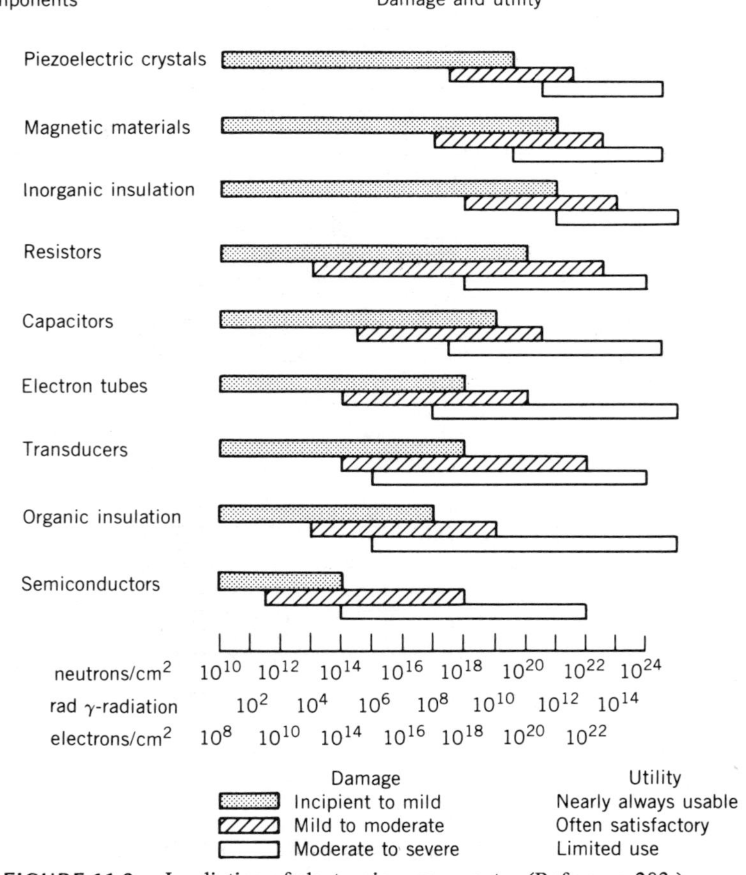

FIGURE 11.3. Irradiation of electronic components. (Reference 203.)

facility in Geneva or at the swimming-pool reactor at Siebersdorf, Austria. The CERN group makes the following statement:

> The results . . . lead to the overall conclusion that the operation of electronic components and circuits is seriously affected by radiation environments with doses in the order of 10^{13} n/cm^2 or 10^4 rad . . . some components and circuits fail completely at doses of 10^{14} n/cm^2 or 10^5 rad.

Earlier summaries[204,205] of radiation damage studies on electronic equipment are available.

TWELVE

NUCLEAR CRITICALITY

12.1 BACKGROUND AND LITERATURE

Because of the many neutron poisons present, and because the Purex process generally removes better than 99% of the uranium and plutonium, nuclear criticality is not ordinarily a problem in handling high-level waste. This, however, is not necessarily true during the processing generating the wastes, or for wastes derived in fuel-fabrication facilities. A minimum amount of basic information is accordingly included here.

The criticality literature appears mostly in the form of articles or reports, although the data for Pu are reviewed in the *Plutonium Handbook*.[206] The most authoritative reference is possibly the "Nuclear Safety Guide."[207] A three-volume reference and training guide has been prepared.[208] The criticality situation for actinides other than the most common fissiles (U-233, U-235, and Pu-239) has been reviewed.[209] Another good source is the report by Paxton,[210] and rather general handling standards have been established by the NRC[211] and by the American Nuclear Society.[212]

12.2 CRITICALITY TERMINOLOGY

Figure 12.1 is presented here as an illustration of a number of terms used in nuclear criticality discussions. It shows the maximum amount of U-233 (after ap-

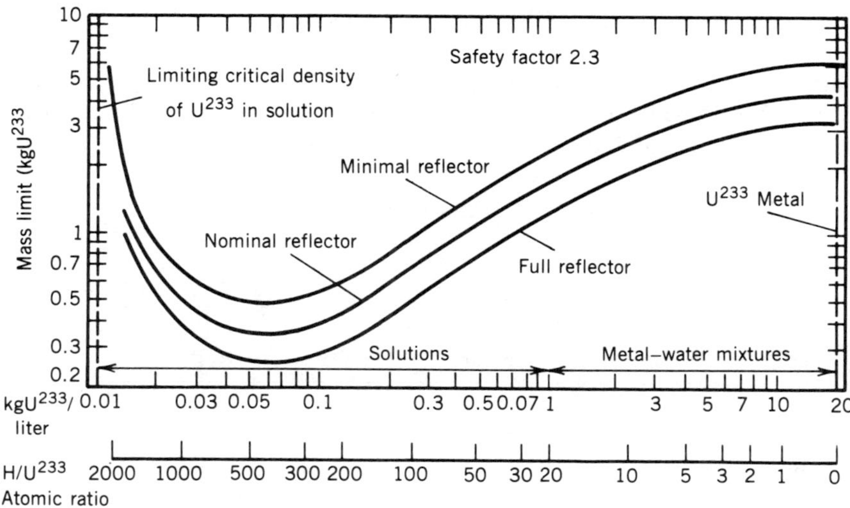

FIGURE 12.1. Critical mass of U-233 and water. (Reference 207.)

plication of a safety factor in this case) that can be handled in different combinations of the pure isotope with water. The data at the far right of the curves are for pure metal, with the admixture of water (the H/X ratio) increasing as it moves across the figure until only a very dilute solution is being considered at the far left.

As one moves from the right to the left, the moderating effect of the water causes a drop in the critical mass, the point where a spontaneous self-sustaining chain fission reaction can occur. At the same time, capture, and thus removal, of neutrons by the hydrogen acts in the other direction. The result of these two opposing effects is to produce minima in the curves, the point of "minimal critical mass." In the figure, this occurs at a concentration of about 60 g/l for U-233 and with an overall limit of about 0.25 kg (fully reflected case).

As one passes the minimum point, increasing losses of neutrons by hydrogen capture allows the mass limit to rise until the "limiting criticality density (or concentration)" is reached. Any solution more dilute than this cannot become critical, no matter how much total material is present.

In the minimal (or bare) reflector case the fissile-water mixtures or solutions are considered to be enclosed in a spherical container of stainless steel or other metal, with $\frac{1}{16}$-$\frac{1}{8}$-inch wall thickness. The pure metal is as a solid sphere. For the nominal reflector case, the spheres are surrounded by a 1-inch thickness of water, or its nuclear equivalent as a neutron reflector. A full reflector is 3 inches of water or its nuclear equivalent.[207]

12.3 CRITICAL MASS DATA

Tables 12.1, 12.2, and 12.3 give critical mass data for U-233, U-235, and Pu-239, respectively. These three are, for practical purposes, the only fissile nuclides currently available in sufficient quantity to produce a criticality accident, although

TABLE 12.1 Criticality Parameters for U-233

	Metal (Density, 18.44)		Water Solution	
Parameter	Fully Reflected	Bare	Fully Reflected	Bare
Minimum critical mass (kg)	6.7	17.0	0.550	1.2
Infinite cylinder (cm)	4.6	8.2	11.5	19.0
Infinite slab (cm)	0.54	4.6	3.0	10.2
Volume (l)	0.407	0.84	3.5	8.7
Limiting critical concentration (g/l)	–	–	10.8	–
Areal density ($g/100\ cm^2$)	–	–	35.0	47.4

SOURCE: References 207, 208, and 211.

TABLE 12.2 Criticality Parameters for U-235

	Metal (Density, 18.44)		Water Solution	
Parameter	Fully Reflected	Bare	Fully Reflected	Bare
Minimum critical mass (kg)	20.1	47.0	0.76	1.40
Infinite cylinder (cm)	7.3	11.4	13.9	21.6
Infinite slab (cm)	1.3	5.6	4.6	11.4
Volume (l)	–	2.7	5.8	14.0
Limiting critical concentration (g/l)	–	–	11.5	–
Areal density ($g/100\ cm^2$)	0.13	–	40.0	56.0

SOURCES: References 207, 208, and 211.

TABLE 12.3 Criticality Parameters for Pu-239

Parameter	Metal (Density, 19.5) Fully Reflected	Metal (Density, 19.5) Bare	Water Solution (N:Pu > 4) Fully Reflected	Water Solution (N:Pu > 4) Bare	Water Solution No NO_3 Fully Reflected
Minimum critical mass (kg)	4.9	10.2	0.51	0.905	0.51
Diameter-infinite cylinder (cm)	4.4	6.1	15.7	23.2	12.5
Thickness-infinite slab (cm)	0.66	2.8	5.8	13.5	3.3
Spherical volume (l)	0.28	0.51	7.7	–	4.5
Limiting critical concentration (g/l)	–	–	7.3	–	7.8
Areal density (g/100 cm^2)	–	–	25.0	–	–

SOURCES: References 207, 208, and 211.

this will not necessarily be so in the future. Pu-241 (which builds up in highly irradiated plutonium) has a minimum critical mass of only 260 g in fully reflected water solution, and some of the values for the higher curiums are even less.[209]

Most of the newer terms in the tables are involved with geometrical considerations, that is, the point where the surface/volume ratio of a particular shape becomes high enough to allow sufficient numbers of neutrons to leak out and be lost, thus making a chain reaction impossible. Figure 12.1 illustrates this for spherical shapes—the minimum critical mass is at the point of minimum critical radius. Similarly, if a cylinder of fissile material has a small enough radius, or a slab, a small enough thickness, there will be enough external surface so that neutron losses by leakage will maintain subcriticality.

THIRTEEN

DECONTAMINATION

13.1 BACKGROUND AND LITERATURE

The criteria established by different countries, and even by organizations in the same country, for an acceptable job of decontamination vary widely,[213] although some official limits appear to be based on an early recommendation of Dunster[214] that cleaning be to a limit of 22,000 dis-min/100 cm^2. Table 13.1 shows the standard that has been used by the NRC for decontamination of decommissioned reactors. The same standard is used by NRC inspectors for other types of facilities, although with considerable flexibility, depending on the planned use of the equipment or space subsequent to cleanup. "Removable" indicates the quantity obtained by a single swipe of the surface with an absorbent tissue or similar material.

An older paper by Stevenson[215] gives a good discussion of the probable mechanisms of contamination and decontamination. Only a few of the numerous meetings on decontamination that have been held and bibliographies available are cited.[216–218] The decommissioning of nuclear facilities is usually associated with major decontamination programs. Reviews are available.[219, 220]

13.2 DECONTAMINATION PROCEDURES

Table 13.2 is a reproduction of a table prepared by Saenger[222] for the AEC. In most cases, the techniques are listed in order of increasing harshness. The milder

TABLE 13.1 NRC Decontamination Standard

Nuclide	(dis/min)/100 cm^2		
	Average	Maximum	Removable
U-nat, ^{235}U, ^{238}U, and decay products	5,000 α	15,000 α	1,000 α
Transuranics, ^{226}Ra, ^{228}Ra, ^{230}Th, ^{228}Th, ^{231}Pa, ^{227}Ac, ^{125}I, ^{129}I	100	300	20
Th-nat, ^{232}Th, ^{90}Sr, ^{223}Ra, ^{224}Ra, ^{232}U, ^{126}I, ^{131}I, ^{133}I	1,000	3,000	200
Beta-gamma emitters (except ^{90}Sr and others noted above)	5,000 $\beta - \gamma$	15,000 $\beta - \gamma$	1,000 $\beta - \gamma$

SOURCE: Condensed from Reference 221.

approaches should be tried first and recourse made to more strenuous treatment only if necessary. The table was also reproduced in a two-volume report by Fitzgerald[223], and much of the material appears in the *Radiological Health Handbook.*[20]

The IAEA recommends[35] either an unbuilt detergent (no admixture with inorganic salts), or a recipe of 5-10% Na_2CO_3, or built detergent plus 1-2% EDTA and 0.5-1.0% citric acid for simple decontamination of glass, stainless steel, copper, aluminum, lead, and rubber. If it is desired to remove some surface as well, chromic acid is used on glass, a 1% H_2SO_4-25% inhibited H_3PO_4 mixture for stainless steel, sodium hydroxide, or citric acid for copper and aluminum, aqua regia for lead, and acetone for rubber.

TABLE 13.2 Decontamination Procedures

Contaminated Area	Decontaminating Agent[a]	Remarks	Maximum Suggested Levels of Contamination
Skin and hands	Mild soap and water or detergent and water.	Wash 2 to 3 min and monitor; do not wash over 3 or 4 times.	Alpha: 150 dis/min/100 cm^2; this is approximately one-half the inhalation level in terms of total dis/min/day. This assumes not more than one-fifth of this material will be inhaled. Additional possible exposure by ingestion is also considered.
	If necessary, follow by scrubbing with soft brush, heavy lather, and tepid water.	Use light pressure with heavy lather; wash for 2 min, 3 times; rinse and monitor; use care not to scratch or erode skin.	
	Lava soap and water.	Apply lanolin or hand cream to prevent chapping.	
	Other procedures		
	A mixture of 50% Tide and 50% corn meal.	Make into a paste; use with additional water with a mild scrubbing action; use care not to scratch or erode the skin.	Beta-gamma; average less than 0.3 mr/hr for each hand surface or 100 cm^2 of skin surface, using Geiger–Mueller instrument calibrated with Ra^{226}.
	A 5% water solution of a mixture of 30% Tide, 65% Calgon, and 5% Carbose (Carboxymethyl cellulose).	Use with water; rub for a minute and rinse.	
	A preparation of 8% Carbose, 3% Tide, 1% Versene, and 88% water homogenized into a cream.	Use without any additional water; rub for 1 min and wipe off; follow with lanolin or hand cream.	
	Chemical procedures (as a last resort)		
	Titanium dioxide paste; prepare paste by mixing precipitated titanium dioxide (a very thick slurry never permitted to dry) with a small amount of lanolin.	Work the paste into affected area for 1 min; rinse and wash with soap, brush, and warm water; monitor.	

	Mix equal volumes of a saturated solution of potassium permanganate and 0.2 *N* sulfuric acid; continue with the next step also. (Saturated solution $KMnO_4$ is 6.4 g/100 ml of water.)	Pour over wet hands, rubbing the surface and using hand brush for not more than 2 min. (Note: will remove a layer of skin if in contact with the skin for more than 2 min.) Rinse with water.	
	Apply a freshly prepared 5% solution of sodium acid sulfite ($NaHSO_3$).	Apply in the same manner as above; apply for not more than 2 min. The above procedure may be repeated; apply lanolin or hand cream when completed.	
	Dissolve in order: (1) Versene (EDTA) 5%	Trivalent metals: Al, Sc, Y, La, Ce, Pr, Nd, Pm, Sm, Eu.	
	(2) Conc. NH_4OH, 3 vol. % (3) Glacial acetic acid 5 vol. %	Rare earths: Ac, Ca, In, Ti, B. Transition metals: Cu, Zn, Fe, Co, Ni, Cd, Sn, Hg, Pb, Th, U, Ag. (Always consider the radioactivity of the cleaning solution when disposing of it.)	
Wounds (cuts and breaks in the skin)	Running tap water; report to Medical Officer and RSO as soon as possible.	Wash the wound with large volumes of running water immediately (within 15 sec); spread the edges of wound to permit flushing action by the water.	Keep wound contamination as low as possible.
Ingestion by swallowing	Immediately induce vomiting; drink large quantities of liquids to dilute the activity.	Urine and fecal analysis will be necessary to determine amount of radionuclides in the body.	
Clothing (see rubber and leather under specific materials)	Wash if levels permit.	Use standard laundering procedures, 3% Versene or citric acid may be added to wash water.	Alpha: 150 dis/min/100 cm^2.
		Wash water must be below the MPL for sewer disposal (see NBS Handbook 69).	Beta-gamma: no area to average more than 0.1 mr/hr; G–M meter Ra^{226} calibrated.

TABLE 13.2 (Continued)

Contaminated Area	Decontaminating Agent[a]	Remarks	Maximum Suggested Levels of Contamination
	Store	To allow for decay if contamination is short-lived.	If clothing is worn 100 hr/week, this will give 1/10 of maximum external dose.
	Disposal	Treat as solid waste if necessary.	
Glassware	Soap or detergent and water.	Monitor wash water and plan disposal of it.	The maximum permissible levels for glassware handled with the bare hands are the same as for the hands and skin.
	Chromic acid cleaning solution or concentrated nitric acid.	Monitor wash water and plan disposal of it.	
	Suggested agents	*Elements removed*	
	Oxalic acid 5% (Caution Poison).	Zr, Nb, Hf.	
	Versene (EDTA) 5% conc. NH_4OH 3%.	Alkaline-earth metals: Be, Mg, Ca, Sr, Ba, Ra, P as PO_4.	
	HCl 10% by volume.	Alkali metals: Na, K, Rb, Cs, and strongly absorbed metals like Po.	
Glass, plastic	See glassware, plastic tools, glass tools.		
Leather	Very difficult to decontaminate.		
Linoleum	CCl_4, kerosene, ammonium citrate, dilute mineral acids.		
Ceramic tile	Mineral acids, ammonium citrate, trisodium phosphate.		
Paint	CCl_4, 10% HCl acid.	Usually best to remove the paint and repaint.	
Brick and concrete	32% HCl acid.	If this is not successful, concrete must be removed.	

Wood	Hot citric acid, remove the wood with a plane or floor chippers and grinders.		
Traps and drains	(1) Flush with water. (2) Scour with rust remover. (3) Soak in a solution of citric acid. (4) Flush again.	Follow all four steps.	
Laboratory tools	Detergents and water steam cleaning.	Use mechanical scrubbing action.	The maximum permissible for tools handled with the bare hands are the same as for the hands and skin.
Metal tools	Dilute nitric acid, 10% solution of sodium citrate or ammonium bifluoride.	As a last resort, use HCl on stainless steel.	
	Metal polish, sandblasting, other abrasives.	Such as brass polish on brass; use caution as these procedures may spread contamination.	
Plastic tools	Ammonium citrates, dilute acids, organic solvents.		
Glass tools	The same as above section on glassware.		
Walls, floors, and benches	Detergents and water with mechanical action.		
	Vacuum cleaning.	The exhaust of the cleaner must be filtered to prevent escape of contamination.	
	Water from high-pressure sources; steam cleaning.	This may spread contamination.	
Specific materials Rubber	Washing or dilute HNO_3.	Short lived contamination may be covered to await decay.	

SOURCE: Reference 222. Also see text.

[a]Ordering is from mild to harsher techniques.

PART FIVE

MISCELLANEOUS DATA

Some generally useful information is given in this Part for convenience. The material is generally available from many other sources, so citations are limited.

FOURTEEN

GENERAL INFORMATION

14.1 RADIOACTIVE DECAY

The rate at which a sample of a radioactive nuclide decays is determined by its half-life, but, internally, there is no way to predict which particular atom is the next to disintegrate. However, the laws of probability can be applied to the total of atoms present, leading to the familiar basic equation for radioactive decay:

$$\frac{N}{N_0} = e^{-\lambda t} \tag{14.1}$$

where N_0 = number of atoms at time zero
N = number of atoms remaining at time t
e = natural logarithmic base, 2.718
λ = the decay constant of the decaying nuclide expressed in reciprocal time units, usually $\sec^{-1}$. Lambda expresses the fraction of N being transformed per unit of time
t = time in seconds

Practically, N and N_0 cannot be measured directly, so the activity A, which will be proportional to λN, is measured. Equation 14.1 is accordingly seen as

$$\frac{A}{A_0} = e^{-\lambda t} \tag{14.2}$$

At the point where the elapsed time is equal to the half-life: $t = T_{1/2}$ and $N/N_0 = \frac{1}{2}$. Then by taking the logarithm of both sides of Equation 14.1

$$\ln \frac{1}{2} = \lambda T_{1/2} \quad \text{or} \quad T_{1/2} = \frac{\ln 2}{\lambda} = \frac{0.693}{\lambda} \tag{14.3}$$

and

$$\lambda = \frac{0.693}{T_{1/2}} \tag{14.4}$$

This being the case, Equation 14.1 can be rewritten:

$$\frac{N}{N_0} = e^{-0.693t/T_{1/2}} \tag{14.5}$$

This equation allows preparation of data such as that shown in Table 14.1. The first column is the ratio of the elapsed time to the half-life, the second, $(e^{-\lambda t})$ is now the numerical value of N/N_0, that is, the fraction of the original sample that has not decayed at time t, and, conversely, $(1 - e^{-\lambda t})$ gives the fraction that has disintegrated. The same type of information can be presented graphically in different ways, an example being Figure 14.1.

TABLE 14.1 Radioactive Decay

$t/T_{1/2}$	$e^{-\lambda t}$	$t/T_{1/2}$	$e^{-\lambda t}$	$t/T_{1/2}$	$e^{-\lambda t}$	$t/T_{1/2}$	$e^{-\lambda t}$
0	1.0000	0.52	0.6974	1.54	0.3439	3.80	0.0718
0.01	0.9931	0.54	0.6878	1.56	0.3391	3.85	0.0693
0.02	0.9862	0.56	0.6783	1.58	0.3345	3.90	0.0670
0.03	0.9794	0.58	0.6690	1.60	0.3299	3.95	0.0647
0.04	0.9726	0.60	0.6597	1.62	0.3253	4.00	0.0625
0.05	0.9659	0.62	0.6507	1.64	0.3209	4.10	0.0583
0.06	0.9593	0.64	0.6417	1.66	0.3164	4.20	0.0544
0.07	0.9526	0.66	0.6329	1.68	0.3121	4.30	0.0508
0.08	0.9461	0.68	0.6242	1.70	0.3078	4.40	0.0474
0.09	0.9395	0.70	0.6156	1.75	0.2973	4.50	0.0442
0.10	0.9330	0.72	0.6071	1.80	0.2872	4.60	0.0412

TABLE 14.1 (Continued)

$t/T_{1/2}$	$e^{-\lambda t}$	$t/T_{1/2}$	$e^{-\lambda t}$	$t/T_{1/2}$	$e^{-\lambda t}$	$t/T_{1/2}$	$e^{-\lambda t}$
0.11	0.9266	0.74	0.5987	1.85	0.2774	4.70	0.0385
0.12	0.9202	0.76	0.5905	1.90	0.2679	4.80	0.0359
0.13	0.9138	0.78	0.5824	1.95	0.2588	4.90	0.0335
0.14	0.9075	0.80	0.5744	2.00	0.2500	5.00	0.0312
0.15	0.9013	0.82	0.5664	2.05	0.2415	5.10	0.0292
0.16	0.8950	0.84	0.5586	2.10	0.2333	5.20	0.0272
0.17	0.8888	0.86	0.5509	2.15	0.2253	5.30	0.0254
0.18	0.8827	0.88	0.5434	2.20	0.2176	5.40	0.0237
0.19	0.8766	0.90	0.5359	2.25	0.2102	5.50	0.0221
0.20	0.8705	0.92	0.5285	2.30	0.2031	5.60	0.0206
0.21	0.8645	0.94	0.5212	2.35	0.1961	5.70	0.0192
0.22	0.8586	0.96	0.5141	2.40	0.1895	5.80	0.0179
0.23	0.8526	0.98	0.5070	2.45	0.1830	5.90	0.0167
0.24	0.8467	1.00	0.5000	2.50	0.1768	6.00	0.0156
0.25	0.8409	1.02	0.4931	2.55	0.1708	6.20	0.0136
0.26	0.8351	1.04	0.4863	2.60	0.1649	6.40	0.0118
0.27	0.8293	1.06	0.4796	2.65	0.1593	6.60	0.0103
0.28	0.8236	1.08	0.4730	2.70	0.1539	6.80	0.0090
0.29	0.8179	1.10	0.4665	2.75	0.1487	7.00	0.0078
0.30	0.8122	1.12	0.4601	2.80	0.1436	7.20	0.0068
0.31	0.8066	1.14	0.4538	2.85	0.1387	7.40	0.0059
0.32	0.8011	1.16	0.4475	2.90	0.1340	7.60	0.0052
0.33	0.7955	1.18	0.4413	2.95	0.1294	7.80	0.0045
0.34	0.7900	1.20	0.4353	3.00	0.1250	8.00	0.0039
0.35	0.7846	1.22	0.4293	3.05	0.1207	8.20	0.0034
0.36	0.7792	1.24	0.4234	3.10	0.1166	8.40	0.0030
0.37	0.7738	1.26	0.4175	3.15	0.1127	8.60	0.0026
0.38	0.7684	1.28	0.4118	3.20	0.1088	8.80	0.0022
0.39	0.7631	1.30	0.4061	3.25	0.1051	9.00	0.0020
0.40	0.7579	1.32	0.4005	3.30	0.1015	9.20	0.0017
0.41	0.7526	1.34	0.3950	3.35	0.0981	9.40	0.0015
0.42	0.7474	1.36	0.3896	3.40	0.0948	9.60	0.0013
0.43	0.7423	1.38	0.3842	3.45	0.0915	9.80	0.0011
0.44	0.7371	1.40	0.3789	3.50	0.0884	10.00	0.0010
0.45	0.7320	1.42	0.3737	3.55	0.0854	10.50	0.0007
0.46	0.7270	1.44	0.3685	3.60	0.0825	11.00	0.0005
0.47	0.7220	1.46	0.3635	3.65	0.0797	11.50	0.0004
0.48	0.7170	1.48	0.3585	3.70	0.0770	12.00	0.0002
0.49	0.7120	1.50	0.3536	3.75	0.0743	13.00	0.0001
0.50	0.7071	1.52	0.3487				

SOURCE: Reference 20.

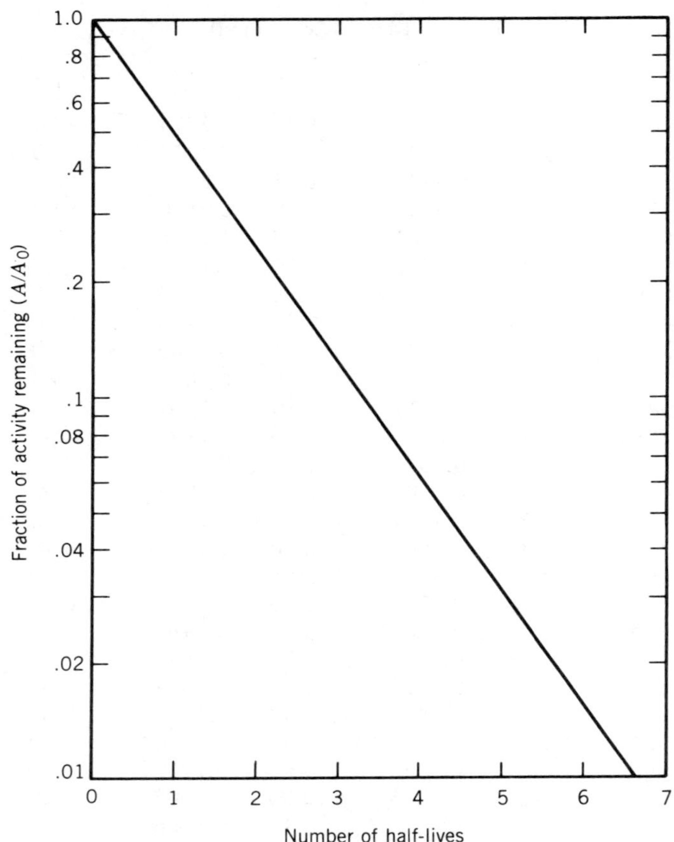

FIGURE 14.1. Radioactive decay. $A/A_0 = (\frac{1}{2})^n$ where A_0 = activity at some original (zero) time; A = activity at any time t; A/A_0 = fraction of activity remaining at time t; n = number of half lives in the time interval between the original time and time t; $n = t/T_{1/2}$. (Reference 20.)

14.2 NEUTRON ACTIVATION

The basic equation for neutron activation in a thin target (one where the neutron flux is not internally reduced by more than the experimental error) is

$$A_\phi = k\sigma_c fN(1 - e^{-\lambda t})\, e^{-\lambda \phi} \tag{14.6}$$

where A_ϕ = measured activity in counts/sec at time ϕ

ϕ = time between end of irradiation and start of counting

k = efficiency of the counter in converting dis/sec to counts/sec

σ_c = capture cross section for the reaction $A(n, \gamma)B$
N = total number of atoms of A in the target
f = neutron flux in n/cm^2-sec
t = time of exposure in the flux

Time units must be consistent, and, in this equation, σ_c is the capture cross section in barns times 10^{-24}.

By bringing in Avogadro's number (which largely cancels out the barn 10^{-24}) and placing the equation on a per minute basis, Equation 14.6 can be further converted to

$$(\text{counts/min})_B = \frac{36.1k\, W_A \theta \sigma_c f}{M_A} (1 - e^{-\lambda t})\, e^{-\lambda \phi} \qquad (14.7)$$

The new terms here are

W_A = weight of the target element in grams
M_A = atomic weight of the target element
θ = isotopic abundance of the target nuclide in the element

The equations above apply only to the simple $A(n, \gamma)B$ case, where A is assumed to be stable and in sufficient quantity so that its abundance is not much changed during the irradiation. If more complicated systems have to be considered (A itself decaying, losses of A or B through side reactions such as fission, etc.), the more involved mathematics illustrated in Figure 3.1 must be used.

In Equation 14.7, the capture cross section used is taken directly from a nuclide chart or other reference. The 10^{-24} multiplier has been incorporated into the equation as shown.

14.3 CONVERSIONS AND EQUALITIES

Both References 9 and 10 present many pages of conversion factors, and such information is available from many other sources. Table 14.2 gives an abbreviated selection of factors likely to be of most interest in radioactive waste management. For the most part, the arrangement is for converting metric to English units. However, as with other similar tables in this book, dividing by the given factor allows conversion of quantities on the right to the units on the left.

Familiar equalities and formulae can become fuzzy in the mind if not used very frequently:

Avogadro's number = 6.0225×10^{23} atoms or molecules/mol
Natural logarithm base = 2.7182

TABLE 14.2 Some Conversion Factors

Multiply	By	To Obtain
Length		
cm	0.3937	in.
	0.03281	ft
m	3.281	ft
	39.37	in.
km	0.6214	miles
	3281	ft
Area		
cm^2	10^{-4}	m^2
	0.1550	$in.^2$
m^2	10.76	ft^2
Volume		
cm^3	10^{-3}	l
	0.06102	$in.^3$
	2.642×10^{-4}	gal (U.S.)
l	61.02	$in.^3$
	0.03531	ft^3
	0.2642	gal (U.S.)
	2.203	lb H_2O (60°F)
m^3	35.31	ft^3
	264.2	gal (U.S.)
Mass		
g	0.03527	oz (avdp)
	0.002205	lb (avdp)
kg	10^{-3}	tonne
	0.001102	ton (short)
	2.205	lb (avdp)
tonne	2205	lb
	1.102	ton (short)
	0.9842	ton (long)
Density		
g/cm^3	0.03613	$lb/in.^3$
	62.43	lb/ft^3
Energy, heat		
MeV	1.602×10^{-6}	ergs
erg	10^{-7}	joules
cal	3.968×10^{-3}	Btu
W	10^7	ergs/s
	3.413	Btu-h

TABLE 14.2 (Continued)

Multiply	By	To Obtain
kW	1.341	horsepower
W/cm^2	3172	$Btu\text{-}h/ft^2$
W/m^3	0.0966	$Btu\text{-}h/ft^3$
Specific heat		
cal/g-°C	1.000	Btu/lb-°F
Thermal conductivity		
W/m-°K	1.730	Btu-h/ft-°F
Temperature		
°C + 273.15	1.000	K
°F + 459.6	1.000	°R′
(°C × 9/5) + 32		°F
°F-32	$\frac{5}{9}$	°C

SOURCE: Primarily Reference 9.

TABLE 14.3 Time Conversions

	Seconds	Minutes
Hour	3600	60
Day	8.64×10^4	1440
Week	6.05×10^5	1.01×10^4
Month (30 d)	2.59×10^6	4.32×10^4
Year (365.26 d)	3.16×10^7	5.26×10^5

TABLE 14.4 Recommended Unit Prefixes

Multiple	Prefix	Symbol	Submultiple	Prefix	Symbol
10^{12}	tera	T	10^{-1}	deci	d
10^9	giga	G	10^{-2}	centi	c
10^6	mega	M	10^{-3}	milli	m
10^3	kilo	k	10^{-6}	micro	μ
10^2	hecto	h	10^{-9}	nano	n
10^1	deka	d	10^{-12}	pico	p
			10^{-15}	femto	f
			10^{-18}	atto	a

SOURCE: Reference 224.

TABLE 14.5 The Greek Alphabet

A	α	Alpha	N	ν	Nu
B	β	Beta	Ξ	ξ	Xi
Γ	γ	Gamma	O	o	Omicron
Δ	δ	Delta	Π	π	Pi
E	ϵ	Epsilon	P	ρ	Rho
Z	ζ	Zeta	Σ	σ	Sigma
H	η	Eta	T	τ	Tau
Θ	θ	Theta	Υ	υ	Upsilon
I	ι	Iota	Φ	ϕ	Phi
K	κ	Kappa	X	χ	Chi
Λ	λ	Lambda	Ψ	ψ	Psi
M	μ	Mu	Ω	ω	Omega

$\log_{10} = 2.3205 \ln_e$

$\pi = 3.1415$

Cylinder volume = $\pi r^2 h$ (where h is the height)

Cylinder surface (no end pieces) = $2\pi rh$

Cylinder surface (with end covers) = $2\pi rh + 2\pi r^2$

Sphere volume = $4/3\,\pi r^3$

Sphere surface = $4\pi r^2$

While there is some duplication with Table 14.2, convenient series of equalities for those handling liquid wastes are

1 gal water (15°C) = 3.782 kg = 8.337 lb = 0.1337 ft^3
= 128 fluid oz = 231 in^3 = 3.7851

1 m^3 water = 1 tonne = 10^3 kg = 10^3 l = 10^6 cm^3 = 264.2 gal (U.S.) = 2205 lb = 1.102 tons (short) = 35.31 ft^3 = 61024 in^3 =
35.31 ft^3 = 61024 in^3

14.4 MISCELLANEOUS DATA

Odd bits of data that might save a bit of investigative time are given in Tables 14.3–14.5.

REFERENCES

1. P. F. Rose and T. W. Burrows, "ENDF/B Fission Product Decay Data" (2 vols). USERDA Report BNL-NCS-5045 (ENDF-245), Aug. 1976.
2. K. F. Flynn and L. E. Glendenin, "Yields of Fission Products for Several Fissionable Nuclides at Various Incident Neutron Energies." USAEC Report ANL-7749, Dec. 1970.
3. W. H. Walker, "Fission Product Data for Thermal Reactors. Part II, Yields." Canadian Report AECL-3037, Part II, April 1973.
4. E. A. C. Crouch, "Chain and Independent Fission Product Yields Adjusted to Conform With Physical Conservation Laws." United Kingdom Report AERE-R-7785, Feb. 1975.
5. M. E. Meek and B. F. Rider, "Compilation of Fission Product Yields, Vallecitos Nuclear Center, 1974." Report NEDO-12154-1, Jan. 1974.
6. K. Wolfsberg, "Estimated Values of Fractional Yields From Low-Energy Fission and a Compilation of Measured Fractional Yields." USAEC Report LA-5553-MS, Feb. 1974.
7. A. F. Voight, "Table of Fission Product Nuclides: Yields, Single Particle States, Observed Spins and Parities, Decay Modes and Half-Lives for Products of Fission of ^{235}U." USAEC Report IS-4052, Dec. 1976.
8. M. J. Fluss, N. D. Dudey, and R. L. Malewicki, "Tritium and Alpha-Particle Yields in Fast and Thermal Neutron Fission of ^{235}U." *Phys. Rev.*, **6C**, 2252–2259 (1972).
9. R. C. Weast, ed., *Handbook of Chemistry and Physics*, 61st ed. CRC Press, Boca Raton, FL, 1980–1981.
   a. R. L. Heath, "Table of the Isotopes," pp. B-258–B-342.
   b. R. L. Heath, "Gamma Energies and Intensities of Radionuclides," pp. B-343–B-399.
   c. J. F. Hunsberger, "Standard Reduction Potentials," pp. D-155–D-160.

10. J. A. Dean, ed., *Lange's Handbook of Chemistry*, 12th Edition. McGraw Hill Publishers, New York, 1979.
   (a) J. R. Peterson, "Table of Nuclides," pp. 3-15–3-116.
11. C. M. Lederer and V. S. Shirley, eds., *Table of Isotopes*, 7th ed. Wiley-Interscience, New York, 1978.
12. V. S. Shirley and C. M. Lederer, eds., Nuclear Wallet Cards. Copies available from National Nuclear Data Center, Brookhaven National Laboratory, Upton, NY.
13. C. W. Kee, "A Revised Light Element Library for the ORIGEN Code." USAEC Report ORNL-TM-4896, May 1975.
14. Anon, "RSIC Data Library Collection, ORYX-E, ORIGEN Yields and Cross Sections–Nuclear Transformations and Decay Data From ENDF/B." USDOE Report DLC-38, Sept. 1979.
15. S. Glasstone and A. Sesonski, *Nuclear Reactor Engineering*, 3rd ed., Van Nostrand-Reinhold, New York, 1981.
16. S. F. Mughabghab and D. I. Garber, "Neutron Cross Sections: Volume 1, Resonance Parameters." USAEC Report BNL-325, 3rd ed., June 1973.
17. F. W. Walker, G. J. Kirouac, and F. M. Rourke, "Chart of the Nuclides," 12th ed., Rev. to April 1977. Knolls Atomic Power Laboratory, Schenectady, New York, June 1973.
18. R. W. Roussin, "RSIC Data Library Collection, PUCOR, 84 Group Neutron Cross Sections for Uranium-Plutonium Cycle LWR and PWR Models in AMPX Master Library Format." USDOE Report DLC-67, Sept. 1979.
19. B. T. Kenna and P. E. Harrison, "Neutron Activation Analysis: Tabulation of Cross Sections, Q-Values, Decay Characteristics and Sensitivities." USERDA Report SAND-76-0254, July 1976.
20. U.S. Department of Health, Education and Welfare (HEW), Public Health Service, *Radiological Health Handbook*, Rev. ed., 1970.
21. R. G. Jaeger, ed.-in-chief, *Compendium on Radiation Shielding*, Vol. I. *Shielding Fundamentals and Methods* (1968); Vol. II. *Shielding Materials* (1975); Vol. III. *Shield Design and Engineering* (1970). Springer-Verlag, Berlin, West Germany.
22. J. W. Healy, "Los Alamos Handbook of Radiation Monitoring, 4th ed." USAEC Report LA-4400, 1970.
23. J. Milsted, E. S. Macias, L. J. Basile, R. W. Anderson, and D. C. Stewart, "Additional Yield Data for Trans-Plutonium Production in Thermal Reactors." USAEC Report ANL-7096, Oct. 1965.
24. W. Schirmer and N. Wächter, "Table of Specific Activities of the Nuclides With Z = 88 to Z = 104." *Actinide Reviews*, **1**, 125–134 (1968).
25. H. W. Kirby, "The Early History of Radiochemistry." USAEC Report MLM-1960, Sept. 1972.
26. H. W. Kirby, "Nuclear Properties and Genetic Relationships of the Naturally Occurring Radioactive Series." USAEC Report MLM-2036, Aug. 1973.
27. H. W. Kirby, Monsanto Research Corporation, Miamisberg, OH. Personal communication.
28. G. A. Cowan, "Migration Paths for Oklo Reactor Products and Applications to the Problem of Geologic Storage of Nuclear Wastes." USERDA Report LA-UR-77-2787, 1977.

29. P. E. Figgins and H. W. Kirby, "Survey of Sources of Ionium (Thorium-230)." USAEC Report MLM-1349, Oct. 1966.

30. Anon, "Nuclear Properties," in *Cf-252, Uses and Market Potential*, E. I. duPont de Nemours and Company, Aiken, SC, 1968, p. 13.

31. J. Blachot, P. Cavallini, A. Ferrieu, and R. Louis, "Rendements de Fission Cumulatifs de ^{252}Cf," *J. Radioanal. Chem.*, **26**, 107–125 (1975).

32. R. M. Harbour, M. Eichor, and D. E. Troutner, "Mass Yields From the Spontaneous Fission of ^{252}Cf," *Radiochim. Acta*, **15**, 146–150 (1971).

33. W. E. Nervik, "Spontaneous Fission Yields of ^{252}Cf," *Phys. Rev.*, **119**, 1685–1690 (1960).

34. International Commission on Radiation Protection, Committee 3, "Data for Protection Against Ionizing Radiation, Suppl. to ICRP Pub. 15," ICRP Publication 21, Pergamon Press, Oxford, 1973.

35. G. J. Appleton and P. N. Krishamoorthy, "Safe Handling of Radioisotopes, Health Physics Addendum," IAEA Safety Series No. 2, 1960.

36. D. C. Stewart, E. P. Horwitz, C. H. Youngquist, and M. A. Wahlgren, "A ^{244}Cm-Be Isotopic Neutron Source," *Nuc. Appl. Tech.*, **9**, 875–878 (1970).

37. M. E. Anderson and M. R. Hertz, "Thick Target Yields for the ^{9}Be(α, n) Reaction," *Nuc. Sci. and Eng.*, **44**, 437–439 (1971).

38. M. E. Anderson and R. A. Neff, "Neutron Energy Spectra of Different Size ^{239}Pu-Be(α, n) Sources," *Nuc. Instr. and Methods*, **99**, 231–235 (1972).

39. O. J. C. Runnals and R. R. Boucher, "Neutron Yields From Actinide-Beryllium Alloys," *Can. J. Phys.*, **34**, 949 (1956).

40. M. E. Anderson, Mound Facility, Monsanto Research Corporation, Miamisberg, OH. Personal communication.

41. H. Bateman, "Solution of a System of Differential Equations Occurring in the Theory of Radioactive Transformations," *Proc. Cambridge Phil. Soc.*, **15**, 423–427 (1910).

42. G. H. Hanson, "Irradiation of Thorium in MTR." USAEC Report IDO-16065, 1956.

43. M. J. Bell, "ORIGEN–The ORNL Isotope Generation and Depletion Code." USAEC Report ORNL-4628, May 1973.

44. Details can be obtained from the Radiation Shielding Information Center (RSIC), Post Office Box X, Oak Ridge, TN 37830.

45. A. G. Croff, M. A. Bjerke, G. W. Morrison, and L. M. Petrie, "Revised Uranium-Plutonium Cycle PWR and BWR Models for the ORIGEN Computer Code." USDOE Report ORNL/TM-6051, Sept. 1978.

46. H. O. Haug, "Calculations and Compilations of Composition, Radioactivity, Thermal Power, Gamma and Neutron Release Rates of Fission Products and Actinides of Spent Power Fuels and Their Reprocessing Rates." Report KFK-1945, 1974.

47. Author's experience in evaluating wastes stored at the Western New York Nuclear Service Center, West Valley, New York.

48. Staff, Oak Ridge National Laboratory, "Siting of Fuel Reprocessing Plants and Waste Management Facilities." USAEC Report ORNL-4451, July 1971.

49. International Union of Pure and Applied Science, "Atomic Weights of the Elements, 1973," *Pure Appl. Chem.*, **4**, 589–603 (1974).

50. J. Kleinberg, W. J. Argensinger, Jr., and Ernest Griswell, *Inorganic Chemistry*, D. C. Heath, Boston, 1960.

51. G. T. Seaborg, *The Transuranium Elements*, Yale University Press, New Haven, 1958.

52. *Gmelin Handbook of Inorganic Chemistry*. 8th ed. "Tc, Technetium," Suppl., Vol. 2, Metal Alloys, Compounds, Chemistry in Solution, Springer-Verlag, Berlin, 1983.

53. W. H. Latimer, *The Oxidation States of the Elements and Their Potentials in Aqueous Solution*, 2nd ed. Prentice-Hall, Englewood Cliffs, NJ, 1952.

54. A. Rose and E. Rose, *Condensed Chemical Dictionary*, 7th ed. Reingold, New York, 1966.

55. C. Keller and B. Kanellakopulos, "Darstellung und Untersuchen eineiger Pertechnetate des Typs $Me^{I}TcO_4$," *Radiochim. Acta*, **1**, 107–108 (1963).

56. J. Bjerrum, G. Schwarzenbach, and L. G. Sillen, *Solubility Constants of Metal Complexes, Part II*, Chem. Soc., London, 1958.

57. Anon, *Chromatography*, *Electrophoresis*, *Immunochemistry*, *Price List I*, Bio-Rad Laboratories, Richmond, CA 1983.

58. Diamond Shamrock Chemical Company, Nopco Chemical Division, 1901 Spring Street, Redwood City, CA.

59. J. R. Wiley, "Decontamination of Savannah River Plant Waste Supernatant," USERDA Report DP-1436, Aug. 1976.

60. U.S. Dept. Energy, "Western New York Nuclear Service Center Study, Final Report for Public Comment," USDOE Report TID-28905-1, Nov 1978.

61. Rohm and Haas Co., Independence Mall, W. Philadelphia, PA.

62. R. V. Panesko, "Recovery of Rhodium, Palladium and Technetium on Strongly Basic Anion Resin," USAEC Report ARH-1279, July 1969.

63. K. Dorfner, *Ion Exchangers, Properties and Applications*, 3rd ed., Ann Arbor Science Publishers, Inc., Ann Arbor, MI, 1972.

64. Dow Chemical, U.S.A., 2020 Dow Center, Midland, MI.

65. W. W. Schulz, Rockwell Hanford Operation, Richland, WA. Personal communication.

66. Union Carbide Corporation, Molecular Sieve Dept., 270 Park Ave, New York.

67. U.S. Dept. Energy, "Draft Environmental Impact Statement, Management of Commercially Generated Radioactive Waste," USDOE Report DOE/EIS-0046D, April 1979.

68. *U.S. Federal Register*, **48**, (#120) 28195–28230 June 21, 1983.

69. L. Garmon, "A Box Within a Box, Within a Box," *Science News*, **120**, 396–399 (1981).

70. U.S. Department of Energy, "Long-Term Management of Liquid High-Level Radioactive Wastes Stored at the Western New York Nuclear Service Center, West Valley, Final Environmental Impact Statement," USDOE Report DOE/EIS-0081, June 1982.

71. National Academy Engineering-National Academy Science, "Solidification of High-Level Radioactive Wastes, Final Report," USNRC Report NUREG/CR-0895, July 1979.

72. U.S. Dept. Energy, "Western New York Nuclear Service Study, Companion Report," USDOE Report TID-28905-2, 1979.

73. U.S. Energy Research Development Administration, "Alternatives for Managing Wastes From Reactors and Post-Fission Operations in the LWR Fuel Cycle," USERDA Report ERDA-76-43, vol. 2, May 1975. In five volumes.

74. U.S. Energy Research Development Administration, "Alternatives for Long-Term Management of Defense High-Level Radioactive Waste, Savannah River Plant," USERDA Report ERDA-77-42, vol. 1, May 1977. In two vols.

75. G. G. Eichholz, *Environmental Aspects of Nuclear Power*, Ann Arbor Science Publishers, Inc., Ann Arbor, MI, 1976.

76. Anon, "High-Level Radioactive Waste Management Alternatives," USAEC Report WASH-1297, 1974.

77. G. J. McCarthy, ed., *Scientific Basis for Nuclear Waste Management*, *Vol. 1.*, Plenum Press, New York, 1979.
   a. W. Lutze, J. Borchardt, and A. K. Dé, "Characterization of Glass and Glass Ceramic Waste Forms," pp. 69–81.
   b. N. E. Brezneva, A. A. Minaev, and S. N. Oziraner, "Vitrification of High Sodium-Aluminum Wastes: Composition Ranges and Properties," pp. 43–50.
   c. C. Houser, I. S. T. Tsong, and W. B. White, "Characterization of Leached Surface Layers on Simulated High-Level Waste Glasses by Sputter-Induced Optical Emission," pp. 131–139.
   d. W. Guber, M. Hussaine, L. Kahl, G. Ondracek, and J. Saidl, "Preparation and Characterization of an Improved High-Level Radioactive Waste (HAW) Borosilicate Glass," pp. 37–42.
   e. J. W. Braithwaite and N. J. Mangani, "Corrosion Considerations for Nuclear Waste Canisters," pp. 283–287.
   f. M. A. Clynne and R. W. Potter II, "P-T-X Relations of Anhydrite and Brine and Their Implications for the Stability of Anhydrite as a Nuclear Waste Repository Medium," pp. 323–328.
   g. D. B. Stewart and R. W. Potter II, "Application of Physical Chemistry of Fluids at Elevated Temperatures and Pressures to Repositories for Radioactive Wastes," pp. 297–311.
   h. B. Allard, H. Kipatsi, B. Torstenfelt, and J. Rydberg, "Nuclide Transport by Groundwater in Swedish Bedrock," pp. 403–410.
   i. B. R. Erdal, W. R. Daniels, D. C. Hoffman, F. O. Lawrence, and K. Wolfsberg, "Sorption and Migration of Radionuclides in Geologic Media," pp. 423–426.

78. U.S. Congress, "West Valley Demonstration Project Act," Public Law 96-368, Oct. 1, 1980.

79. C. J. M. Northrup, Jr., ed., *Scientific Basis for Nuclear Waste Management, Vol. 2*, Plenum Press, New York, 1980.
   a. J. L. Crandall, "Development of Solid Radioactive Waste Forms in the United States," pp. 39–51.
   b. W. S. Aaron, T. C. Quinby, and E. H. Kobisk, "Development and Characterization of Cermet Forms for Radioactive Wastes," pp. 315–322.
   c. R. Odoj, E. Merz, and R. Wolters, "Effect of Denitration on Ruthenium Volatility," pp. 911–917.

80. J. A. Stone, S. T. Goforth, Jr., and P. K. Smith, "Preliminary Evaluation of Alternative Forms for Immobilization of Savannah River Plant High-Level Waste," USDOE Report DP-1545, Dec. 1979.

81. J. E. Mendel and R. D. Nelson, eds., "State-of-the-Art Review of Materials Properties of Nuclear Waste Forms," USDOE Report PNL-3802, April 1981.

82. U.S. Department of Energy, "Long-Term Management of Defense High-Level Radioactive Wastes, Savannah River Plant, Final Environmental Impact Statement," USDOE Report DOE/EIS-0023, Nov. 1979.

83. Anon, "A Cost Comparison of Solidification Processes," prepared by TERA Corp., Berkeley, CA, for Lawrence Livermore Laboratory, USERDA Report UCRL-13740, Rev. 1, Aug. 1977.

84. G. S. Barney, "Fixation of Radioactive Waste by Hydrothermal Reaction With Clays," USAEC Report ARH-ST-110A, Nov. 1974.

85. R. O. Lokken, "A Review of Radioactive Waste Immobilization in Concrete," USDOE Report PNL-2654, June 1978.

86. J. A. Stone, "Evaluation of Concrete as a Matrix for Solidification of Savannah Plant Waste," USERDA Report DP-1448, June 1977.

87. Anon, "Management of Intermediate-Level Radioactive Waste at Oak Ridge National Laboratory. Final Environmental Impact Statement," USERDA Report ERDA-1553, Sept. 1977.

88. D. M. Roy and G. R. Gouda, "High-Level Radioactive Waste Incorporation Into (Special) Concretes," *Nucl. Tech.*, **40**, 214–219 (1978).

89. L. R. Dole, G. C. Rogers, M. T. Morgan, D. P. Stinton, J. H. Kessler, S. M. Robinson, and J. G. Moore, "Cement-Based Radioactive Waste Hosts Formed Under Elevated Temperatures and Pressures (FUETAP Concretes) for Savannah River Plant High-Level Defense Waste," USDOE Report ORNL/TM-3579, March 1983.

90. K. J. Schneider, "Status of Technology in the United States for the Solidification of Highly Radioactive Wastes," USAEC Report BNWL-820, Oct. 1968.

91. W. F. Bonner, H. T. Blair, and L. S. Romero, "Spray Solidification of Nuclear Waste," USERDA Report BNWL-2059, Aug. 1976.

92. Anon, "Alternatives for Long-Term Management of Defense High-Level Wastes, Idaho Processing Plant," USERDA Report 77-43, Sept. 1977.

93. K. M. Lamb and H. S. Cole, "Development of a Pelletized Waste Form for High-Level ICCP Zirconia Wastes," USDOE Report ICP-1185, 1979.

94. I. Kostantinovich, "Features of a Process for Vitrifying Radioactive Waste Without Precalcination and Radionuclide Behavior in the Process," in *Proceedings of Symposium on Management of Radioactive Waste from the Nuclear Fuel Cycle*, IAEA, Vienna (1976).

95. J. E. Mendel, "High-Level Waste Glass," *Nucl. Tech.*, **32**, 72–88 (1977).

96. W. A. Ross, "Development of Glass Formulations Containing High-Level Nuclear Wastes," USDOE Report PNL-2481, Feb. 1978.

97. J. L. Buelt and C. C. Chapman, "Liquid-Fed Ceramic Melter: A General Description Report," USDOE Report PNL-2735, Oct. 1978.

98. J. H. Simmons, P. B. Macedo, A. Barkatt, and T. A. Litovitz, "Fixation of Radioactive Waste in High-Silica Glass," *Nature*, **278**, 729–731 (1979).

99. G. J. McCarthy, "High-Level Waste Ceramics: Materials Considerations, Process Simulation and Product Characterization," *Nucl. Tech.*, **32**, 92–105 (1977).

100. G. J. McCarthy, "Crystalline Ceramics From Defense High-Level Waste," *Nucl. Tech.*, **44**, 451 (1979).

101. D. Gombert, H. S. Cole, and J. R. Berrett, "Vitrification of High-Level ICPP Calcined Wastes," USERDA Report ICP-1177, 1979.

102. D. M. Strachnan, "Crystalline Materials for the Long Term Storage of Hanford Nuclear Defense Wastes," USDOE Report RHO-SA-13, May 1978.

103. A. E. Ringwood, S. E. Kesson, N. G. Ware, W. Hibberson, and A. Major, "Immobilization of High-Level Nuclear Reactor Wastes in SYNROC," *Nature*, **278**, 219–223 (1979).

104. A. E. Ringwood, "Immobilization of Radioactive Waste in SYNROC," *Amer. Sci.* **70**, 201–207 (1982).

105. R. L. Schwoebel and J. K. Johnstone, "The Sandia Titanate: A Brief Overview," in D. W. Readey and C. R. Cooley, eds., *Ceramic and Glass Radioactive Waste Forms*, USERDA Report CONF-77102-1, 1977, pp. 101–109.

106. L. J. Jardine and M. J. Steindler, "A Review of Metal-Matrix Encapsulation of Solidified Radioactive High-Level Waste," USDOE Report ANL-78-19, May 1978.

107. J. M. Rusin, W. J. Gray, and J. W. Wald, "Multi-Barrier Waste Forms, Part II: Characterization and Evaluation," USDOE Report PNL-2668-2, 1979.

108. K. M. Lamb, "Final Report: Development of a Metal Matrix for Incorporating High-Level Commercial Waste," USDOE Report ICP-1144, 1979.

109. W. S. Aaron, T. C. Quinby, and E. H. Kobisk, "Cermet High-Level Waste Forms, a Progress Report," USDOE Report ORNL/TM-6404, 1978.

110. J. M. Rusin, R. O. Lokken, K. R. Sump, M. F. Browning, and G. J. McCarthy, "Multibarrier Waste Forms, Part I: Development," USDOE Report PNL-2668-1, 1978.

111. W. J. Lackey, P. Angelini, F. L. Layton, D. P. Stinton, and J. S. Varuska, "Sol-Gel Technology Applied to Glass and Crystalline Ceramics," *Proc. Symp. Waste Mngmt.*, **2**, 391–417 (1980) (CONF 800313).

112. Organization for Economic Cooperation and Development, *Proceedings of Symposium on Management of Radioactive Wastes from Fuel Reprocessing*, Paris, March 1973.
    a. L. T. Lakey and B. R. Wheeler, "Solidification of High-Level Radioactive Waste at Idaho Processing Plant," pp. 593–612.
    b. G. Rudolph, J. Saidl, S. Brobnik, W. Guber, W. Hild, H. Krause, and W. Miller, "Lab-scale R&D Work on Fission Product Solidification by Vitrification and Thermite Processes," pp. 655–681.
    c. W. Bocola, A. Donato, and G. Sgalambro, "Survey of the Present State of Studies on the Solidification of Fission Product Solutions in Italy," pp. 449–487.

113. J. vanGeel, H. Eshrich, H. Humerl, and P. Grziwa, "Solidification of High-Level Liquid Wastes to Phosphate Glass-Metal Matrix Blocks," in *Management of Radioactive Wastes From the Nuclear Fuel Cycle*, IAEA, Vienna, Paper IAEA-SM-207/83, 1976.

114. J. R. Grover and B. E. Chidley, "Glasses Suitable for the Long Term Storage of Fission Products," UKAEA Report AERE-R-3178, 1960.

115. Materials Characterization Center (Battelle Pacific Northwest Laboratory, Richland, WA), "Nuclear Waste Materials Handbook–Waste Form Test Methods," USDOE Report DOE/TID-11400, 1980.

116. Schott Optical Glass, Inc., Duryea, PA, "Radiation Resistant Optical Glass (Cerium Stabilized)," Brochure 7600e, 1975.

117. J. A. Stone, "An Experimental Comparison of Alternative Solid Forms for Savannah River High-Level Wastes," in S. V. Topp, ed., *Scientific Basis for Nuclear Waste Management*, Proceedings of Conference 811122. North-Holland, New York, 1982, pp. 1–8.

118. E. D. Hespe, "Leach Testing of Immobilized Radioactive Waste Solids, A Proposal for a Standard Method," *Atomic Energy Review*, **9**(1), 195–207 (1971).

119. G. S. Barney, "Leachability Criteria for Immobilized Radioactive Waste," Quarterly Report., Waste Management and Transportation Technical Development, Atlantic Richfield Hanford Co., USAEC Report ARH-ST-110B, Feb. 1975, p. 63.

120. R. M. Wallace, H. L. Hull, and R. Bradley, "Solid Forms for Savannah River Plant High Level Waste," USAEC Report DP-1335, Dec. 1973.

121. J. H. Goode, C. L. Fitzgerald, and V. C. A. Vaughn, "The Dissolution of Unirradiated and Irradiated (U, Pu)O_2 in Nitric Acid," USAEC Report ORNL-5015, Feb. 1975.

122. M. N. Elliot, R. Gayler, J. R. Groves, and W. H. Handwick, "The Fixation of Radioactive Wastes in Glass: Part III, the Removal of Ruthenium and Dust From Nitric Acid Vapors." UKAEA Report AERE-R-4098, 1962.

123. R. B. Keeley, W. F. Bonner, and D. E. Larson, "Technology Status of Spray Calcination/Vitrification of High-Level Liquid Waste for Full-Scale Application," talk presented 70th Annual Meeting, AICE, New York, Nov. 13–17, 1977.

124. R. H. Burns, "Solidification of Low- and Intermediate-Level Wastes," *Atomic Energy Review*, **9**, 547–599 (1971).

125. Anon, *Management of Low- and Intermediate-Level Wastes*, IAEA, Vienna, STI/PUB/264, 1970.

126. National Academy of Science-National Research Council, "The Shallow Land Burial of Low-Level Radioactively Contaminated Solid Waste," 1976.

127. U.S. Congress, "Marine Protection, Research and Sanctuaries Act, Title I," Public Law 92-532, 1972.

128. U.S. Dept. Energy, "Nuclear Waste Management Program, Summary Document, FY 1980," USDOE Report DOE/ET-0094/D, April 1979.

129. U.S. Dept. Energy, "Report of Task Force of Nuclear Waste Management," USDOE Report DOE/ER-0094/D, Feb. 1978.

130. U.S. Congress, "Low-Level Waste Policy Act," Public Law 97-35, Dec. 22, 1980.

131. U.S. Nuclear Regulatory Commission, "Licensing Requirements for Land Disposal of Radioactive Waste," 10CFR, Part 61.

132. U.S. Nuclear Regulatory Commission, "Licensing Requirements for Land Disposal of Radioactive Waste, Final Environmental Impact Statement," USNRC Report NUREG-0945, 1982.

133. *U.S. Federal Register*, **39**, 32921–32923 (Sept. 12, 1974).

134. L. J. Carter, "WIPP Goes Ahead, Amid Controversy," *Science* **222**, 1004–1006 (1983).

135. IAEA, "Standardization of Radioactive Waste Categories," IAEA Technical Report No. 101, STI/DOC/10/101, 1970.

136. T. B. Mullarkey, T. L. Jentz, J. M. Connelly, and J. P. Kane, "A Survey and Evaluation of Handling and Disposing of Solid Low-Level Nuclear Fuel Cycle Waste," Report AIF/NESP-0008, Atomic Industry Forum, Washington, D.C., 1976.

137. U.S. Department of Transportation, "A Review of the Department of Transportation (DOT) Regulations for Transportation of Radioactive Materials," DOT Office Hazardous Materials, 1974.

138. International Atomic Energy Authority, "Regulations for Safe Transport of Radioactive Materials, 1967 Edition," IAEA Safety Series No. 6, 1967.

139. International Atomic Energy Authority, "Bitumenization of Radioactive Wastes," IAEA Tech. Series No. 116, STI/DOC/10/116, 1970.

140. C. L. Fitzgerald, H. W. Godbee, R. E. Blanco, and W. Davis, Jr., "The Feasibility of Incorporating Radioactive Wastes in Asphalt or Polyethylene, *Nucl. Appl. Technol.*, **9**, 821–829 (1970).

141. F. Kreith, *Principles of Heat Transfer*, 2nd ed., International Textbook, Scranton, PA, 1965.

142. E. P. Blizard, "Nuclear Radiation Shielding," in H. E. Etherington, ed., *Nuclear Engineering Handbook*, McGraw-Hill, New York, 1958 Section 7.3.

143. E. P. Blizard, "Analytical Methods of Shield Design," in E. P. Blizard and L. S. Abbott, eds., *Reactor Handbook*, Vol. III, 2nd ed., Part B. Interscience, New York, 1962 Chap. 11.

144. E. D. Arnold and B. F. Maskewitz, "Radiation Shielding Information Center Code Package CCC-60/SDC, A Shielding-Design Calculation Code for Fuel Handling Facilities," USAEC Report ORNL-3041, March 1966.

145. U.S. Congress, "Nuclear Policy Act of 1982," Public Law 97-425, Jan. 7, 1983.

146. C. Norman, "High-Level Politics Over Low-Level Wastes," *Science* **223**, 258–260 (1984).

147. *U.S. Federal Register*, **46**, 35280–35296 (July 8, 1981).

148. *U.S. Federal Register*, **48**, 28195–28230 (June 21, 1983).

149. *U.S. Federal Register*, **47**, 58204–58206 (Dec. 29, 1982).

150. W. D. Weast, "A Bedded Salt Repository for Defense Radioactive Wastes in Southeastern New Mexico," in S. Fried, ed., *Radioactive Waste in Geologic Storage*, American Chemical Society Symposium Series 100, 1979.

151. D. Isherwood, "Geoscience Data Base Handbook for Modeling a Nuclear Waste Repository," USNRC Report NUREG/CR-0912, 1981. In two volumes.

152. C. Klingsberg and J. Duguid, "Isolating Radioactive Waste," *Amer. Sci.* **70**, 182–190 (1982).

153. S. Gonzales, "Host Rocks for Radioactive Waste Disposal," *Amer. Sci.*, **70**, 191–200 (1982).

154. D. A. Morris and A. I. Johnson, "Summary of Hydrologic and Physical Properties of Rock and Soil Materials as Analyzed by the Hydrologic Laboratory of the USGS," U.S. Geol. Survey Paper 1839-0, 1967.

155. Office of Nuclear Waste Isolation (ONWI), "Technical Support for GEIS: Radioactive Waste Isolation in Geologic Formations," USDOE Report Y/OWI/TM-36, 1978.

156. H. H. Hess and A. Polderwaart, eds., *Basalts*, Wiley-Interscience, New York, 1967. In two volumes.

157. H. Van Olphen and J. J. Fripiat, eds., *Data Handbook for Clay Materials and Other Non-Metallic Minerals*, Pergamon Press, Oxford, 1979.

158. D. Hill, B. L. Pierce, W. C. Metz, M. D. Rowe, E. T. Haefle, F. C. Bryant, and E. J. Tuthill, "Management of High-Level Waste Repository Siting," *Science* **218**, 859–864 (1982).

159. K. J. Schneider and A. M. Platt, "High-Level Radioactive Waste Management Alternatives," USAEC Report BNWL-1900, 1974.

160. E. R. Ekren, "Geologic and Hydrologic Considerations for Various Concepts of High-Level Radioactive Waste Disposal in Conterminous U.S.," U.S. Geological Survey Open File Report 74-158, 1974.

161. V. V. Mirkovich, "Experimental Study Relating Thermal Conductivity to Thermal Piercing of Rocks," *Int. J. Mech. Min. Sci.*, **5**, 205–211 (1968).

162. D. P. Linroth and W. G. Krawza, "Heat Content and Specific Heat of Six Rock Types at Temperatures to 1000°C," U.S. Bureau of Mines Investigation 7503, 1971.

163. G. B. Clark, et al., "Rock Properties Related to Rapid Excavation: Final Report to the Office of High Speed Ground Transportation (Contract 3-0142)," 1969.

164. D. T. Griggs, F. J. Turner and H. C. Heard, "Deformation of Rocks at 500° to 800°C," in *Rock Deformation*, Geological Society of America Memoir 79, 1960.

165. Office of Nuclear Waste Isolation (ONWI), "Granitic Briefing Book (Draft)," Feb. 1981.

166. J. M. Cleveland, T. F. Rees, and K. L. Nash, "Neptunium and Americium Speciation in Selected Basalt, Granite, Shale and Tuff Groundwaters," *Science* **221**, 271–273 (1983).

167. M. G. Seitz, P. G. Rickert, S. M. Fried, A. M. Friedman, and M. J. Steindler, "Studies of Nuclear Waste Migration in Geologic Media," USDOE Report ANL-78-8, 1978.

168. R. Sallach, "Model for Estimating the Distribution of Cations in Multicomponent Ion Exchange Reactions," USAEC Report SC-RR-67-861, 1967.

169. J. Relyea and R. Serne, "Controlled Sample Program Publication No. 2: Interlaboratory Comparison of Batch K_d Values," USDOE Report PNL-2872, 1979.

170. J. H. Westsik, Jr., L. A. Bray, F. N. Hodges, and E. J. Wheelwright, "Permeability, Swelling and Radionuclide Retardation Properties of Candidate Backfill Materials," in S. V. Topp, ed., *Scientific Basis for Nuclear Waste Management*, North Holland Publications, New York, 1982, pp. 329–336.

171. T. Rockwell, III, ed., "Handbook of Radiation Shielding Data," USAEC Report TID-7004, 1956. Also published in hardback by Van Nostrand, New York, 1956.

172. W. H. Steigelmann, "Radioisotope Shielding Design Manual," USAEC Report NYO-10721, 1963.

173. E. D. Arnold, "Handbook of Shielding Requirements and Radiation Characteristics of Isotopic Power Sources for Terrestrial, Marine and Space Applications," USAEC Report ORNL-3576, 1964.

174. N. M. Schaeffer, ed., "Reactor Shielding for Nuclear Engineers," USAEC Report TID 2595, 1973.

175. J. C. Courtney, ed., "Handbook of Radiation Shielding Data," American Nuclear Society Report ANS/SD-76/14, 1976.

176. D. C. Stewart, *Handling Radioactivity*, Wiley-Interscience, New York, 1981.

177. Stanford Research Institute, "The Industrial Uses of Radioactive Fission Products," SRI Report 361 (1951); also published as USAEC Report AECU 1673 (1951).

178. J. Moteff, "Tenth-value Thicknesses for Gamma-ray Absorption," *Nucleonics*, **13**(7), 24 (1955).

179. Mailing addresses for both the NCRP and ICRU are 7910 Woodmount Ave., Washington, D. C. 20014.

180. K. Z. Morgan and J. E. Turner, eds., *Principles of Radiation Protection, A Textbook of Health Physics*, Wiley, New York, 1967.

181. H. F. Henry, *Fundamentals of Radiation Protection*, Wiley, New York, 1969.

182. L. S. Taylor, *Radiation Protection Standards*, Chemical Rubber Co., Cleveland, OH, 1971.

183. J. Shapiro, *Radiation Protection, A Guide for Scientists and Physicians*, Harvard University Press, Cambridge, MA, 1972.

184. I. G. Wilms and C. E. Moss, "Bookshelf on Radiological Health," *Am. J. Public Health*, **65**, 1231–1237 (1975).

185. International Atomic Energy Agency, P.O. Box 100, A-1400, Vienna, Austria or UNIPUB, P.O. Box 433, Murray Hill Station, New York, NY 10016.

186. International Commission on Radiation Units and Measurements, "Radiation Quantities and Units," ICRU Report 19 (1971).

187. International Atomic Energy Authority, "Radiation Protection Procedures," IAEA Safety Series No. 38 (1973).

188. International Atomic Energy Authority, "Safe Handling of Radionuclides, 1973 Edition," IAEA Safety Series No. 1 (1973).

189. W. A. Jennings, "SI Units in Radiation Measurement," *Brit. J. Radiol.*, **45**, 784-785 (1972).

190. National Council for Radiation Protection, "Basic Radiation Protection Criteria," NCRP Report 39, 1971.

191. International Commission for Radiation Protection, Committee 2, *Recommendations on Permissible Doses for Internal Radiation, 1959*, Pergamon Press, Oxford, 1960. Reproduced in *Health Phys.*, **3**, 1-380 (1960).

192. National Council for Radiation Protection, "Radiation Protection in Educational Institutions," NCRP Report No. 32, 1966.

193. N. J. Rudie, *Principles and Techniques of Radiation Hardening* (3 vols), Western Periodicals, North Hollywood, CA, 1976.

194. H. Schönbacher and A. Stolarz-Izycha, "Compilation of Damage Test Data: Part I: Cable Insulating Materials," Report CERN 79-04, 1979; "Part II: Thermosetting and Thermoplastic Resins," Report CERN 79-08, 1979.

195. J. F. Kirchner and R. Bowman, *Effects of Radiation on Materials and Components*, Reinhold, New York, 1964.

196. A. J. Swallow, *Radiation Chemistry, An Introduction*, Wiley, New York, 1973.

197. R. J. Chaffin, *Microwave Semiconductor Devices. Fundamentals and Radiation Damage*, Wiley, New York, 1973.

198. Anon, "Radiation Effects Data Book," USAEC Report AGC 2277, Vol. 2 (Rev), 1973.

199. Anon, "How Radiation Affects Materials, A Special Report," *Nucleonics*, **14**(9), 53-88 (1956).

200. W. F. Sheely, ed., *Radiation Effects*, Gordon and Breach, New York.

201. G. D. Calkins and P. Shall, "Radiation Damage—Miscellaneous Materials," in C. R. Tipton, ed., *Reactor Handbook*, Vol. I, 2nd Ed., pp. 74-83. Wiley-Interscience, New York, 1960.

202. R. W. King, N. J. Broadway, and S. Palinchak, "The Effect of Nuclear Radiations on Elastomers and Plastic Components and Materials," USAEC REIC Report 21, 1961; N. J. Broadway and S. Palinchak, (same title), REIC Report 21 (Add.), 1964.

203. S. Battisti, R. Bossart, H. Schönbacher, and M. Van de Voorde, "Radiation Damage to Electronic Components," Report CERN 75-18, 1975.

204. E. F. Laine, "Radiation Effects on Electronic Components," USAEC Report UCID 4544, 1962.

205. "Special Issue on the Effects and Uses of Energetic Radiation on Electronic Materials," *Proc. IEEE*, **62**(9), 1187-1277 (1974).

206. O. J. Wick, ed., *Plutonium Handbook, A Guide to the Technology* (2 vols.), Gordon and Breach, New York, 1967.

207. J. T. Thomas, ed., "Nuclear Safety Guide, TID 7016, Revision 2," USNRC Report NUREG/CR-0095, 1978.

208. R. D. Carter, K. R. Ridgway, G. R. Kiel, and W. A. Blykert, "Criticality Handbook," USAEC Report ARH-600, Vol. I (Rev.), 1968; Vol. II (Rev.), 1969; Vol. III, 1971.

209. E. D. Clayton and S. R. Bierman, "Criticality Problems of Actinide Elements," *Actinides Rev.*, **1**, 409–439 (1971).

210. H. D. Paxton, "Criticality Control in Operations With Fissile Materials," USAEC Report LA-3366 (Rev.), 1972.

211. U.S. Nuclear Regulatory Commission, "Nuclear Criticality Safety in Operations With Fissionable Materials Outside Reactors," NRC Reg. Guide 3.4, Feb. 1978.

212. American Nuclear Society, Subcommittee ANS-8, "American National Standard for Nuclear Criticality Safety in Operations With Fissionable Materials Outside Reactors," Standard ANSI N 16.1-1975, 1975.

213. G. D. Schmidt, "Limits for Radioactive Surface Contamination," in N. V. Steere, ed., *Handbook of Laboratory Safety*, 2nd ed., Chemical Rubber Co., Cleveland, 1971, pp. 477–481.

214. H. J. Dunster, "Surface Contamination Measurements as an Index of Control of Radioactive Materials," *Health Phys.*, **8**, 353–356 (1962).

215. D. G. Stevenson, "Radiological Decontamination: Theoretical and Practical Aspects," *Research* (London), **13**, 383–389 (1960).

216. USERDA, "Proceedings of the Conference on Decontamination and Decommissioning (D&D) of ERDA Facilities," CONF 750827, Idaho Falls, ID, August 19–21, 1975.

217. H. Phillip and E. Wichmann, comps., "Bibliographies in Nuclear Science and Technology, Section 13, Decontamination, Number 10," Report AED-C-13-10, 1974.

218. W. E. Sande, H. D. Freeman, M. S. Hanson, and R. I. McKeever, "Decontamination and Decommissioning of Nuclear Facilities, A Literature Search," USERDA Report BNWL 1917, 1975.

219. G. R. Bainbridge, P. A. Bonhote, G. H. Daly, E. D. Detilleux, and H. Krause, "Decommissioning of Nuclear Facilities, a Review of Status," *At. Energy Rev.*, **12**(1), 146–160 (1974).

220. K. J. Schneider and C. E. Jenkins, coords., "Technology, Safety and Costs of Decommissioning a Reference Nuclear Fuel Reprocessing Plant," USNRC Report NUREG 0278, 1977. In two vols.

221. U.S. Nuclear Regulatory Commission, "Termination of Operating Licenses for Nuclear Reactors," *NRC Regulatory Guide*, 1.86 (6–74), 1974.

222. E. L. Saenger, "Medical Aspects of Radiation Accidents," USAEC Report TID 18867, 1963.

223. J. J. Fitzgerald, *Applied Radiation Protection and Control*, Gordon and Breach, New York, 1970. In two vols.

224. U.S. National Bureau of Standards, "NBS Miscellaneous Publication 253," Rev. May 1969.

INDEX

Acids, properties of, 134, 135
Actinides:
  buildup in ORIGEN reactor, 103-105
  composition in HLLW, 113, 144, 146
  cross sections of, 26, 72
  properties, radiometric, 62-64
Actinium series, 79, 83, 85
Activation, nuclear, 276
  by alpha particles, 74, 88-93
  by fast neutrons, 51, 72, 80
  by gamma rays, 54, 55
  by resonance neutrons, 35, 38, 39, 43, 72, 80, 84
  by slow neutrons, 35, 38, 39, 43, 72, 80, 84
  see also Cross sections
Activation products, 42
  buildup in ORIGEN reactor, 105-109
  composition in HLLW, 114
  properties, radiometric, 38, 42
Alpha particles:
  penetration of skin, 56
  shielding against, 232
Alpha-n reaction, 88
  neutron yields from, 91
Americium-241, production of, 59, 61
Aqueous silicate process, 149, 150
Atomic Energy Commission (U.S.), 145, 187, 194, 264

Basalts, 215, 218, 219
Bases, properties of, 134, 135
Bequerel, definition of, 247
Beta-emitters, long lived, 108
Beta particles:
  dose from, 246
  penetration of skin, 56
  range of, 56, 233
  shielding against, 232
Borosilicate glass, *see* Wastes, solidification of high-level
Bremsstrahlung, 233, 234
Brines, composition of, 222
British rising-glass process, 166
Buildup chains for transuranics, 57
Buildup factors for shielding, 232, 236-239

Calcines, *see* Wastes, solidification of high-level
Californium-252, 63, 87-89
  dose rates from, 88, 90

Californium-252 (*Continued*)
fission product yields from, 88, 89
neutron emission from, 75
properties of, 63, 87
shielding for, 88, 242
Ceramics, *see* Wastes, solidification of high-level
Clays, *see* Wastes, solidification of high-level
Code of Federal Regulations (U.S.), 145, 188, 191, 194, 210, 244
Commercial wastes, definition of, 142
Concretes, *see* Wastes, solidification of high-level
Conversion equations, 40, 97
Conversion factors, 277-279
for compound weights, 120, 124-129
for heat and energy units, 201
for mass units, 103
for radiation doses, 248
for time units, 278
Cooling time, effect on HLLW, 108, 112-114
Continuous ceramic melter, 168
Criticality, *see* Nuclear criticality
Cross sections:
of activation products, 38
fission, 72-74
of fission products, 35
of heavy nuclides (Z>88), 72-74
of naturally occurring nuclides, 80, 84
reactor, 51
of selected active nuclides, 39
of stable isotopes, 43, 51
see also Activation, nuclear
Curium-242, production of, 59, 61

Decontamination, 264-269
standards for, 264
Defense wastes, definition of, 141
Department of Energy (U.S.), 172, 187, 210
Department of Transportation (U.S.), 194
Dose equivalent, definition of, 245, 247

Elastomers, radiation damage to, 256
Electronics, radiation damage to, 258
Elements, the, 117
heavy, radiometric properties of, 57-92
light, radiometric properties, of, 1-56
naturally occurring radioactive, 75-87
volatility of, 179
Environmental Protection Agency (U.S.), 210
Equalities, 272
Extractants, properties of, 134, 136

Federal Register (U.S.), 210, 211
FINGAL process, 166
Fission products:
buildup in ORIGEN reactor, 97-103
decay chains, 3-25
decay in wastes, 108, 112
radiometric properties, 27, 35-37
Fission product yields, 26, 28-34
from Cf-252, 88
from even-even nuclides, 27, 32
from major fissiles, 26, 28
miscellaneous data, 27, 34
Fuel assemblies, *see* Reactor fuel assembiles
FUETAP process, 153

Gamma rays:
dose conversion factors, 206, 248
dose from, 56, 247
shielding against, 235
Glasses, *see* Wastes, Solidification of high-level
Granites, 215, 220, 221
Gray, definition of, 247
Greek alphabet, 280
Groundwaters, composition of, 223

Half-value layer, definition of, 232
HARVEST process, 166
Health Physics, 244-253
literature, 244
protection standards, 248
terms, 245
Helium, production from alphas, 63, 81, 86
High-level wastes, *see* Wastes, radioactive high-level
High Temperature Gas Reactor, *see* Reactors, nuclear
Hydroxides, properties of, 121, 131

International Atomic Energy Agency, 174, 193, 195, 265

International Commission of Radiation Protection, 244
International Commission on Radiation Units and Measurements, 244, 245, 247
Inverse square law, 56
Ion exchangers, properties of, 136

Leachability, 174-179
  calculation of, 174
  of cesium, 179
  rates of, 179, 181
  of uranium, 179
Light water reactors, *see* Reactors, nuclear
Limiting critical density, definition of, 261
Liquid metal fast breeder reactor, *see* Reactors, nuclear
LOTES process, 161
Low-level Waste Policy Act of 1980, 188
Lubricants, radiation damage to, 255

Maximum permissisble concentrations, 175, 191, 193, 249
Materials Characterization Center, 172, 174
Minimum critical mass, definition of, 261
Multibarrier waste disposal forms, *see* Wastes, solidification of high-level

National Commission on Radiation Protection (U.S.), 244, 248
National Nuclear Data Center (U.S.), 3, 43
Naturally occurring radionuclides, 75-87
  cross sections of, 80, 84
  decay chains, 77, 80, 81, 83
  helium production from, 81, 86
  in high-level wastes, 75
  historical names of, 77
  Q values for, 80, 84, 86
  radiometric properties of, 77, 79
  specific activities of, 77, 83
Neptunium series, 78, 81, 84
Neptunium-237, production of, 61
Neutron activation, 43, 51, 272. *See also* Activation, nuclear; Cross sections
Neutrons:
  dose conversion factors, 248
  shielding against, 237
Neutron sources:
  alpha-n, 88
  Californium-252, 87
  gamma-n, 54, 55
Nitrates, properties of, 115, 121, 130
Nuclear criticality, 260-263
Nuclear Policy Act of 1982, 210
Nuclear reactions, calculations for, 94
Nuclear Regulatory Commission (U.S.), 145, 188, 194, 210, 249, 264
Nuclide migration through rocks, 221
Nu-values, 75, 76

Oak Ridge National Laboratory, 96, 153, 193
Oils, radiation damage to, 255
Oklo natural reactor, 87
Operations, data for, 229-268
ORIGEN computer code, 40, 43, 94-114
  buildup of actinides, 103
  buildup of activation products, 105
  buildup of fission products, 97
  computational problem, 94
  description of, 96
ORIGEN reactor model:
  description of, 97
  effect of cooling on wastes from, 108, 112, 113, 114
Oxidation potentials, standard, 117, 122
Oxides, properties of, 121, 131

Pacific Northwest Laboratories, 141, 161, 166, 168, 172, 184
Phosphate glass, 149, 162
Photoneutron reaction, 54, 55
Plastics, radiation damage to, 256
Plutonium:
  absorption on rocks, 225, 226
  buildup in ORIGEN reactor, 103, 106
Plutonium-236, production of, 62
Plutonium-238, production of, 61
Plutonium-239:
  criticality data for, 262, 263
  fission yields from, 26, 28
Plutonium-241, fission yields from, 26, 28
Process chemicals, properties of, 134
Protection standards, 248
Purex extraction process, 75, 115, 260

Quality factor, definition of, 245

Q values:
of actinides, 68
of activation products, 38
of fission products, 35
of miscellaneous radionuclides, 39
of naturally occurring isotopes, 84
of natural thorium and uranium, 86

Rad, definition of, 245
Radiation damage, 254-259
to elastomers and plastics, 256
to electronics, 258
to oils and lubicants, 255
Radiation Shielding Information Center, 40, 51, 209, 231
Radioactive decay, 273
Radionuclides, miscellaneous, radiometric properties of, 27, 39
Reactor fuel assemblies:
fuel composition, 111
mass distribution in, 110
metals in, 106, 109
physical characteristics, 111
Reactors, nuclear:
HTGR, wastes from, 142, 146
LMFBR, wastes from, 142, 146
LWR's, wastes from, 139, 140, 142, 146
Repositories, high-level waste:
backfill materials for, 212, 226
data for, 210-227
Rocks, properties of, 212-227
ion-exchange characteristics, 221
permeability of, 212-217
shear strengths of, 215, 218
thermal characteristics of, 212, 214
see also Basalts; Granites; Tuff, volcanic
Roentgen, definition of, 245
Rules of thumb, 55

Savannah River Plant, 141, 144, 145, 148, 150, 172, 178
Shale fracture disposal of wastes, 153, 193
Shielding, 231-238
of alphas, 232
of betas, 232
buildup factors for, 232
of gammas, 88, 235
materials for, 243
of neutrons, 237
terminology for, 231
Sievert, definition of, 247
SI units, 42, 246
Soils, permeability of, 213
Solubility constants, 121, 133
Solvents, properties of, 134, 136
Spontaneous fission, 63
half lives, 64
neutrons from, 75, 76
Supercalcine, 154, 157, 158
SYNROC, 155, 156, 158

Technicium compounds, solubility of, 121
Tenth value layer, definition of, 232, 235
Thorium, natural:
ionium in, 87
Q value of, 83
specific activity of, 83
Thorium series, 77, 80, 84
Transuranium elements, *see* Antinides
Tuffs, volcanic, 215

Unit prefixes, 279
Uranium, natural:
burnup in ORIGEN reactor, 103
isotopic composition of, 87
Q value for, 83
specific activity of, 83
Uranium series, 78, 82, 84
Uranium-232, production of, 61
Uranium-233:
criticality data for, 260, 262
fission yields from, 26, 28
production of, 61
Uranium-235:
criticality data for, 260, 262
fission yields from, 26, 28

Vitromet, 156, 162

Waste canisters:
capacity data, 207, 209
corrosion of, 208, 209
Waste Isolation Pilot Plant (WIPP), 192, 212
Wastes:
high-level liquid (HLLW), 141-185
composition of, 94-115, 142-147

definition of, 193
types of, 141, 142
see also ORIGEN reactor model; Reactors, nuclear
intermediate-level, 193-194
definition of, 193
disposal of, 194
low-level, 186-191
burial sites for, 187
classification of, 188
disposal, regulations for, 186, 188, 191
immobilization of, 194
non-high-level, 186-198
immobilization of, 194-198
packaged, 199-209
container materials for, 208, 209
container sizes, 207, 209
exposure calculations for, 201-209
temperature calculations for, 199-201
radioactive, 139-209
sources of, 140
repositories for, 210-227
host rock properties, 212-221
nuclide migration from, 221-227
regulatory aspects of, 210
solidification of high-level, 145-174
borosilicate glass, 149, 166-173
calcines, 149, 150, 155, 158-162
ceramics, 149, 154
clays, 149, 150
concretes, 149, 153
forms for, comparison of, 148, 150-152
forms for, summary of, 149
glasses, 149, 162-174
multibarrier, 149, 155
see also Element volatility; Leachability
TRU, 186, 191
Weight conversion factors, 121, 124-129
Western New York Nuclear Service Center, 142, 147, 194

Zeolites, 138